Carbon Allotropes and Composites

Scrivener Publishing
100 Cummings Center, Suite 541J
Beverly, MA 01915-6106

Publishers at Scrivener
Martin Scrivener (martin@scrivenerpublishing.com)
Phillip Carmical (pcarmical@scrivenerpublishing.com)

Carbon Allotropes and Composites

Materials for Environment Protection and Remediation

Edited by

Chandrabhan Verma

Interdisciplinary Research Center for Advanced Materials, King Fahd University of Petroleum and Minerals, Dhahran, Saudi Arabia

and

Chaudhery Mustansar Hussain

Department of Chemistry and Environmental Science, New Jersey Institute of Technology, Newark, USA

Scrivener Publishing

WILEY

This edition first published 2023 by John Wiley & Sons, Inc., 111 River Street, Hoboken, NJ 07030, USA and Scrivener Publishing LLC, 100 Cummings Center, Suite 541J, Beverly, MA 01915, USA

For more information about Scrivener publications please visit www.scrivenerpublishing.com.

Wiley Global Headquarters
111 River Street, Hoboken, NJ 07030, USA

For details of our global editorial offices, customer services, and more information about Wiley products visit us at www.wiley.com.

Library of Congress Cataloging-in-Publication Data

ISBN 978-1-394-16650-3

Cover image: Pixabay.com
Cover design by Russell Richardson

Set in size of 11pt and Minion Pro by Manila Typesetting Company, Makati, Philippines

Printed in the USA

10 9 8 7 6 5 4 3 2 1

Contents

Preface

Due to their huge surface area and numerous other distinguishing characteristics, nanostructure materials are widely used in a variety of applications. The production of substrates for better environmental protection and cleanup has been prompted by these qualities. They offer a superior surface for the adsorption of impurities and pollutants that contaminate industrial influents, wastewater, air, and soil as contaminants. Those examples all include a variety of harmful environmental substances, such as toxic metals, phenolic compounds, dyes, and other substances that must be treated appropriately before being released into the environment. There have been numerous earlier initiatives for environmental protection and restoration. However, composites made of highly efficient and relatively noble carbon allotropes are attracting significant attention. The use of carbon allotropes offers many benefits, including low cost, low toxicity, simple manufacture, and high efficiency. They are also ideal replacements for previously established materials. This book discusses the most recent developments and trends in the use of carbon allotropes and their composites for environmental restoration and protection. The synthesis, characterization, and application of carbon allotropes and their composites in environmental preservation and cleanup are also covered in the book.

There are fourteen chapters in the book. Chapters 1 and 2 cover the creation and characterization of carbon allotropes, as well as their fundamental characteristics. Chapter 3 addresses how carbon allotropes can be functionalized or modified in covalent and noncovalent ways to improve their ability to maintain and repair the environment. The application of carbon allotropes and their composites for specialized environmental protection and cleanup is explored in the following chapters (4–14). Lead (Pb) and nickel (Ni) decontamination are covered in Chapters 4, and 5, respectively. Chapter 6 describes the ability of molybdenum-modified-carbon allotropes in wastewater treatment. The purification of other common elements, such as zinc (Zn), copper (Cu), iron (Fe), and others, is covered in Chapter 7. The capacity of carbon allotropes to remove phenolic compounds is discussed in Chapter 8. The ability of carbon allotropes in carbon dioxide (CO_2) to capture and purify the air is addressed in Chapters 9 and 10, respectively. Waste breakdown and the management potential of carbon allotropes is covered in Chapter 11. The topic of carbon allotropes and green and sustainable development is explored in Chapters 12–14.

We editors, Drs. Chandrabhan Verma and Chaudhery Mustansar Hussain would like to thank all contributors for their great efforts. On behalf of Scrivener Publishing (Wiley-Scrivener imprint), we are very thankful to the authors of all chapters for their amazing and passionate efforts in producing this book. Special thanks to Martin Scrivener (President of Scrivener Publishing) for his dedicated support and help during this project. Our final thanks go to Scrivener Publishing (Wiley-Scrivener imprint) for publishing the book.

Chandrabhan Verma, PhD
Chaudhery Mustansar Hussain, PhD
(Editors)

1

Preparation of Carbon Allotropes Using Different Methods

Omar Dagdag[1*], Rajesh Haldhar[2], Seong-Cheol Kim[2], Elyor Berdimurodov[3†], Sheerin Masroor[4], Ekemini D. Akpan[1] and Eno E. Ebenso[1‡]

[1]*Centre for Materials Science, College of Science, Engineering and Technology, University of South Africa, Johannesburg, South Africa*
[2]*School of Chemical Engineering, Yeungnam University, Gyeongsan, Republic of Korea*
[3]*Faculty of Chemistry, National University of Uzbekistan, Tashkent, Uzbekistan*
[4]*Department of Chemistry, A.N. College, Patliputra University, Patna, Bihar, India*

Abstract
Carbon-containing substances have long been employed as sources of energy, and carbon is crucial to contemporary industry. Consequently, understanding carbon allotropes are essential for creating novel materials. Due to their distinctive characteristics, which make them ideal for a wide range of prospective uses, carbon nanotubes (CNTs) have been the subject of scientific study for more than fifteen years. The fields of nanoscience and nanotechnology continue to advance their research in order to create CNTs with adequate characteristics for applications in the future. Recently, a new type of nanocarbon material known as carbon quantum dots (CQDs) has attracted a lot of attention, particularly in solar cells, bioimaging, electrocatalysis, nanomedicine, and chemical sensors, as well as light-emitting diode (LED). The preparatory processes for CNTs and CQDs are the main topic of this chapter. The appropriate examples were used to discuss the complementary arc discharge, laser ablation, acid oxidation, and further carbon allotropes manufacturing processes. This chapter has also covered the benefits and downsides of each technique. New carbon allotropes might be created using the information in this chapter.

Keywords: Carbon, energy, science, technology, CNTs, CQDs and synthesis methods

**Corresponding author*: dagdao@unisa.ac.za
†*Corresponding author*: elyor170690@gmail.com
‡*Corresponding author*: ebensee@unisa.ac.za
Equal contribution by all the authors

Chandrabhan Verma and Chaudhery Mustansar Hussain (eds.) Carbon Allotropes and Composites: Materials for Environment Protection and Remediation, (1–16) © 2023 Scrivener Publishing LLC

Abbreviations

DMF:	dimethylformamide
NaHS:	Sodium hydrosulfide
NaHSe:	Sodium selenide
FC:	Floating catalyst
SWNTs:	Single-walled carbon nanotubes
0D:	Zero dimension
3D:	Three dimensions
1D:	One-dimensional
2D:	Two-dimensional
PL:	Photoluminescence
QY:	Quantum yield
SAC:	single-atom catalysts
PEG200:	Poly(ethylene glycol)

1.1 Introduction

Carbon is a very important element due to its multifunctional binding nature. It has the atomic number six, which means it has 6 e^- that can occupy $1s^2$, $2s^2$, and $2p^2$ atomic orbitals. Among them, 4 e^- are valence e^- that can be hybridized as sp, sp^2, or sp^3. Carbon can create many different forms at the macro and nano scales. It can take various allotropic forms depending on the hybridization and has a wide range of properties. The most common carbon allotropes are soft and conductive, such as graphite (sp^2), and hard and insulating, such as diamond (sp^3) [1]. Recently, many new allotropes have been developed, such as fullerenes, graphene, and carbon nanotubes. These allotropes are not only a very interesting and broad area of research but also have many applications due to their unique properties. Carbon is the only element that can be allotropic from 0D to 3D [2]. 0D structures include nanoclusters and quantum dots; 1D includes nanofibers, nanorods, nanowires, and nanotubes; 3D consists of thin layers and nano-coatings; 3D includes powders and bulk materials [1].

These are usually 8 allotropic carbon atoms [3] (diamond, graphite, lonsdaleite, C60, C540, C70, carbon nanotubes and amorphous carbon). Carbon has been studied for many years. In this section, it is indicated the current scope of global research on carbon. Figure 1.1 shows the 8 allotropes of carbon and their compounds.

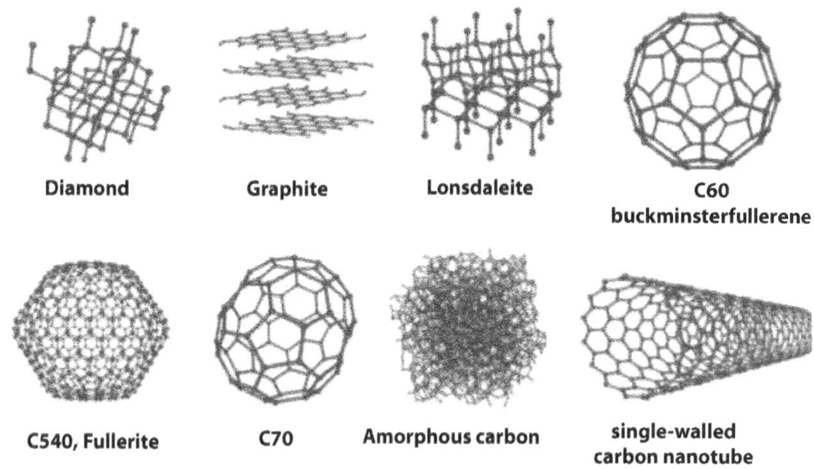

Figure 1.1 The structure of 8 kinds of carbon allotropes [4]. Reproduced from Ref. [4], [http://dx.doi.org/10.5714/CL.2014.15.4.219], under the terms of the CC BY 4.0 license.

1.2 Synthesis Methods

1.2.1 Synthesis of CNTs

1.2.1.1 Arc Discharge Method

The arc discharge method was first used in the synthesis of Buckminster fullerenes in 1985 [5]. In 1991, Iijima applied the original arc design method to CNT production. The MWNT was placed on a carbon black graphite electrode with a current of 100 A. It was originally thought to produce fullerenes [6]. This synthesis way is based on the explosion of electric current. Figure 1.2 shows the representative illustration of an arc discharge system. In this method, the SWCNTs and MWCNTs carbon allotropes are formed with high efficiency, as well as simple and easy production of high-quality nanotubes [7]. In this method, the anode and cathode electrodes (graphite) were used in the syntheses of carbon allotropes, such as nanotubes, fullerene, C60, C54, C70, and so on (Figure 1.2). In this electrical cell, there are two electrodes: anode and cathode, both are made from the graphite. The cathode was made from pure graphite while the anode was made with activated carbon, which is the source of the catalytic effect. Therefore, the formation of SWCNTs and MWCNTs carbon allotropes in the arc discharge method is done under the catalytic effect of graphite in the electric cell with a high voltage (20 V) and a high temperature (above 1700 °C). It gives a fixed time of 50-100 A because

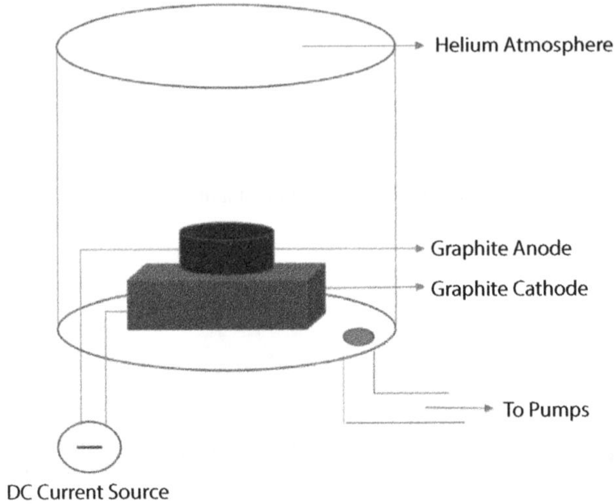

Figure 1.2 Representative illustration of an arc discharge system [1]. Reproduced from Ref. [1], [10.1007/s42823-019-00068-2], under the terms of the CC BY 4.0 license.

fewer structural problems arise during the production of CNTs than other methods [8, 9]. One of the most important prerequisites for maintaining the arc is to maintain a permanent link between the anode and cathode is 1 mm, placed in a building, usually filled in an inert gas (e.g., He or Ar) under low pressure. When the arc strikes the electrodes, plasma is formed containing inert gases, carbon, and catalytic steam.

However, the syntheses of carbon allotropes in the electrochemical cell have some technical problems. For example, the graphite anode was slowly destroyed under high temperature and pressure. The production of carbon allotropes in the electrical cell was covered or accumulated on the surface of the cathode electrode, this accumulation destroyed the cathode electrode. The oscillating arc causes plasma instability, which affects the quality of the final product [10]. Some doping agents (Co, Fe, and other metals) are doped on the anode electrode to increase the catalytic effects of the anode, these doped agents influence the formation of nanotubes in the form of soot. The evolution of H_2 from electrical cells has been shown to guarantee the best synthesis of MWCNT with high crystallinity and a small number of synchronized carbon nanoparticles [11, 12].

1.2.1.2 Laser Ablation Method

It first used the laser ablation technique in 1995 to create single-walled carbon nanotubes. Laser ablation techniques consist of vaporizing particles

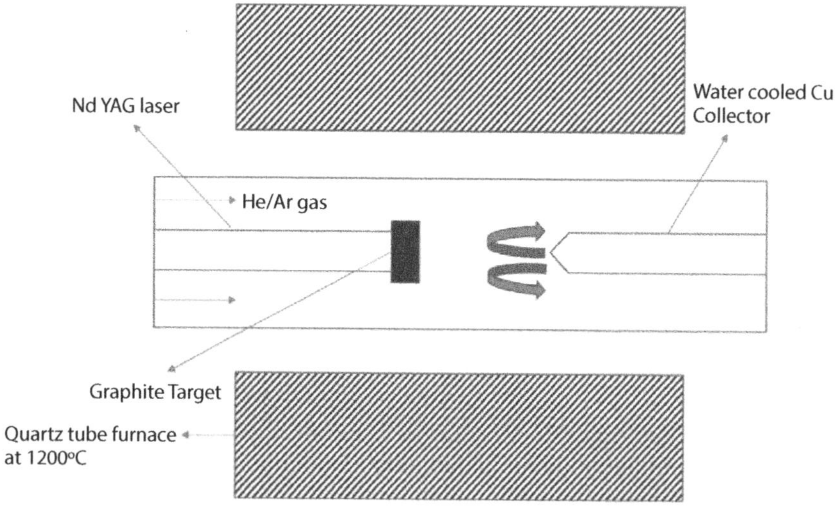

Figure 1.3 Representative graph of laser ablation device [1]. Reproduced from Ref. [1], [10.1007/s42823-019-00068-2], under the terms of the CC BY 4.0 license.

from a solid object [13]. Figure 1.3 shows the main mechanism of the laser ablation device, in which the formation of carbon nanomaterials is formed between two quartz tube furnaces. The temperature of these quartz tube furnaces is 1200°C. They catalytically affect the formation of carbon nanomaterials. The reaction chamber was located between two quartz tube furnaces and it was filled with the inert gas (He or Ar) under 500 Torr (Figure 1.3) [14]. The laser source is continuous or pulses lasers, which are performed to vaporize the target graphite. The distinction between the continuous (CW) or pulse lasers is the super light intensity that must be used for pulsed lasers (100 kW/cm^2 for pulsed lasers versus 12 kW/cm^2 for CW lasers). Arepalli *et al.* [15] fabricated the SWCNTs using spiral laser evaporation and focused on length and aspect ratio. It was found in the obtained results that the individual long nanotubes (thousands of microns) formed near the target and then clustered into spheres.

1.2.1.3 Chemical Vapor Deposition (CVD)

It is indicated that the carbon allotropes were synthesized from the chemical reaction of organic compounds. This method is more convenient in carbon nanomaterials preparation, because, this method require low pressure, normal temperatures, dot required expensive apparatuses and specially designed laboratories [16]. The use of carbon monoxide and hydrogen vapor in Fe was first described in 1959 [17]. In 1993, CNTs were obtained by this method using acetylene in Fe at a temperature of 700°C [18]. In 1996, CVD was introduced as a method for using large CNTs [19]. Arc and laser-cultured CNTs are more crystalline than CVD-cultured CNTs. The next benefit of the CVD method is that the carbon nanomaterials were prepared in CVD with pure and higher-quality degrees. Figure 1.4 shows the representative graph of the chemical vapor deposition system [1]. In this CVD schema, the tubular reactor was designed with a metallic catalyst, such as Fe, Co, and Ni, that affects the carbonization reaction as catalytically effect at 600°C to 1200°C. The hydrocarbon vapors are passed through the catalyst. The carbonization reaction was done on the surface of the catalyst. As a result, the carbon nanomaterials are accumulated on the surface of the catalyst. After finish reaction, the reactor is cooled and the carbon nanomaterials product was separated from the catalyst surface [19]. Depending on the location, the CVD process can be divided into combined [20] and FC-CVD [21]. In excited CVD, the catalyst is activated and reduced to SWNTs, which complicates the whole process. In addition, the preparation is complicated by the interaction between components and ventilators [22]. In comparison, FC-CVD is one step and uses the entire process on SWNTs in a gaseous environment [23]. FC-CVD also produces highly pure and uncontaminated SWNTs. Proper selection of functional groups has been shown to play a main role in controlling the chirality and morphology of SWNTs [24].

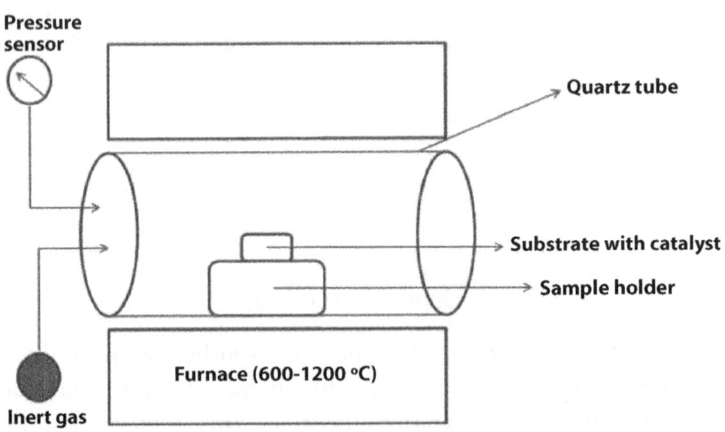

Figure 1.4 Representative graph of chemical vapor deposition system [1]. Reproduced from Ref. [1] [10.1007/s42823-019-00068-2], under the terms of the CC BY 4.0 license.

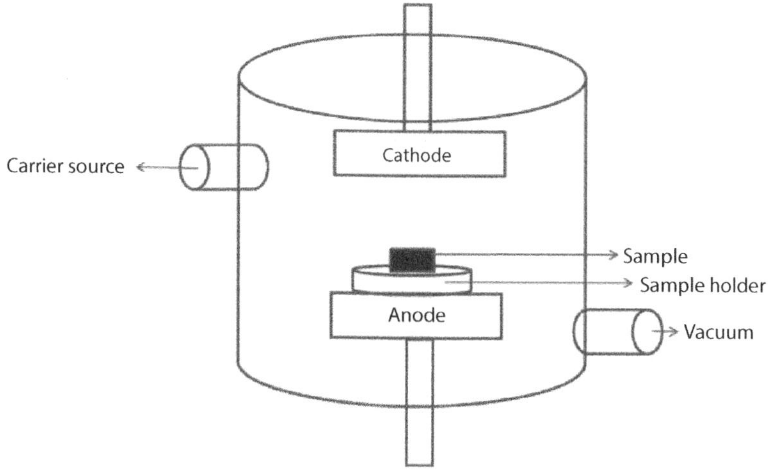

Figure 1.5 Representative figure of plasma-enhanced CVD [1]. Reproduced from Ref. [1], [10.1007/s42823-019-00068-2], under the terms of the CC BY 4.0 license.

1.2.1.4 Plasma-Enhanced CVD (PE-CVD)

In recent years, CVD is mostly applying in the synthesis of carbon nanotubes. However, the ability of PE-CVD to produce vertically aligned nanotubes and its new and more productive explored method. This method is a new way to build CNT composites and change their properties [25]. PE-CVD provides another way to lower the temperature, known as ambient temperature, for many processes, and therefore, it becomes an important factor in the production of composite materials [26]. A representative figure of plasma-enhanced CVD was indicated in Figure 1.5. Therefore, the glow discharge is made of a high voltage that is used for both electrodes. The targeted chemicals are placed on the electric ground. The reaction gas is distributed over several plates, which creates a uniform film. Transfer transition metals (Ni, Fe, and Co) deposited on a substrate (glass, Si, or SiO_2) by hot CVD or sputtering [27]. After the formation of small nanosized metals, the carbonaceous reaction gas (CH_4, C_2H_2, C_2H_4, C_2H_6, and CO) is introduced into the deposition chamber, and the carbon nanotubes are attached to the metal parts of the substrate are released from high-frequency [12].

1.2.2 Synthesis of CQDs

Since the invention of CQDs, many methods of CQD synthesis have been developed [28, 29]. In general, CQD assembly methods can be divided into two groups: bottom-up and top-down methods (Figure 1.6). In the

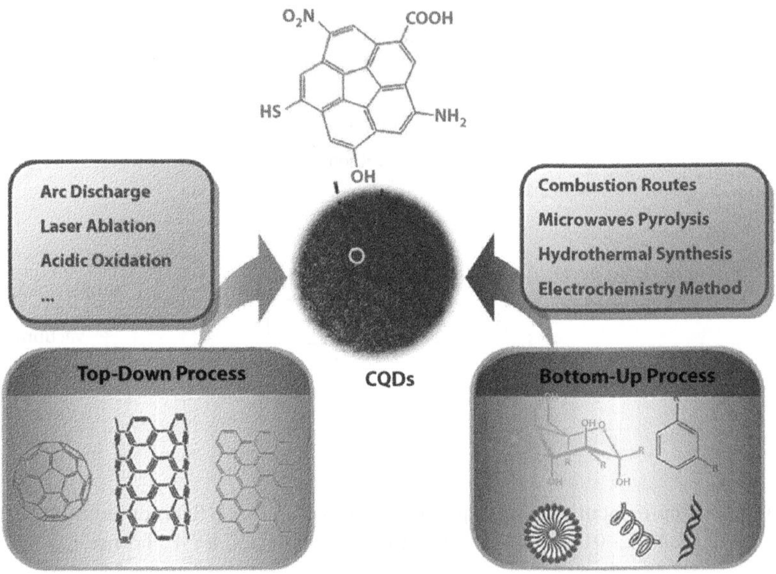

Figure 1.6 Synthetic routes of CQDs [30]. Reproduced from Ref. [30], [https://doi.org/10.3389/fchem.2019.0067], under the terms of the CC BY 4.0 license.

above process, macromolecules are chemically or physically separated or dispersed into small CQDs. It is clear from the obtained results that the CQDs were formed by the carbonization and polymerization reactions of organic compounds.

1.2.2.1 Arc Discharge

Arora and Sharma reported that the arc discharge method [31] required the separation techniques of carbon nanomaterials. All reactions were done in the closed reactor under forming high energy gas plasma, which is formed at 4000 K and higher electrical current. In this method, the carbon particles are separated from the reaction product to an anodic electrode. In the cathode, there is a concentration of carbon vapor to form CQDs. Fabrication of CQDs by arc discharge was introduced in 2004 [32]. Xu *et al.* spontaneously produced three types of carbon nanoparticles with different molecular weights and optical properties in the preparation of SWNTs by the arc discharge method. CQDs can emit blue-green, yellow, and orange light at 365 nm. Further studies showed that the CQD surface was bound to hydrophilic carboxyl groups. The water solubility of the CQDs is obtained in this way with a good yield, but the particle dispersion is usually high because carbon particles of different sizes are produced during the removal process. Additionally, the surface size of the catalyst was decreased after the carbon reactions; as a result, the number of catalytically active regions decreased. These processes may influence the reaction efficiency in this method.

1.2.2.2 Laser Ablation

The use of laser ablation method [33–35] uses the energy of laser pulses to ablate the surface in a thermodynamic state that generates high temperature and pressure, which quickly heat and melt into a plasma state, and the vapor crystallizes and forms nanoparticles [36]. Li *et al.* [37] proposed a simple method for synthesizing CQDs using laser irradiation of a carbon precursor dispersed in a normal solvent. The resulting CQD shows visible and flexible PL. In addition, Hu *et al.* [38] demonstrated that the condition of the CQD surface can be improved by choosing an appropriate organic solvent during the laser irradiation to match the PL properties of the CQD preparation. Laser ablation is a good method for synthesizing CQDs with a narrow optical distribution, good water solubility, and fluorescent properties. However, its use is limited by the complexity of handling and the high cost.

1.2.2.3 Acidic Oxidation

Chemical treatment (acidic oxidation) is often used to transport and degrade carbon into nanoparticles containing hydrophilic groups, such as -OH or C=0 groups to obtain CQDs [39, 40]. Yang *et al.* [41] synthesized new CQDs by the hydrothermal and acid oxidation methods with a high yield and purity. First, carbon nanoparticles were oxidized in H_2SO_4, HNO_3, and $NaClO_3$ solutions. The oxidized CQDs were treated with DMF, NaHS, and NaHSe as N source, S source, and Se source, respectively. The results of N-CQD, SCQD and Se-CQD showed better PL performance, higher QY and longer fluorescence lifetime than pure CQD. Experimental results show that the doped heteroatom can affect the PL characteristics, which corresponds to the electronegativity of N, S, and Se. The strong heteroatoms in CQDs control the electronic structure of CQDs and therefore provide excellent electrocatalytic activity when used as an electrocatalyst. On the other hand, as shown in this study, these heavy CQDs can bind to transition metal ions, N-CQD, S-CQD and Se-CQD can bind to other metal ions as Fe^{3+}, Co^{2+} and Ni^{2+} to form SAC [41].

1.2.2.4 Combustion/Thermal Routes

In modern times, it is important to note enhancing a production system for CQD because of the simplicity of the method, ease of manufacture, first order, low cost, and quality [42–44]. The use of combustion/thermal to produce CQDs was first described by Xu *et al.* and followed by many researchers. Li *et al.* [45] obtained modified GQDs using carboxyl groups by burning citric acid and then adding acetic acid residues at low temperatures. The resulting GQD has a small gap of 8.5 nm and more carboxyl groups in the GQD. Its presence favors the useful use of H_2O in the electrocatalytic process in aqueous solutions.

1.2.2.5 Microwave Pyrolysis

Among the simple methods, the microwave heating pyrolysis technique was found to be effective due to its rapid production and availability [46, 47]. Zhu *et al.* PEG200 and fructose reported a simple microwave combustion method for the synthesis of CQDs by combining sugars, such as glucose with H_2O, to form a pure drug after microwave heating [48]. The obtained CQDs exhibit target-dependent PL properties. A simple, fast, and ecological way to enrich the CQDs of existing clusters. It becomes the site of the fusion of metal ions and forms the carbon base of the four electrodes.

1.2.2.6 Electrochemistry Method

Electrochemical technology is a modest and easy manufacturing process that can be carried out at high temperatures. The production of CQDs by electrochemical methods has become widespread due to the easy control of the multifunction and PL efficiency of the obtained CQDs [49, 50]. Hou *et al.* [51] prepared the blue-emitting CQDs by the electrochemical carbonization method; these CQDs are a super detector for the mercury ions in the aquatic system. The electrochemical catalytic method is also efficient and widely used for the production of high-quality electronic materials; however, the nature of electrochemical catalysts and electrochemical catalytic mechanisms are not deeply studied. Therefore, the synthesis of CQD in electrochemical ways is a modern topic, and it would be more dominant in future research.

1.2.2.7 Hydrothermal/Solvothermal Synthesis

The hydrothermal method is a widely used method for the synthesis of CQDs [52, 53], because the structure is simple, and the product has almost the same size and high yield. In this synthesis way, the organic compounds and polymers are mixed in an aquatic solution. Then, the formed mixture was heated in a closed autoclave at 100°C to 300°C. The size and properties of targeted carbon dots depend on the reaction temperature and times [54]. Zhu *et al.* [55] prepared the CQDs with a high yield (over 80%) by the hydrothermal methods. In this preparation, citric acid and ethylenediamine as sources of C and N under strong hydrothermal treatment. The prepared CQDs are efficient biosensors for iron determination.

Hola *et al.* [56] prepared the overall color and fluorescence of the finished CQDs at different wavelengths that can be tuned by adjusting the amount of N-graphite in hot water. In addition, Lu *et al.* [57] found carbon- and nitrogen-rich biomolecules that can be used to coordinate the internal molecules of CQDs during hydrodynamic condensation. The simplicity of the method and application of heteroatom doping represents a promising way to design and fabricate electronic devices with novel doping and electronic structures.

1.3 Conclusions

This chapter presents and discusses the processes used to create carbon quantum dots (CQDs), and carbon nanotubes (CNTs). The creation of carbon nanomaterials greatly benefits from the explanation of synthetically carbon allotropes. This is because they might be used in a variety of sectors. In this article, we evaluated the various CNT synthesis techniques, such as plasma-enhanced CVD (PE-CVD), chemical vapor deposition (CVD), and arc discharge, laser ablation. Due to its easy controllability of composition and structure through precursor optimization, the hydrothermal technique is a good choice for the production of CQDs used as electrocatalysts. Additionally, electrodeposition of CQDs is a preferable option that can result in CQDs with homogeneous particle size, which is more crucial, as it makes it possible for CQDs to work together with other conventional electrocatalysts in a single pot during a green chemistry manufacturing process. Therefore, scientists and engineers are interested in carbon quantum dots and carbon nanotubes.

References

1. Gupta, N., Gupta, S.M., Sharma, S., Carbon nanotubes: Synthesis, properties and engineering applications. *Carbon Lett.*, 29, 419–447, 2019.

2. Selvam, A., Sharma, R., Sutradhar, S., Chakrabarti, S., Synthesis of carbon allotropes in nanoscale regime, in: *Carbon Nanomaterial Electronics: Devices and Applications*, pp. 9–46, Springer, Singapore, 2021.
3. Hirsch, A., The era of carbon allotropes. *Nat. Mater.*, 9, 868–871, 2010.
4. Karthik, P. and Singh, S.P., Carbon-allotropes: Synthesis methods, applications and future perspectives. *Carbon Lett.*, 15, 219–237, 2014.
5. Kroto, H.W., Heath, J.R., O'Brien, S.C., Curl, R.F., Smalley, R.E., C60: Buckminsterfullerene. *Nature*, 318, 162–163, 1985.
6. Iijima, S., Helical microtubules of graphitic carbon. *Nature*, 354, 56–58, 1991.
7. Arora, N. and Sharma, N., Arc discharge synthesis of carbon nanotubes: Comprehensive review. *Diam. Relat. Mater.*, 50, 135–150, 2014.
8. Purohit, R., Purohit, K., Rana, S., Patel, V., Carbon nanotubes and their growth methods. *Proc. Mater. Sci.*, 6, 716–728, 2014.
9. Farhat, S. and Scott, C.D., Review of the arc process modeling for fullerene and nanotube production. *J. Nanosci. Nanotechnol.*, 6, 1189–1210, 2006.
10. Gang, X., Shenli, J., Jian, X., Zongqian, S., Analysis of the carbon nano-structures formation in liquid arcing. *Plasma Sci. Technol.*, 9, 770, 2007.
11. Ando, Y., Carbon nanotube: The inside story. *J. Nanosci. Nanotechnol.*, 10, 3726–3738, 2010.
12. Koziol, K., Boskovic, B.O., Yahya, N., Synthesis of carbon nanostructures by CVD method, in: *Carbon and Oxide Nanostructures*, pp. 23–49, Springer, Berlin, Heidelberg, 2010.
13. Arepalli, S., Laser ablation process for single-walled carbon nanotube production. *J. Nanosci. Nanotechnol.*, 4, 317–325, 2004.
14. Scott, C.D., Arepalli, S., Nikolaev, P., Smalley, R.E., Growth mechanisms for single-wall carbon nanotubes in a laser-ablation process. *Appl. Phys. A*, 72, 573–580, 2001.
15. Arepalli, S., Nikolaev, P., Holmes, W., Files, B.S., Production and measurements of individual single-wall nanotubes and small ropes of carbon. *Appl. Phys. Lett.*, 78, 1610–1612, 2001.
16. Shah, K.A. and Tali, B.A., Synthesis of carbon nanotubes by catalytic chemical vapour deposition: A review on carbon sources, catalysts and substrates. *Mater. Sci. Semicond. Process.*, 41, 67–82, 2016.
17. Walker Jr., P.L., Rakszawski, J., Imperial, G., Carbon formation from carbon monoxide-hydrogen mixtures over iron catalysts. I. Properties of carbon formed. *J. Phys. Chem.*, 63, 133–140, 1959.
18. José-Yacamán, M., Miki-Yoshida, M., Rendon, L., Santiesteban, J., Catalytic growth of carbon microtubules with fullerene structure. *Appl. Phys. Lett.*, 62, 657–659, 1993.
19. Popov, V.N., Carbon nanotubes: Properties and application. *Mater. Sci. Eng. R Rep.*, 43, 61–102, 2004.
20. Yang, F., Wang, X., Si, J., Zhao, X., Qi, K., Jin, C. *et al.*, Water-assisted preparation of high-purity semiconducting (14, 4) carbon nanotubes. *ACS Nano*, 11, 186–193, 2017.

21. Ding, E.-X., Jiang, H., Zhang, Q., Tian, Y., Laiho, P., Hussain, A. *et al.*, Highly conductive and transparent single-walled carbon nanotube thin films from ethanol by floating catalyst chemical vapor deposition. *Nanoscale*, 9, 17601–17609, 2017.
22. Zhou, W., Ding, L., Liu, J., Role of catalysts in the surface synthesis of single-walled carbon nanotubes. *Nano Res.*, 2, 593–598, 2009.
23. Nasibulin, A.G., Moisala, A., Brown, D.P., Jiang, H., Kauppinen, E., II, A novel aerosol method for single walled carbon nanotube synthesis. *Chem. Phys. Lett.*, 402, 227–232, 2005.
24. Ahmad, S., Liao, Y., Hussain, A., Zhang, Q., Ding, E.-X., Jiang, H. *et al.*, Systematic investigation of the catalyst composition effects on single-walled carbon nanotubes synthesis in floating-catalyst CVD. *Carbon*, 149, 318–327, 2019.
25. Prasek, J., Drbohlavova, J., Chomoucka, J., Hubalek, J., Jasek, O., Adam, V. *et al.*, Methods for carbon nanotubes synthesis. *J. Mater. Chem.*, 21, 15872–15884, 2011.
26. Meyyappan, M., Delzeit, L., Cassell, A., Hash, D., Carbon nanotube growth by PECVD: A review. *Plasma Sources Sci. Technol.*, 12, 205, 2003.
27. Chen, M., Chen, C.-M., Chen, C.-F., Preparation of high yield multi-walled carbon nanotubes by microwave plasma chemical vapor deposition at low temperature. *J. Mater. Sci.*, 37, 3561–3567, 2002.
28. Mosconi, D., Mazzier, D., Silvestrini, S., Privitera, A., Marega, C., Franco, L. *et al.*, Synthesis and photochemical applications of processable polymers enclosing photoluminescent carbon quantum dots. *ACS Nano*, 9, 4156–4164, 2015.
29. Lu, S. and Yang, B., One step synthesis of efficient orange-red emissive polymer carbon nanodots displaying unexpect two photon fluorescence. *Acta Polym. Sin.*, 7, 1200–1206, 2017.
30. Wang, X., Feng, Y., Dong, P., Huang, J., A mini review on carbon quantum dots: Preparation, properties, and electrocatalytic application. *Front. Chem.*, 7, 671, 2019.
31. Yatom, S., Bak, J., Khrabryi, A., Raitses, Y., Detection of nanoparticles in carbon arc discharge with laser-induced incandescence. *Carbon*, 117, 154–162, 2017.
32. Xu, X., Ray, R., Gu, Y., Ploehn, H.J., Gearheart, L., Raker, K. *et al.*, Electrophoretic analysis and purification of fluorescent single-walled carbon nanotube fragments. *J. Am. Chem. Soc.*, 126, 12736–12737, 2004.
33. Kuzmin, P.G., Shafeev, G.A., Bukin, V.V., Garnov, S.V., Farcau, C., Carles, R. *et al.*, Silicon nanoparticles produced by femtosecond laser ablation in ethanol: Size control, structural characterization, and optical properties. *J. Phys. Chem. C*, 114, 15266–15273, 2010.
34. Liu, C., Zhao, Z., Zhang, R., Yang, L., Wang, Z., Yang, J. *et al.*, Strong infrared laser ablation produces white-light-emitting materials via the formation of

silicon and carbon dots in silica nanoparticles. *J. Phys. Chem. C*, 119, 8266–8272, 2015.
35. Xiao, J., Liu, P., Wang, C., Yang, G., External field-assisted laser ablation in liquid: An efficient strategy for nanocrystal synthesis and nanostructure assembly. *Prog. Mater. Sci.*, 87, 140–220, 2017.
36. Sun, Y.-P., Zhou, B., Lin, Y., Wang, W., Fernando, K.S., Pathak, P. *et al.*, Quantum-sized carbon dots for bright and colorful photoluminescence. *J. Am. Chem. Soc.*, 128, 7756–7757, 2006.
37. Li, X., Wang, H., Shimizu, Y., Pyatenko, A., Kawaguchi, K., Koshizaki, N., Preparation of carbon quantum dots with tunable photoluminescence by rapid laser passivation in ordinary organic solvents. *Chem. Commun.*, 47, 932–934, 2010.
38. Hu, S.-L., Niu, K.-Y., Sun, J., Yang, J., Zhao, N.-Q., Du, X.-W., One-step synthesis of fluorescent carbon nanoparticles by laser irradiation. *J. Mater. Chem.*, 19, 484–488, 2009.
39. Shen, P. and Xia, Y., Synthesis-modification integration: One-step fabrication of boronic acid functionalized carbon dots for fluorescent blood sugar sensing. *Anal. Chem.*, 86, 5323–5329, 2014.
40. Zhang, Q., Sun, X., Ruan, H., Yin, K., Li, H., Production of yellow-emitting carbon quantum dots from fullerene carbon soot. *Sci. China Mater.*, 60, 141–150, 2017.
41. Yang, S., Sun, J., Li, X., Zhou, W., Wang, Z., He, P. *et al.*, Large-scale fabrication of heavy doped carbon quantum dots with tunable-photoluminescence and sensitive fluorescence detection. *J. Mater. Chem. A*, 2, 8660–8667, 2014.
42. Li, H., Kang, Z., Liu, Y., Lee, S.-T., Carbon nanodots: Synthesis, properties and applications. *J. Mater. Chem.*, 22, 24230–24253, 2012.
43. Guo, Y., Zhang, L., Cao, F., Leng, Y., Thermal treatment of hair for the synthesis of sustainable carbon quantum dots and the applications for sensing Hg2+. *Sci. Rep.*, 6, 1–7, 2016.
44. Rai, S., Singh, B.K., Bhartiya, P., Singh, A., Kumar, H., Dutta, P. *et al.*, Lignin derived reduced fluorescence carbon dots with theranostic approaches: nano-drug-carrier and bioimaging. *J. Lumin.*, 190, 492–503, 2017.
45. Li, S., Zhou, S., Li, Y., Li, X., Zhu, J., Fan, L. *et al.*, Exceptionally high payload of the IR780 iodide on folic acid-functionalized graphene quantum dots for targeted photothermal therapy. *ACS Appl. Mater. Interfaces*, 9, 22332–22341, 2017.
46. Schwenke, A.M., Hoeppener, S., Schubert, U.S., Synthesis and modification of carbon nanomaterials utilizing microwave heating. *Adv. Mater.*, 27, 4113–4141, 2015.
47. Choi, Y., Jo, S., Chae, A., Kim, Y.K., Park, J.E., Lim, D. *et al.*, Simple microwave-assisted synthesis of amphiphilic carbon quantum dots from A3/B2 polyamidation monomer set. *ACS Appl. Mater. Interfaces*, 9, 27883–27893, 2017.

48. Zhu, H., Wang, X., Li, Y., Wang, Z., Yang, F., Yang, X., Microwave synthesis of fluorescent carbon nanoparticles with electrochemiluminescence properties. *Chem. Commun.*, 34, 5118–5120, 2009.
49. Deng, J., Lu, Q., Mi, N., Li, H., Liu, M., Xu, M. *et al.*, Electrochemical synthesis of carbon nanodots directly from alcohols. *Chem. Eur. J.*, 20, 4993–4999, 2014.
50. Ahirwar, S., Mallick, S., Bahadur, D., Electrochemical method to prepare graphene quantum dots and graphene oxide quantum dots. *ACS Omega*, 2, 8343–8353, 2017.
51. Hou, Y., Lu, Q., Deng, J., Li, H., Zhang, Y., One-pot electrochemical synthesis of functionalized fluorescent carbon dots and their selective sensing for mercury ion. *Anal. Chim. Acta*, 866, 69–74, 2015.
52. Shen, J., Zhu, Y., Yang, X., Zong, J., Zhang, J., Li, C., One-pot hydrothermal synthesis of graphene quantum dots surface-passivated by polyethylene glycol and their photoelectric conversion under near-infrared light. *New J. Chem.*, 36, 97–101, 2012.
53. Liu, J., Li, D., Zhang, K., Yang, M., Sun, H., Yang, B., One-step hydrothermal synthesis of nitrogen-doped conjugated carbonized polymer dots with 31% efficient red emission for in vivo imaging. *Small*, 14, 1703919, 2018.
54. Anwar, S., Ding, H., Xu, M., Hu, X., Li, Z., Wang, J. *et al.*, Recent advances in synthesis, optical properties, and biomedical applications of carbon dots. *ACS Appl. Bio Mater.*, 2, 2317–2338, 2019.
55. Zhu, S., Meng, Q., Wang, L., Zhang, J., Song, Y., Jin, H. *et al.*, Highly photoluminescent carbon dots for multicolor patterning, sensors, and bioimaging. *Angew. Chem.*, 125, 4045–4049, 2013.
56. Hola, K., Sudolská, M., Kalytchuk, S., Nachtigallová, D., Rogach, A.L., Otyepka, M. *et al.*, Graphitic nitrogen triggers red fluorescence in carbon dots. *ACS Nano*, 11, 12402–12410, 2017.
57. Lu, S., Sui, L., Wu, M., Zhu, S., Yong, X., Yang, B., Graphitic nitrogen and high-crystalline triggered strong photoluminescence and room-temperature ferromagnetism in carbonized polymer dots. *Adv. Sci.*, 6, 1801192, 2019.

2

Carbon Allotrope Composites: Basics, Properties, and Applications

Sheerin Masroor

Department of Chemistry, A. N. College, Patliputra University, Patna, Bihar, India

Abstract

This chapter envelopes the fundamentals, properties, and applications of carbon allotropes and their composites. The ongoing recognition of multiple forms of carbon nanostructures has inspired research in different fields. The first section emphasizes the fundamentals of carbon and allotropes. The ambidexterity of the different arrangements of carbon atoms leads to the formation of different allotropes and multiple phases, which causes various unique properties. To enhance the potential of these compounds to be applied in different industries, they may often be combined with other materials to achieve the next level of properties. The resultant composites have significantly improved properties.

Keywords: Carbon allotropes, composites, graphene, carbon-nanotubes

2.1 Introduction

The development of carbon chemistry and technology happened in the last 20th century with the specific development in carbon materials. The ultimate property of carbon can develop a specific structure in bulk in addition to the nano range. Around 95% are called carbon-dependant compounds. This all happens due to the presence of valence electrons which are four (04) in number and helps to make bonds in single, double, or triple bonds. For this, carbon can also react to give a stable compound with extra electronegative and electropositive elements present in the periodic table. By getting so much diversity in carbon compounds, multiple nanostructures

Email: masroor.sheerin@gmail.com

Chandrabhan Verma and Chaudhery Mustansar Hussain (eds.) Carbon Allotropes and Composites: Materials for Environment Protection and Remediation, (17–30) © 2023 Scrivener Publishing LLC

have been accompanied by mismatched biological, chemical, and physical properties. All this literature markedly showed that carbon can be considered the most experimented element in research, science, and technology [1–6]. Here different allotropic parts depend on the hybridization which happens with different properties. The element carbon can form multiple structures in volume, as well as in the nanometre scale range. Different allotropic forms are formed based on their hybridization, which shows a great range of stuff.

Graphite (sp^2) and diamond (sp^3) are the most easily found carbon allotropes that are conductive in nature. Later on, other allotropes have also been discovered in addition to graphene and diamond, such as fullerene/buckyball and a vast variety of carbon nanotubes.

It is the only element whose different allotropes can exist in multiple forms, such as zero dimension (0D) to three dimensions (3D). The different structures include the following:

i. Zero-dimensional structures: fullerene, CNT, quantum dots, and nanoclusters;
ii. One-dimensional: nanotubes, nanofibers, nanowires, and nanorods;
iii. Two-dimensional: graphene, thin films, and nanocoatings;
iv. Three-dimensional: powders and bulk materials.

Furthermore, the diversity of carbon arrangements to make multiple allotropes makes compounds that are extremely useful in almost all kinds of industries [7].

2.2 Allotropes of Carbon

There are around eight allotropes of carbon that are mainly found easily, and these are:

a. Diamond,
b. Graphite,
c. Lonsdaleite,
c. C_{60} buckminsterfullerene,
d. C_{540} fullerite,
e. C_{70} fullerene,
f. Amorphous carbon,
g. Zig-zag single-walled carbon nanotube.

a. Diamond

It is a well-known allotrope of carbon and the hardest known natural mineral, which is very hard, has an extremely high refractive index, and high dispersion of visible light. It all makes it beneficial for industrial applications and jewellery. This property forms an excellent abrasive effect and makes it a very good polishing and lustrous effect.

Eight atoms make up each unit cell in the face-centred cubic lattice crystal structure of a diamond. This results in a cubic diamond structure. The four other carbon atoms form a tetrahedral geometry with all carbon atoms covalently connected to them. This results in a chair-shaped three-dimensional network of six-membered carbon rings that has no bond angle strain. C–C bonds are formed through sp3 hybridized orbitals, resulting in a 154-pm bond length [8, 9]. Figure 2.1 shows how a diamond is made up.

b. Graphite

If the carbon forms a trigonal planar structure, it is called graphite [10, 11] (Figure 2.2). Graphene is another name for these distinct layers. The carbon atoms in a particular layer are arranged in a honeycomb lattice with a bond length of 0.142 nm and a spacing of 0.335 nm between various planes [12].

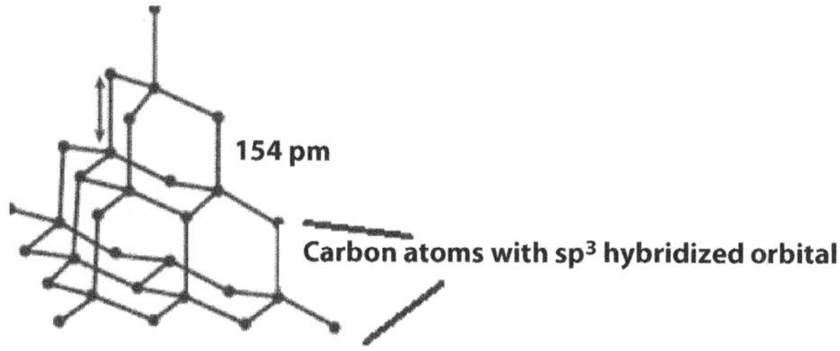

Figure 2.1 Structure of diamond.

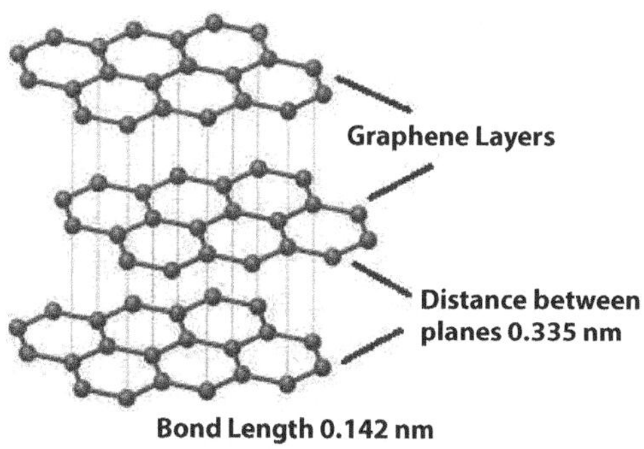

Figure 2.2 Sheets of graphene.

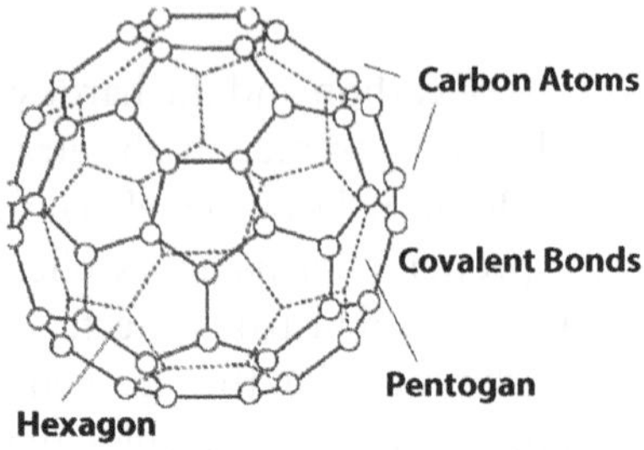

Figure 2.3 Pictorial presentation of buckminsterfullerene.

The van der Waals force, which refers to the comparatively weak connection between the layers, makes it possible for layers that resemble graphene to detach and move past one another [13].

c. Lonsdaleite

It is a kind of diamond with a hexagonal crystallographic structure and is considered a natural substance [14, 15] (Figure 2.3). It has great mechanical properties, which makes it attractive for multiple uses, [16, 17]. It is synthesized just like a diamond, i.e., in presence of high static pressure and temperature [18].

In the laboratory, lonsdaleite may be produced by chemical vapor deposition [19–21], in addition to the thermal decomposition of a polymer such as poly(hydridocarbyne), under atmospheric pressure, in presence of argon atmosphere, at 1,000°C/1,832°F [22]. C_{60} buckminsterfullerene.

As an icosahedron with 60 vertices and 32 faces (20 hexagons and 12 pentagons, with no pentagon sharing a vertice), buckminsterfullerene has a carbon atom at each of its polygonal vertices and a unique bond running down each of its edges. A C60 molecule's van der Waals diameter in this situation is around 1.01 nm. A C60 molecule has a nucleus-to-nucleus diameter of 0.71 nm. The C60 molecule has two types of bonds. Bonds typically have a length of 0.14 nm. Each carbon atom in the structure is joined by a covalent link to three other carbon atoms.

d. C_{540} fullerite

These are solid-state structures made of fullerene molecules that are found naturally within interstellar gas clouds or they are bulk solid forms of pure or mixed fullerenes, which are called fullerite (Figure 2.4). Fullerites are known for their unique structural properties that may point out as helpful to humankind. They have the best application in technological sectors, such as electronics and engineering, as well as the development of heat-resistant weapon systems and ultra-hard metal alloys.

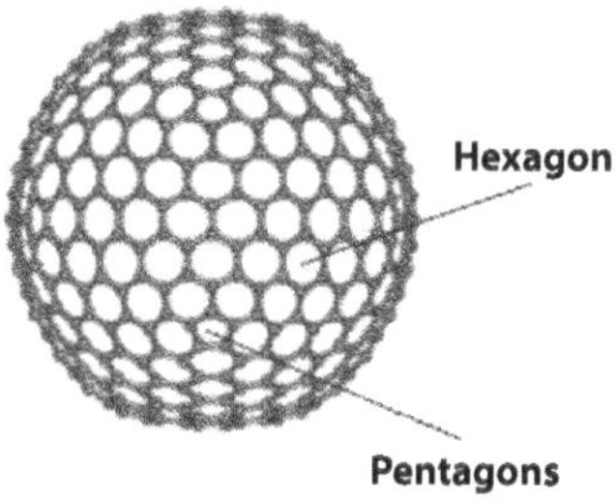

Figure 2.4 Pictorial presentation of fullerites.

e. C_{70} fullerene
C_{70} fullerene is a molecule of fullerene having 70 carbon atoms (Figure 2.5). The combined carbon atoms form a fused ring structure that resembles a rugby ball and is composed of 25 hexagons and 12 pentagons, with a carbon atom located at each polygon's vertex and a bond running down each edge.

f. Amorphous carbon
It is freely found in nature and is reactive carbon that has no crystalline structure (Figure 2.6). These carbon molecules may be stabilized by winding up dangling-π bonds with hydrogen. This kind of carbon is generally abbreviated as general amorphous carbon.

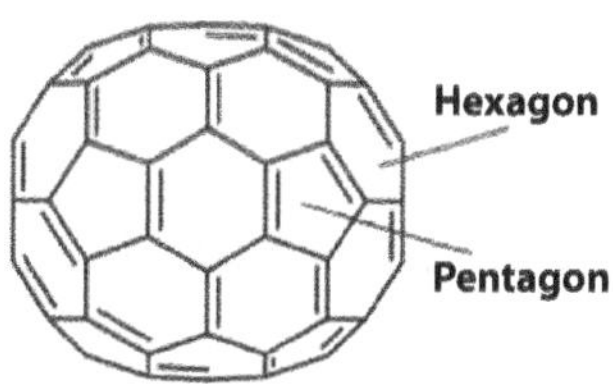

Figure 2.5 Pictorial presentation of fullerene.

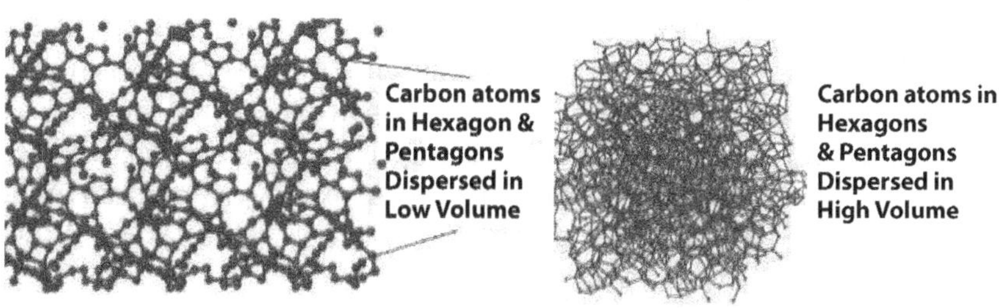

Figure 2.6 Structure of amorphous carbon.

g. Zig-zag single-walled carbon nanotube
These are single-walled carbon nanotubes with (n,m) types of indices which are equal to (n,0) or (0,m). The carbon atoms present in Zigzag carbon nanotubes have a chiral angle of 0° and can be either metallic or semiconducting.

2.3 Basics of Carbon Allotrope Composites and Their Properties

Carbon-carbon allotropic hybrids make up the majority of the carbon allotrope composites. Graphite, graphene, graphene oxide, carbon nanotubes, carbon nanofibers, carbon metal complexes, carbyne chains, graphene quantum dots, and carbon nanodots are typically present in these composites in more complicated forms. In contrast to their counterparts, the majority of these composites feature three-dimensional structures with any link, such as a covalent bond or van der Waals interactions, present between carbon atoms. There are numerous procedures involved in creating carbon composites, which are described here [23–28].

Synthesis processes involving already existing carbon allotropes.

a. Synthesis processes involving the in-situ making of carbon allotropes, such as pyrolysis, redox reactions, ultrasonication, chemical vapor deposition, and solvothermal techniques, in liquid-phase methods, frequently including redox steps.

2.4 Composites of Graphite or Graphite Oxide (GO)

Graphite oxide (GO) formerly also known as graphitic oxide or graphitic acid. The chemical composition includes atoms of carbon, hydrogen, and oxygen, which are present in different ratios, mainly obtained by reacting graphite with various strong oxidizers and acids for resolving extra metals. The highest oxidized bulk material is a yellow mass of carbon/oxygen ratio between 2.1 and 2.9. This allows the formation of the layered structure of graphite. But the structure formed is having an irregular spacing between atoms [29, 30]. The structural model of graphite oxide was proposed in the year 1998 [31] (Figure 2.7).

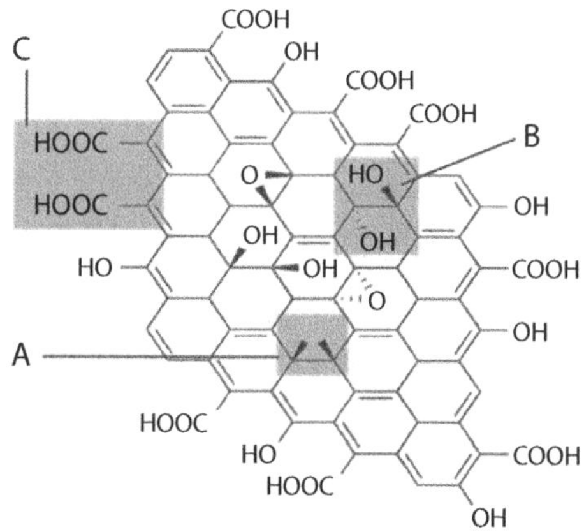

Figure 2.7 Structure of graphite oxide.

Here in the molecular structure of graphite oxide, the presence of functional groups is highlighted as:

A: Epoxy bridges,
B: Hydroxyl groups,
C: Pairwise carboxyl groups.

The composites of graphite are mainly found in association with carbon nanotubes (Graphite-CNT) composite. These can be synthesized by a different route including chemical vapor deposition (CVD). The making of composites with single-walled carbon nanotubes along with graphene and graphite on the bed of nickel foam via CVD, where the reactants were acetylene gas and carbon precursor in equation 1 [32].

$$3H\text{-}C{\equiv}C\text{-}H \rightarrow 6C + 3H_2$$

Another composite made from CNTs along with coated graphite was extracted from the pyrolysis of CNT/polyaniline composites at the temperature of 1500°C. In this formed composite the graphene layers and CNT, both make an angle of 110°C. This causes the alignment of graphene layers, which in turn makes carbon nanotubes high in performance [33].

2.4.1 Applications of Graphite Oxide

The important applications of graphite oxide are given here as follows:

1. These act as an insulator, or possibly a semiconductor, which is having differential conductivity values between 1 and 5×10^{-3} S/cm at a bias voltage of 10 V [34].
2. It helps in the making of graphene, when the graphite oxide disperses quickly in water, by breaking up into macroscopic flakes, in the form of layers. The chemical reduction of these formed flakes yields a suspension of graphene flakes. This was first experimentally obtained by Hanns-Peter Boehm in 1962 [35].
3. It is used in the desalination of water via the reverse osmosis process in the 1960s [36].

2.5 Composites of Graphene

The graphene oxides are the resultants when graphite is chemically exfoliated or oxidized naturally or artificially [37]. The synthetic methods include hydrothermal treatment and chemical vapor deposition (CVD) technique [38–40].

When compared to other 2-dimensional carbon molecules, graphene oxide has a single or multilayered structure that can be divided by functional groups bound to carbon atoms [41]. Much earlier than graphene was discovered, Benjamin Brody wrote a study on graphene oxide in 1859 [42]. According to Brodie, Staudenmaier, and Hoffman, one of the alternative methods for making graphene oxide involves reacting potent acids like sulfuric or nitric with potassium chlorate. While Hummers-Offerman uses KMnO4 and a concentrated combination of H2SO4 and NaNO3. In comparison to earlier methods used for the manufacture of graphene oxide, this process provides a good yield and takes less time [43].

2.5.1 Applications of Graphene Oxide

1. Graphene oxide is primarily used as transparent protective and conductive coatings in the form of films, which are multi-atomic layered thick [44–46].
2. Large application in solar cells in the form of electrodes, which are based on organic materials, leading to the use of

a large area of reduced graphene oxides in transparent electrodes which can become indispensable perovskite-based solar cells [47].

3. It is used in photocatalysis which is one of the intensively emerging technologies applied in particular for environmental protection [48].

2.6 Composite of Graphite-Carbon Nanotube (Gr-CNT)/Polythene or Silicon

Any polymer can be loaded using this kind of composite material. For instance, a composite made of polyethene and a hybrid of graphite and carbon nanotubes (Gr-CNT) [49] (Figure 2.8). Here in this composite, the Gr-CNT disperse themselves in between the interfaces of polyethylene.

The next example comes with the silicon as an external molecule in between Gr-CNT and present in the system as Si/Gr/CNTs composite material.

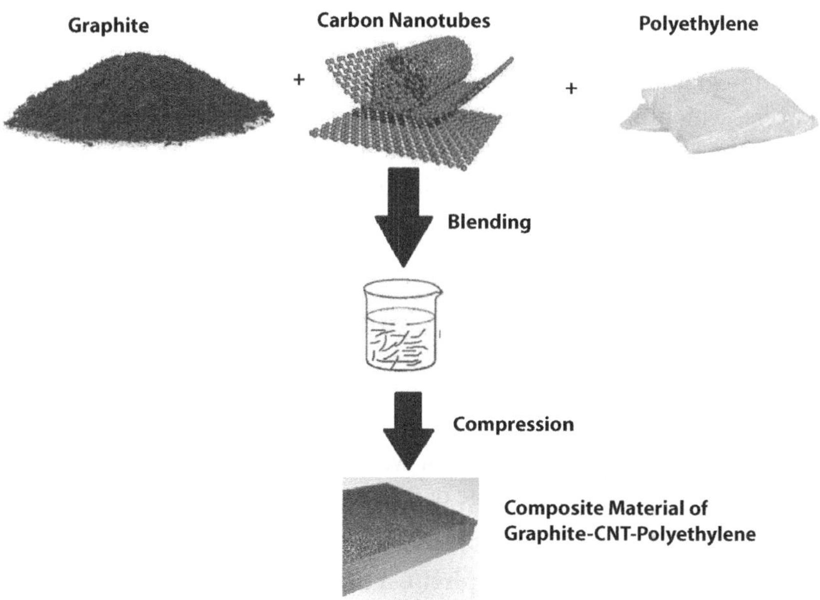

Figure 2.8 The given picture shows the outline of the composite making.

2.6.1 Applications of Graphite-Carbon Nanotube (Gr-CNT)/Polythene or Silicon [50]

1. Gr-'CNT/Polythene system is explored as a highly conductive network that served to develop shielding materials.
2. The increased electrochemical properties of the Si/Gr/CNT composite material (silicon, graphite, and carbon nanotubes) were attributed to the homogeneous dispersion of carbon nanotubes in the interior surface of silicon and graphite phases.

2.7 Graphene (or Graphene Oxide)–Carbon Nanofiber (CNF) Composites

The carbon nanofibers are known to be very minute cylindrical nano-meter scaled nanostructures having different layers of graphene [51].

Although having similarities with CNTs, the CNFs have lesser attention from researchers. In addition, various composites of CNFs were found with graphene/graphene oxide/graphite oxide.

2.7.1 Applications of CNF Composites

It is used in electrochemical device applications, such as batteries [52]. In the impregnation of CNF in a reduced GO network, the CNFs act as nano-spacers, which increases the volumetric electrochemical performance of this supercapacitor to many folds. The next generation of commercially available portable electronics benefit greatly from the flexible paper constructed from graphene-carbon fiber, which acts as a lateral heat spreader.

2.8 Graphene-Fullerene Composites

The composite made from graphene and fullerene molecules is called Graphene Nanobuds [53]. They have a high tensile strength of value around 50 GPa, whereas the elastic moduli were observed to become a little less but remain high too in the range of 0.43 to 0.77 TPa [54].

2.8.1 Applications of Graphene-Fullerene Composites

It is best used in supercapacitors as having specific capacitance. So these electrodes applied in supercapacitors are having chemical inertness. Some researchers found great structural featured stability in graphene/fullerene composites of C_{60} with distinct sizes.

2.9 Conclusion

In this chapter, some common forms of carbon composites have been reviewed. The literature survey revealed the fact that they have been synthesized by high-temperature techniques, such as Chemical Vapor Deposition or pyrolysis or carbonization. This paper goes into great length about some of the characteristics of carbon nanoparticles, nanolayers, and other nanostructures, as well as their composites in polymeric and other materials. These carbon nanostructure-based composite materials have numerous uses in engineering, electronics, and sensors. Recent advancements in the realm of carbon materials, such as quenched carbon and activated graphene, indicate that more discoveries with a wide range of potential uses are still feasible.

References

1. Hirsch, A., The era of carbon allotropes. *Nat. Mater.*, 9, 868–871, 2010.
2. Skoda, M., Dudek, I., Jarosz, A., Szukiewicz, D., Graphene: One material, many possibilities-application difficulties in biological systems. *J. Nanomater.*, 2014, 190, 2014.
3. Harris, P.J.F., New Perspectives on the Structure of Graphitic Carbons. *Crit. Rev. Solid*, 30, 235–253, 2005.
4. Torres, L.E.F., Roche, S., Charlier, J.-C., Introduction to carbon-based nanostructures, 2nd ed., pp. 1–10, Cambridge University Press, Cambridge, UK, 2020.
5. Khalaj, Z., Monajjemi, M., Diudea, M.V., Main allotropes of carbon: A brief review, in: *Sustainable Nanosystems Development, Properties, and Applications*, M.V. Putz, and M.C. Mirica, (Eds.), pp. 185–213, IGI Global, Hershey, PA, USA, 2017.
6. Slepicka, P., Slepickova Kasalkova, N., Siegel, J., Kolska, Z., Bacakova, L., Svorcik, V., Nano-structured and functionalized surfaces for cytocompatibility improvement and bactericidal action. *Biotechnol. Adv.*, 33, 1120–1129, 2015.
7. Torres, L.E.F., Roche, S., Charlier, J.-C., Introduction to carbon-based nanostructures, 2nd ed., pp. 1–1, Cambridge University Press, Cambridge, UK, 2020.
8. Delhaes, P., Polymorphism of carbon, in: *Graphite and Precursors*, P. Delhaes, (Ed.), pp. 1–24, Gordon & Breach, Philadelphia, PA, US, 2000.

9. Pierson, H.O., *Handbook of Carbon, Graphite, Diamond, and Fullerenes: Properties, Processing, and Applications*, pp. 40–41, Noyes Publications, Saddle River, NJ, US, 2012.
10. Delhaes, P., Polymorphism of carbon, in: *Graphite and Precursors*, P. Delhaes, (Ed.), pp. 1–24, Gordon & Breach, Philadelphia, PA, US, 2000.
11. Pierson, H.O., *Handbook of Carbon, Graphite, Diamond, and Fullerenes: Properties, Processing, and Applications*, pp. 40–41, Noyes Publications, Saddle River, NJ, US, 2012.
12. Delhaes, P., *Graphite and Precursors*, CRC Press, Boca Raton, FL, USA, 2001.
13. Chung, D.D.L., Review graphite. *J. Mater. Sci.*, 37, 8, 1475–1489, 2002.
14. Dubinchuk, V.T., Simakov, S.K., Pechnikov, V.A., Lonsdaleite in diamond-bearing metamorphic rocks of the Kokchetav Massif. *Dokl. Earth Sci.*, 430, 40, 2010.
15. Bundy, F.P. and Kasper, J.S., Hexagonal diamond—A new form of carbon. *J. Chem. Phys.*, 46, 3437, 1967.
16. Chen, H., Zhang, W.Y., Wang, Z.L., Comparative studies on photonic band structures of diamond and hexagonal diamond using the multiple scattering method. *J. Phys. Condes. Matter*, 16, 741, 2004.
17. Pan, Z.C. *et al.*, Harder than diamond: Superior indentation strength of wurtzite BN and lonsdaleite. *Phys. Rev. Lett.*, 102, 055503, 2009.
18. He, H.K., Sekine, T., Kobayashi, T., Direct transformation of cubic diamond to hexagonal diamond. *Appl. Phys. Lett.*, 81, 610, 2002.
19. Monkhorst, H.J. and Pack, J.D., Special points for Brillouin-zone integrations. *Phys. Rev. B*, 13, 5188, 1976.
20. Hebbache, M., First-principles calculations of the bulk modulus of diamond. *Solid State Commun.*, 110, 559, 1999.
21. Wang, S.Q. and Ye, H.Q., *Ab initio* elastic constants for the lonsdaleite phases of C, Si and Ge. *J. Phys. Condes. Matter*, 15, L197, 2003.
22. Yu, M.F. *et al.*, Strength and breaking mechanism of multiwalled carbon nanotubes under tensile load. *Science*, 287, 637, 2000.
23. Hahm, M.G., Reddy, A.L.M., Cole, D.P., Ajayan, P.M., Vajtai, P.M., Carbon nanotube-nanocup hybrid structures for high power supercapacitor applications. *Nano Lett.*, 12, 5616, 2012.
24. Zhu, Z., Fthenakis, Z.G., Tománek, D., Electronic structure and transport in graphene/haeckelite hybrids: An *ab initio* study. *2D Mater.*, 2, 035001, 2015.
25. Hahm, M.G., Lee, J.H., Hart, A.H.C., Song, S.M., Nam, J., Jung, H.Y., Ajayan, P.M., Carbon nanotube core graphitic shell hybrid fibers. *ACS Nano*, 7, 10971, 2013.
26. Lv, R., Cruz-Silva, E., Terrones, M., Building complex hybrid carbon architectures by covalent interconnections: Graphene–nanotube hybrids and more. *ACS Nano*, 8, 4061, 2014.
27. Gan, X., Lv, R., Bai, J., Zhang, Z., Wei, J., Huang, Z.-H. *et al.*, Efficient photovoltaic conversion of graphene–carbon nanotube hybrid films grown from solid precursors. *2D Mater.*, 2, 034003, 2015.

28. Zhang, Y., Shang, J., Fu, W., Zeng, L., Tang, T., Cai, Y., A sp^2+sp^3 hybridized carbon allotrope transformed from AB stacking graphyne and THD-graphene. *AIP Adv.*, 8, 015028, 2018.
29. Hummers, W.S. and Offeman, R.E., Preparation of graphitic oxide. *J. Am. Chem. Soc.*, 80, 6, 1339, 1958.
30. Sadri, R., Experimental study on thermo-physical and rheological properties of stable and green reduced graphene oxide nanofluids: Hydrothermal assisted technique. *J. Dispers. Sci. Technol.*, 38, 9, 1302–1310, 2017.
31. He, H., Klinowski, J., Forster, M., Lerf, A., A new structural model for graphite oxide. *Chem. Phys. Lett.*, 287, 1, 53, 1998.
32. Chen, H.-C., Wang, Y.-C., Huang, K.-T., Selective synthesis of carbon-nanotubes/graphite or carbon-nanotubes/multi-graphene composites on 3-D nickel foam prepared with different nickel catalysts and pre-treatment. *Synth. Met.*, 219, 124, 2016.
33. Hoon Nam, D., Cha, S., Jin Jeong, Y., Hyung Hong, S., Enhanced graphitization of carbon around carbon nanotubes during the formation of carbon nanotube/graphite composites by pyrolysis of carbon nanotube/polyaniline composites. *J. Nanosci. Nanotechnol.*, 13, 7365, 2013.
34. Gómez-Navarro, C., Weitz, R.T., Bittner, A.M., Scolari, M., Mews, A., Burghard, M., Kern, K., Electronic transport properties of individual chemically reduced graphene oxide sheets. *Nano Lett.*, 7, 11, 3499–3503, 2007.
35. Sprinkle, M., *Boehm's 1961 Isolation of Graphene*, Graphene Times. London, UK, Archived from the original on 2010-10-08, Dec. 07, 2009.
36. Bober, E.S., *Final Report on Reverse Osmosis Membranes Containing Graphitic Oxide*, p. 116, U.S. Dept. of the Interior, Washington, DC, 1970.
37. Park, S. and Ruoff, R.S., Chemical methods for the production of graphenes. *Nat. Nanotechnol.*, 5, 217–24, 2010.
38. Chen, D., Feng, H., Li, J., Graphene oxide: Preparation, functionalization, and electrochemical applications. *Chem. Rev.*, 112, 6027–53, 2012.
39. Tang, L., Li, X., Ji, X., Teng, K.S., Tai, G., Ye, J. *et al.*, Bottom-up synthesis of large-scale graphene oxide nanosheets. *J. Mater. Chem.*, 22, 5676–83, 2012.
40. Liu, Y. and Chen, Y., Synthesis of large scale graphene oxide using plasma enhanced chemical vapor deposition method and its application in humidity sensing. *J. Appl. Phys.*, 119, 103301, 2016.
41. Bianco, A., Cheng, H.-M., Enoki, T., Gogotsi, Y., Hurt, R.H., Koratkar, N. *et al.*, All in the graphene family-a recommended nomenclature for two-dimensional carbon materials. *Carbon*, 65, 1–6, 2013.
42. Geim, A.K., Graphene prehistory. *Phys. Scr.*, T146, 014003, 2012.
43. Chen, W., Yan, L., Bangal, P.R., Preparation of graphene by the rapid and mild thermal reduction of graphene oxide induced by microwaves. *Carbon*, 48, 1146–52, 2010.
44. Novoselov, K.S., Falko, V.I., Colombo, L., Gellert, P.R., Schwab, M.G., Kim, K., A roadmap for graphene. *Nature*, 490, 192–200, 2012.

45. Wu, S.X., Yin, Z.Y., He, Q.Y., Huang, X., Zhou, X.Z., Zhang, H., Electrochemical deposition of semiconductor oxides on reduced graphene oxide-based flexible, transparent and conductive electrodes. *J. Phys. Chem. C*, 114, 11816, 2010.
46. Eda, G., Fanchini, G., Chhowalla, M., Large-area ultrathin films of reduced graphene oxide as a transparent and flexible electronic material. *Nat. Nanotechnol.*, 3, 270–4, 2008.
47. Assadi, M.K., Bakhoda, S., Saidur, R., Hanaei, H., Recent progress in perovskite solar cells. *Renew. Sust. Energ. Rev.*, 81, 2812–22, 2018.
48. Ong, C.B., Ng, L.Y., Mohammad, A.W., A review of ZnO nanoparticles as solar photocatalysts: Synthesis, mechanisms, and applications. *Renew. Sust. Energ. Rev.*, 81, 536–51, 2018.
49. Jia, L.C., Yan, D.X., Jiang, X. *et al.*, Synergistic effect of graphite and carbon nanotubes on improved electromagnetic interference shielding performance in segregated composites. *Ind. Eng. Chem. Res.*, 57, 11929, 2018.
50. Li, X., Zhang, G., Zhang, L., Zhong, M., Yuan., X., Silicon/graphite/carbon nanotubes composite as anode for lithium ion battery. *Int. J. Electrochem. Sci.*, 10, 2802, 2015.
51. Radushkevich, L.V. and Lukyanovich, B.M., About the structure of carbon formed by the thermal decomposition of carbon oxide on the iron contact. *J. Phys. Chem. (Moscow)*, 26, 88, 1952.
52. Graphene-coated carbon nanofiber/sulphur composite material, preparation and application thereof, CN104269538A, 2014.
53. Zheng, Y., Xu, L., Fan, Z. *et al.*, Mechanical properties of graphene nanobuds: A molecular dynamics study. *Curr. Nanosci.*, 8, 1, 89, 89–96, 2012.
54. Ma, J., Guo, Q., Gao, H.L., Qin, X., Synthesis of C60/graphene composite as electrode in supercapacitors. *Fuller. Nanotub. Carbon Nanostructures*.

3

Activation of Carbon Allotropes Through Covalent and Noncovalent Functionalization: Attempts in Modifying Properties for Enhanced Performance

Richika Ganjoo[1], Shveta Sharma[1] and Ashish Kumar[1,2*]

[1]Department of Chemistry, School of Chemical Engineering and Physical Sciences, Punjab, India
[2]NCE, Bihar Engineering University, Department of Science and Technology, Government of Bihar, Bihar, India

Abstract

Since a few hundred years ago, the element carbon has been employed as a source of energy; in the current technological period, carbon has assumed a substantial and prominent role in practically all scientific and technological sectors. We should thus be aware of its varied forms of life as a way to honour this wonderful element. Because of its valency, carbon has the fascinating property of being able to create several allotropes. Graphene, fullerenes, and carbon nanotubes (CNTs) are only a few of the carbon allotropes and morphologies that have been created in recent years (GR). In addition to discussing current research in these areas, this chapter will also include a few additional intriguing new applications. We will concentrate primarily on the characteristics and the methods for synthesizing these carbon nanomaterials to customize them for particular uses, notably in the fields of biomedicine, waste management, pollutant removal, anti-corrosion, and other related fields.

***Keywords*:** Carbon allotropes, fullerene, graphene, applications, carbon nanotubes

**Corresponding author*: drashishchemlpu@gmail.com

Chandrabhan Verma and Chaudhery Mustansar Hussain (eds.) Carbon Allotropes and Composites: Materials for Environment Protection and Remediation, (31–50) © 2023 Scrivener Publishing LLC

3.1 Introduction

3.1.1 Carbon Allotropes: Fundamentals and Properties

An element with the chemical symbol C and atomic number 6, carbon (from the Latin carbo, "coal") is a mineral. It is tetravalent, meaning that four of its atom's electrons are accessible to create covalent chemical connections. It is also nonmetallic. In the periodic table, it belongs to group 14 [1–4]. In the crust of the Earth, carbon is the 17th most prevalent element. It is the sixth most frequent element in the universe, the fourth most frequent in the solar system after hydrogen, helium, and oxygen, and the seventeenth most frequent element in the crust of the Earth. Carbon is a unique and necessary element on our planet. Because of its abundance, capacity to form organic compounds and extraordinary capability to form polymeric compounds, carbon may be considered a common component of all known life due to its abundance, and it forms around 18.5% of the body mass. There are several ways in which the carbon atoms may bond, giving rise to different allotropes of carbon, most common examples of allotropes of carbon are amorphous carbon, diamond, graphite, etc. [5, 6]. Different allotropic forms of carbon have different physical characteristics. For instance, whereas diamond is a very clear and hardest known mineral, while graphite is opaque, black, and soft and may produce a spot on paper. Even the electrical conductivity varies in the case of diamond and graphite. The materials with the greatest known thermal conductivities under normal circumstances include diamond, carbon nanotubes, and graphene. At normal temperatures and pressures, all carbon allotropes are solids, and out of them, graphite is the most thermodynamically stable form [7]. Carbon allotropes require high temperatures to react with different chemicals. In inorganic compounds, the most prevalent oxidation state for carbon is +4, whereas transition metal carbonyl complexes and carbon monoxide are found in +2. Significant amounts, however, may be discovered in biological deposits of coal and methane clathrates etc. The main sources of inorganic carbon include limestones, dolomites, and carbon dioxide [8]. More than any other element, carbon can be found in approximately ten million different compounds; however, this figure only represents a small portion of the molecules that are theoretically feasible under normal circumstances. Because of this, carbon is often considered the "king of the elements."

Because atomic carbon has a relatively limited life span, it is stabilized in a variety of multiatomic forms known as allotropes that have different molecular arrangements. Amorphous carbon, graphite, and diamond are the most common allotropes of carbon. Fullerenes, which include carbon nanotubes, buckyballs, carbon nanobuds, and nanofibers, were formerly regarded as exotic materials, but they are now frequently manufactured and employed in research (Figure 3.1). There have also been findings of several other rare allotropes, including glassy carbon, carbon nanofoam, etc. The two naturally occurring crystalline allotropic forms of carbon known as graphite and diamond have been known to exist for a very long time. But more significantly, the crystal structures and characteristics of these two compounds' chemistry are different. According to chemistry, the valence of carbon atoms may lead to the formation of several more allotropes [9]. When one carbon atom joins with another carbon atom, covalent bonds are created. Allotropes are chemically similar elements with radically different physical characteristics to make this phenomenon easier to explain. As a result, an allotrope is created if the atoms of a material that solely contains one kind of atom are arranged differently. Because of this, several other carbon allotropes and forms, including graphene, buckminsterfullerene, carbon nanotubes, etc., were found, making carbon the material with the greatest number of allotropes that have been detected.

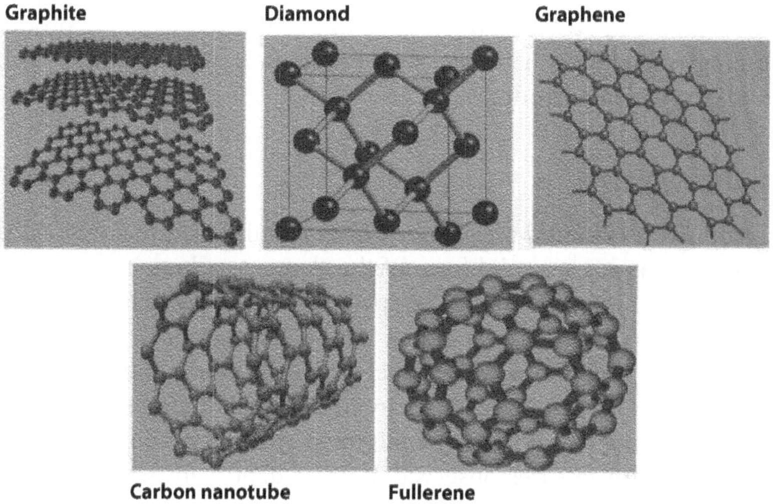

Figure 3.1 Structures of several 0-, 1-, 2-, and 3-dimensional carbon nanomaterials showing sp2 and sp3 hybridization allotropes in various crystallographic forms [10]. Reproduced from Ref. [10], [https://doi.org/10.3390/ma8063068], under the terms of the CC BY 4.0 license.

3.1.1.1 *Graphite*

The crystalline form of the element carbon is called graphite. Graphene is arranged in layered layers throughout. The most stable form of carbon under normal circumstances is graphite, which occurs naturally. Due to in-plane metallic bonding, which makes it an excellent electrical and thermal conductor inside the layers and a poor conductor perpendicular to the layers, graphite is anisotropic (because the van der Waals forces between the layers are weak) [11]. Within each layer of graphite's layer structure, the atoms are organized in a hexagonal arrangement, and the layers are layered in the AB pattern. As a consequence, the hexagonal unit cell has the following measurements: c = 6.71 A° and a = 2.46 A°. Carbon sheets with trigonal planar structures make up graphite. Graphene is the name of the individual layers. Each layer's carbon atoms are organized in a honeycomb lattice with 0.142-nm-long bonds and 0.335-nm-wide gaps between the planes. The comparatively weak van der Waals connections and frequent gas vacancies in the layer bonds enable the graphene-like sheets to glide past one another and be readily separated [12]. Alpha (hexagonal) and beta(rhombohedral) are the two names for graphite. They have a lot of comparable qualities. They vary in how the graphene layers are stacked: ABA stacks in alpha graphite whereas ABC stacks in energetically less stable and less prevalent beta graphite. When heated to a temperature over 1300 °C, the beta form may be mechanically changed from the alpha form to the beta form, and vice versa. Graphite is utilized in many products, including refractories, batteries, steel production, brake linings, pencils, headphones, and electrodes.

3.1.1.2 *Diamond*

The atoms of a diamond are grouped in a crystal structure known as diamond cubic, and it is a solid form of the element carbon. The chemically stable form of carbon at ambient temperature and pressure is another solid form of carbon known as graphite, although diamond transforms into it gradually [13, 14]. Due to its superior hardness and thermal conductivity, diamond is employed in many important industrial applications, including cutting and polishing equipment. They also allow materials to be subjected to pressures observed at great depths in the Earth in diamond anvil cells. Few forms of impurities may contaminate the diamond because of how rigidly the atoms are arranged in it, but boron and nitrogen are two exceptional cases. About one impurity or flaw in a million lattice atoms gives the diamond its many colors, including blue, yellow, brown, and green from boron, nitrogen, defects and radiation respectively. Additionally, the optical dispersion and refractive index of diamond are also quite high. A solid form of pure carbon with its atoms organized in a crystal is called a diamond. Depending on the sort of chemical connection, solid carbon may take on several shapes known as allotropes [15]. Diamond and graphite are the two most prevalent allotropes of unadulterated carbon. In graphite, the atoms form planes and are bonded to their three closest neighbors at a distance of 120 degrees. The bonds are sp^2 orbital hybrids. In a diamond, the atoms by bonding with neighboring four atoms form tetrahedra. The bonds are strong, and the diamond contains the most atoms per unit volume of any known material, making it the hardest and least compressible.

3.1.1.3 Graphene

With its exceptional optical, electrical, thermal, chemical, and mechanical capabilities, graphene is a two-dimensional material that is one atom's thickness thick. It is a desirable material for use in many different applications, such as composite materials, energy-related systems, sensors, actuators, and electronics. Graphene must regularly undergo chemical modification to meet the criteria in actual applications. First, pure graphene decomposes before melting, is insoluble, and is difficult to work with. As a result, it cannot be shaped into desired forms using standard material processing methods [16]. The physical stabilization of graphene layers is secondly limited to solid supports. Free-standing graphene layers often develop creases or pile together due to π bonding and hydrophobic interactions. Thirdly, graphene has a zero bandgap. To use graphene in electronics or optoelectronics, the bandgap of the material must be opened. Fourth, clean graphene often has weak interactions with other small molecules or polymers and low catalytic activity, which restricts its use in composites, sensors, and catalysis. The surfaces and electrical structures of graphene sheets have been altered using a variety of chemical techniques to solve these problems [17]. The structure and characteristics of graphene may be changed effectively by chemical functionalization. Graphene sheets may be uniformly distributed in aqueous and/or organic liquids by attaching functions to certain surfaces of the materials. The switch of carbon hybridization from sp^2 to sp^3 caused by the covalent attachment of chemical groups, on the other hand, may have opened a tunable bandgap in graphene. The optical, chemical and mechanical characteristics of graphene materials may also be modified via chemical functionalization.

3.1.1.4 Activated Carbon

Activated carbon, sometimes known as activated charcoal, is a kind of carbon that is often used for a variety of purposes, including the filtration of impurities from water and air. It is treated (activated) to have tiny, low-volume pores, increasing the surface area accessible for chemical reactions or adsorption (which is different from absorption) [18, 19]. The process of activation is comparable to creating popcorn from dried corn kernels since the finished product is lighter, fluffier, and has a bigger surface area than the kernels do. Sometimes, active is used instead of activated. Because of a large number of pores surface area of activated charcoal is very high and about one gram of it has a surface area of more than 3,000 m^2 (32,000 sq ft). Before activation, the specific surface area of charcoal is between 2.0 and 5.0 m^2/g. This large surface area is responsible for an activation level high enough for meaningful use. Adsorption can be further enhanced by chemical treatment. Waste from paper mills has been investigated as a source of activated carbon, which is generally made from waste materials like coconut husks [20]. To be "activated," these bulk sources are first turned into charcoal. Activated coal is what it's called when it's made from coal.

3.1.1.5 Carbon Nanotubes and Fullerene

A carbon tube with a diameter commonly measured in nanometers is known as a carbon nanotube (CNT). Single-wall carbon nanotubes have dimensions of the order of nanometer and are situated in between fullerene cages and flat graphene [21, 22]. Single-wall carbon nanotubes may be envisioned as cuttings from a two-dimensional hexagonal lattice of carbon atoms that have been rolled up along one of the Bravais lattice vectors to create a hollow cylinder, even though they are not manufactured in this manner. A helical lattice of flawlessly connected carbon atoms is produced on the cylinder surface in this construction by the imposition of periodic boundary constraints throughout the length of this roll-up vector. Single-wall carbon nanotubes are nestled together to form multi-wall carbon nanotubes (MWCNTs), which have a ring-like shape and are only loosely connected by van der Waals interactions [23, 24]. These tubes resemble the long straight and parallel carbon layers cylindrically stacked around a hollow tube proposed by Oberlin, Endo, and Koyama, though not identically. The phrase "carbon nanotubes" are the tubes with carbon in the structure with sizes under 100 nm. Radushkevich and Lukyanovich discovered these tubes in 1952. Although it is often significantly bigger than its diameter, the length created by standard manufacturing techniques is seldom recorded.

Thus, end effects are disregarded and the length of carbon nanotubes is considered to be unlimited for many uses. Some carbon nanotubes are semiconductors, while others may display astounding electrical conductivity. The strong bonds between the carbon atoms in their nanostructure and their extraordinary tensile strength and thermal conductivity give them these qualities as well. Additionally, they are subject to chemical modification. Because of these properties, CNTs are considered to be beneficial to a wide variety of technological fields and other applications of materials science. Fullerenes are molecular allotropes of carbon that display a variety of fascinating occurrences because of their nature as -electron molecules that are simple to work with chemically. The fullerene family is a versatile building block of materials with important applications in physics, chemistry, and biology due to the unusually high degree of curvature in their conjugated electron systems. We provide a short overview of the family of fullerenes, describe its most prevalent and extensively investigated member, C_{60}, in its molecular and crystal structure, and make brief mentions of its derivatives, heterofullerenes, and fullerene polymers [25–27]. Additionally, the current hot spots for applications are emphasized. A novel family of carbon-only compounds known as fullerenes (formerly buckminsterfullerenes) includes 60 carbon atoms (C_{60}) organized in a soccer ball configuration. In 1985, fullerenes' first instance was found. Since their discovery, fullerenes, a huge class of carbon allotropes, have garnered a great deal of interest across several scientific disciplines. Carbon atoms make up fullerene molecules, which may take the form of a hollow sphere, ellipsoid, or tube. Buckyballs are another name for spherical fullerenes. They are carbon clusters, with any number of hexagons and 12 pentagons making up their surface.

3.1.2 Functionalization of Carbon Allotropes: Synthesis and Characterization

The fundamental problem with the vast number of synthetic methods that are widely used is that they produce samples with a wide range of diameters and chirality of nanotubes, and these samples are frequently corrupted with impurities and these may of metallic nature or amorphous [28, 29]. To enhance the beneficial electronic and mechanical properties of these materials, various methods for the purification and separation of tubes are the method to get desired tubes. The efficiency of CNTs has been severely hampered by the weak van der Waals interaction at between CNTs and the polymer framework. Because CNTs have a high aspect ratio (>1000), tiny diameter (nanometer scale), and huge surface area, they have a dispersion issue that is somewhat different from that of other typical fillers, like spherical particles and carbon fibres. Additionally, the offered marketed CNTs come in the shape of intricate bundles, which makes dispersion challenging by nature.

3.1.2.1 Covalent Functionalization of Carbon Allotropes: Synthesis and Characterization

Comparatively to the side walls, the end caps of nanotubes often consist of hemispheres that resemble fullerenes and are extremely curved. Defect sites including pentagon-heptagon pairings known as Stone-Walls defects, sp^3-hybridized defects, and vacancies in the nanotube lattice may be found on the sidewalls themselves. The carbon form of CNTs' covalent connection with functional groups serves as the foundation for chemical functionalization [30]. It may be carried out at the various defects-filled sidewalls of nanotubes or their end caps. Direct covalent sidewall functionalization is linked to a switch from sp^2 to sp^3 hybridization and a concurrent loss of the p-conjugation system on the graphene layer. By reacting with a few molecules that have a high chemical reactivity, this process may be created. Because the sidewalls of CNTs are anticipated to remain inert, fluorination of CNTs has grown in popularity for early exploration of covalent functionalization. SWCNT that has been covalently functionalized with polytyrosine were synthesized and characterized by Egulaz *et al.* They also studied the given experimental conditions critically to know about the effective dispersion of the modified carbon nanotubes and performance of modified glassy carbon electrodes modified for the total polyphenolic content of tea extracts was successfully determined using the suggested sensor [3]. By combining the phytochemicals curcumin (CUR), glycyrrhizin (GLY), and rutin (RUT), Ohadi *et al.* devised a quick and effective approach for noncovalent functionalization of CNTs (RUT). Different techniques were used, and the resulting nanoparticles were assessed using a zeta analyzer, EDX, TEM, and size analysis. The acquired findings demonstrated that noncovalent functionalization with CUR, GLY, or RUT produced stable CNT suspension in aqueous media. CNTs have been surface functionalized has been proven with the increased value of size and zeta potentials. According to cytotoxicity studies, the IC 50 value of CNTs may be increased by functionalizing them with these phytochemicals, going from 301 g/mL to a high of 4088 g/mL in RUT functionalized CNTs. Overall, the findings demonstrated that functionalizing CNTs using this practical technique is possible and that they may be used to create an intriguing, less hazardous drug delivery system [31]. By mixing modified MWNTs with specific polymer, composite film, conductive in nature with outstanding mechanical strength was created. The homogeneous dispersion of MWNTs in the polymer as a result of their surface functionalization, which was accomplished using ureidopyrimidinone (upy) and pyrene-upy through covalent and noncovalent techniques, respectively, considerably enhanced the electrical and mechanical characteristics of its nanocomposite [32]. By adding chemical functional groups to the sidewalls of CNTs, it is possible to covalently functionalize them, creating, for example, carboxylated CNTs etc. There may be an interaction between various functional groups present on CNTs and on the other part of the composite to produce a covalent link that helps the biomolecule adhere to (be immobilized on) the surface of the CNT. For instance, carboxylated CNTs may interact directly with the primary amines in biomolecules by utilizing a carbodiimide chemical to activate the carboxyl groups on the CNTs. N-ethyl-N'-(3-dimethylaminopropyl) carbodiimide hydrochloride is the most often used carbodiimide that is water soluble (EDC). A biomolecule's principal amine may quickly replace the intermediate o-acylisourea ester created by EDC's reaction with carboxyl groups. Figure 3.2 depicts the crosslinking chemistry strategy for EDCs. As a result, the main amine and carboxyl groups create an amide connection. For instance, Zhang *et al.* described immobilising horseradish peroxidase utilising EDC and poly-L-lysine as cross-linkers [33]. To speed up the coupling process, N-hydroxysuccinimide (NHS) or sulfo-NHS is often used. Since o-acylisourea ester is unstable in an aqueous solution, the NHS ester produced by the EDC-NHS coupling technique is more stable than the intermediate.

3.1.2.2 Noncovalent Functionalization of Carbon Allotropes: Synthesis and Characterization

Noncovalent functionalization has the advantage of not altering the final structural characteristics of the CNT walls. An alternate technique for adjusting the interfacial characteristics of nanotubes is noncovalent functionalization. Noncovalently, functionalization is performed by aromatic chemicals, surfactants, and polymers, mostly by π-π stacking or hydrophobic interactions. In these methods, CNTs' noncovalent alterations may significantly improve their solubilities while preserving their desirable features.

The sidewalls of CNTs may and do interact with aromatic molecules through stacking interactions, such as pyrene, porphyrin, and its derivatives, resulting in the noncovalent functionalization of CNTs. According

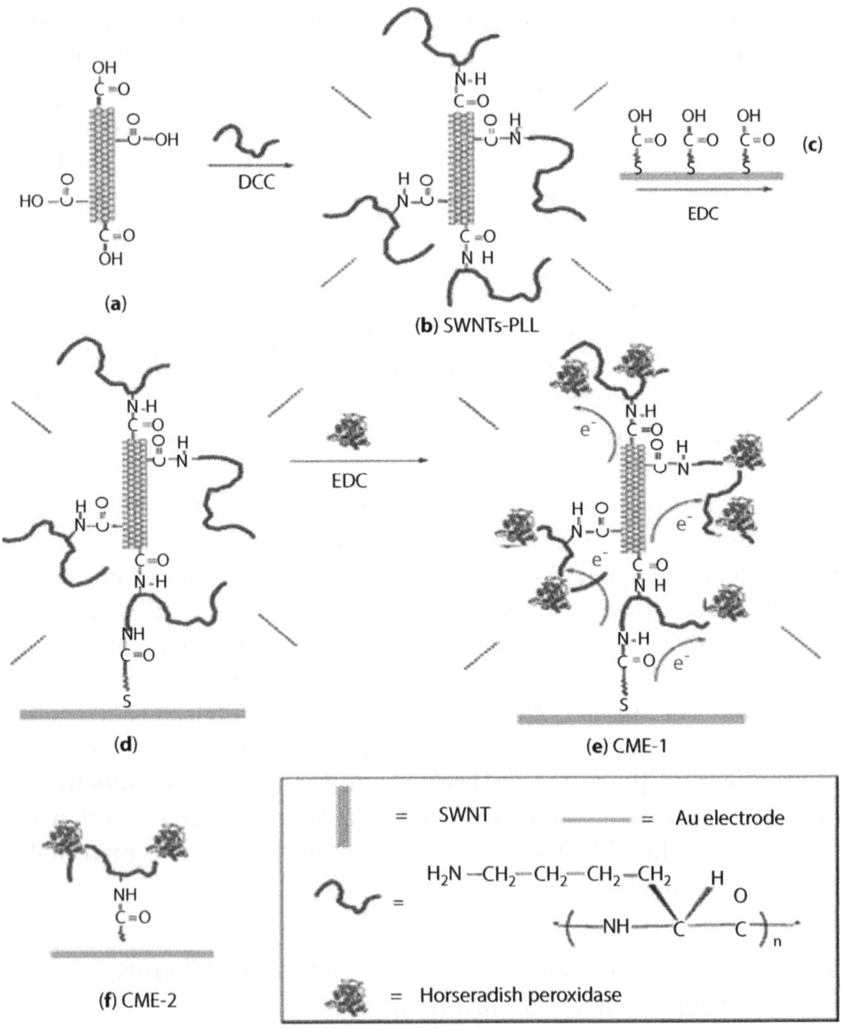

Figure 3.2 Covalent crosslinking of a biomolecule with a carbon nanotube using the EDC and EDC-NHS methods [33]. Reproduced from Ref. [33] with permission, Copyright@ The American Chemical Society, 2004.

to Hong S. C., the polymer wrapping procedure was effective in functionalizing MWNTs with water-soluble poly(2-ethyl-2-oxazoline) (PEOX). Functionalized MWNTs (DMF) showed excellent dispersion in various organic solvents along with water [34].

CNTs are often dissolved using soluble polymers or amphiphilic polymers. By lowering the entropic cost of micelle generation, polymers have several advantages over tiny molecule surfactants. Additionally, certain conjugated polymers interact with nanotubes at substantially greater energy than tiny molecules do. The fundamental issue with polymers is that they may be forced into energetically unfavorable conformations by interactions with mechanically hard SWNTs. It has been proposed that certain polymers may round nanotubes in a helical pattern to reduce strain in various conformations [35]. By noncovalently joining SWNTs to a variety of polymers, which are linear in nature, SWNTs were reversibly solubilized in water in the g/L concentration range, according to Smalley *et al.* (PSS). It is possible to imagine PVP, a polymer with a hydrophobic alkyl backbone and hydrophilic pendant groups, coiling around the nanotube such that its backbone is in touch with the nanotube surface and its pyrrolidone groups are exposed to water [35]. To improve CNTs' solubility and processability, tiny chemicals and polymers are often used to covalently and noncovalently modify their surfaces. Polymer grafting has the advantage of increasing the affinity of the CNTs-g-polymer in polymer matrices and based on the natural state of the grafted polymer, even at slight functionalization, making CNTs soluble in a variety of solvents [36]. The optimal composition of 0.5 wt. % MWNTs and 2.0 wt. % SDS about water was used by Gao *et al.* to achieve homogenous dispersion of MWNTs. CNT dispersion is improved by the steric repulsion and greater negative surface charge brought on by the interaction between CNTs and SDS via the hydrophobic region [37].

Additionally, CNTs have been functionalized using surfactant polymers. Surfactant was physically adsorbed onto the surface of CNTs, which reduced their surface tension and essentially stopped them from aggregating. This method's effectiveness was highly reliant on the characteristics of the surfactants, medium chemistry, and polymer matrix. The effectiveness of surfactant interactions with SWNTs is also strongly influenced by the length and form of the alkyl chain: alkyl groups that are longer and more branched are preferred over those that are linear and straight, correspondingly [38, 39].

Due to their tiny size, cylindrical form, high surface-to-volume ratio, high conductance value, and favorable absorbency in living beings, carbon nanotubes have received a great deal of attention and have been employed for the creation of electrochemical biosensors. CNTs are used in electrochemical biosensors for a dual function: they assist the freezing the movement of biomolecules and offer electrical conductivity. This biofunctionalization (used to join biomolecules to CNTs) is a key factor in determining a recognition molecule's capacity to detect the analyte. Physical adsorption, covalent cross-linking, polymer encapsulation, and other biofunctionalization techniques have all been investigated and published. The benefits and drawbacks of each approach are different. The creation of electrochemical biosensors uses the noncovalent functionalization of carbon nanotubes with a range of biomolecules, which is covered in detail in this article. Developing bioelectrodes with enzymes for biosensor and biofuel cell applications is adopting this approach of immobilization more often.

Pyrene has been shown to have a very strong affinity for the surface of nanotubes. It belongs to a group with a complex aromatic system. The efficiency of the pyrene moiety in noncovalently functionalizing CNTs is well acknowledged [40].

3.2 Applications of Functionalized Carbon Allotropes

Carbon allotropes are innovative and interesting nanomaterials for a variety of applications because of their distinctive physical, chemical, thermal, and mechanical characteristics [41]. However, the insolubility of these prospective uses in solvents owing to intense intertube van der Waals interactions often limits their possible use. Solubilization of CNTs requires strategic methods, which is why CNT applications depend on it.

3.2.1 Biomedical

Active carbon (AC), carbon nanotubes (CNT), graphene, etc., are all examples of carbon allotropes that have found applications in biomedicine thanks to their excellent thermal conductivity, structural rigidity, amenability to modification and functionalization, high surface-to-volume-ratio, and high biocompatibility. Nanomaterials for carrying drugs have been studied recently to enhance medication delivery methods for cancer therapy. However, there is a persistent need for more useful therapeutic materials that provide quick clearance, a strong ability to reduce systemic toxicity through the targeted tumor, and improved drug solubility. Bioimaging, targeting, chemical sensing, therapies, delivery, catalysis, and energy harvesting have all been used for carbon nanomaterials in the biomedical area. To take advantage of the desired attribute of the nanoparticles in each application, a customized surface functionalization is necessary [42]. Their potential for usage, particularly in biomedical applications, is increased when carbon nanotubes are used as scaffolding for biomolecules. Since CNTs have a high specific surface area, they have been considered to be excellent candidates for immobilizing proteins and enzymes. Since protein/enzyme immobilization is a crucial step in the creation of biosensors, which may address many of the difficulties in the healthcare system, the development of innovative CNT-protein/enzyme conjugates is crucial for advancing biomedical research and applications. Additionally, the distinct optical and electrochemical qualities of CNTs make them perfect for medicinal, imaging, sensing, and energy applications.

3.2.2 Waste Treatment

Due to their huge specific surface area and developed pore structure, carbon atoms' tubular cylinders in carbon nanotubes (CNTs) have been shown in research to be capable of adsorbing and removing personal care products (PPCPs) and endocrine-disrupting substances (EDCs). Additionally, they have strong mechanical properties and great photocatalytic activity. CNTs exhibit good PPCP and EDC removal when paired with membrane filtration, with removal rates of up to 95% under ideal experimental circumstances. In the elimination of triclosan, acetaminophen, and ibuprofen, nanocomposite membranes incorporating CNTs have shown good results. The breaking down of chemical compounds in the presence of light is another use for CNTs in addition to their shown promise in adsorption and membrane filtering. Reactive oxygen species are produced by the use of CNT as a photocatalyst, and they are capable of oxidizing pollutants into CO_2 and H_2O [43]. The carbon nanotube-based composite membrane is the greatest option when it comes to a popular kind of separation membrane for water treatment since it combines the superb performance of the standard membrane material with the astounding performance of the carbon nanotubes. The use of carbon nanotube-based composite membranes has been extensively studied for certain emerging water treatment applications and the growing contaminants in water.

3.2.3 Pollutants Decontamination

One of the most crucial and fundamental decontamination techniques is adsorption. Its simple operation and comparably cheap investment costs give it a significant edge over many other technologies. Adsorption is the process by which contaminants are physically and/or chemically attracted to the surface of a substance (adsorbent). Applications using carbon allotropes have shown considerable potential together with the advent and growth of nanotechnologies, particularly the derivatives of graphene. Applications using carbon nanomaterials, particularly the variants of graphene, have shown considerable potential together with the development and growth of nanotechnologies. As was already noted, GO has a two-dimensional, layered structure with a significant amount of surface area and is also abundant in different oxygen-containing functionalities. Unlike graphene, GO has excellent water-borne solubility because it is mono-dispersed in water, begins to unravel its 2D structure, and ends up creating a homogenous colloidal suspension. Furthermore, it is capable of interacting favorably with a variety of contaminants in ionic or molecule form through processes, most notably electrostatic interaction, pi-pi interaction, and hydrophobic interaction. GO has thus gained a lot of interest as a nanosized adsorbent [44]. The CNTs are then used as an absorbent in the cleanup of environmental contamination. One technological process that includes the absorption of gaseous, liquid, and organic contaminants is called absorption. In addition, CNTs may be used for hybrid catalysis, membrane filtration, disinfection, sensing, and monitoring. Its many applications have led to the development of new, environmentally friendly goods that lessen and regulate pollution. As a result, this study will go into more detail about the mechanism of carbon nanotubes and how they may be used to remove developing contaminants including estrogen, pesticides, and heavy metals via adsorption, membrane filtration, disinfection, hybrid catalysis, as well as sensing and monitoring [45].

3.2.4 Anticorrosive

Organic substances have been found to effectively suppress metallic corrosion. However, there has been a significant increase in the use of functionalized carbon allotropes as anticorrosive agents, notably carbon dots, carbon nanotubes, and graphene. Because of their exceptional inhibitive efficiency, biocompatibility, mechanical qualities, chemical stability, solubility, affordability, and environmental friendliness, they have won the hearts of corrosion scientists and engineers [46, 47]. According to studies, CNTs (functionalized) stick to metallic samples and prevent hostile solutions from interacting with them. According to reports, carbon dots have good protective capabilities ranging from 84% to 99%, carbon nanotubes are primarily used in coating systems, and graphene-based materials have excelled as coatings and inhibitors in a variety of media.

3.2.5 Tribological

In the subject of tribology, carbon allotropes are of particular interest due to their unusual structures and characteristics. Graphene nanoplatelets (GNPs) and carbon nanotubes (CNTs) are appealing reinforcements for creating light, strong, wear-resistant metal matrix composites with great mechanical and tribological performance because of their remarkable

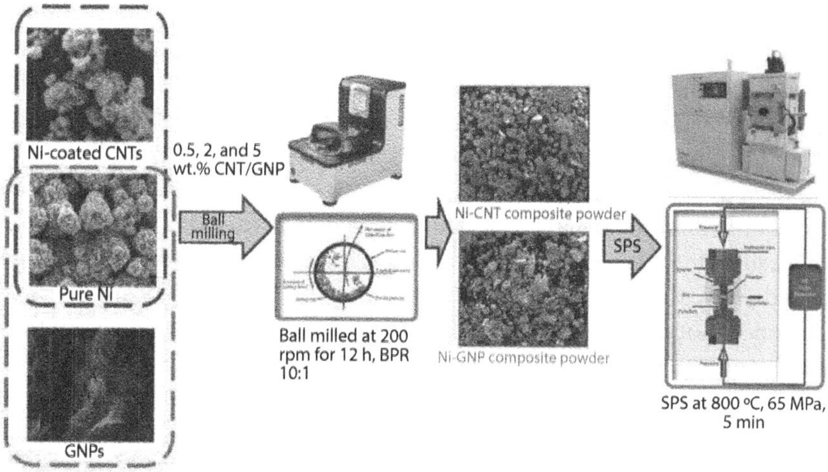

Figure 3.3 Diagrammatical presentation of nickel-carbon and nickel-graphene nanotubes composite processing [48]. Reproduced from Ref. [48], [https://doi.org/10.3390/ma14133536], under the terms of the CC BY 4.0 license.

mechanical, chemical, etc. characteristics [48]. Parfenov *et al.* investigated VM-CNS model systems using rheological and tribological techniques [49].

Patil *et al.* developed nickel-carbon nanotube composites and nickel-graphene nanoplatelet composites [48]. To study the impact of reinforcement concentration and its dispersion on the nickel microstructure, the Ni-CNT/GNP composites with varied reinforcement concentrations (0.5, 2, and 5 wt%) were ball milled for twelve hours. Figure 3.3 displays the processing schematics for composites made of Ni-CNT and GNP. The findings showed that adding CNTs/GNPs to the nickel matrix significantly improved the microhardness of these composites and refined the grain structure. In a sliding wear test, both the CNTs and the GNPs were successful in creating a lubricant layer that increased wear resistance and decreased the coefficient of friction, as compared to the pure nickel equivalent.

3.2.6 Catalytic

Since the development of carbon nanotubes (CNTs) and graphene, carbon-based composites have been a major application area to take advantage of the perfect nanostructures and exceptional material characteristics of graphitic carbons [50]. Following their large-scale fabrication and subsequent discovery, nanometer-scale carbon tubules are currently receiving interest for their possible utility in several areas of materials research, including superconductivity, catalysis, and so on. Numerous carbon materials have been used in the area of heterogeneous catalysis to stabilize and scatter metallic particles. With highly intriguing potential in catalysis, carbon nanotubes have evolved as a distinct carbon allotrope. Most often, they are used as substrates for inorganic metal catalysts, such as molecular catalysts or even more sophisticated hierarchical hybrids [51]. To make a precise estimate of the lifespan of this substance as a catalyst for oxidative dehydrogenations, Frank *et al.* looked into the oxidation of multiwalled carbon nanotubes (CNTs) [52]. Shi *et al.* proposed a unique, straightforward, and one-pot method for synthesizing graphene-encapsulated Fe_3C that is embedded in carbon nanotubes by the direct pyrolysis of renewable biomass [53]. The high-quality CNTs formed, according to research on the effects of different pyrolysis temperatures. Characterization results demonstrated that significant numbers of CNTs have grown using Fe3C-encapsulated N-doped graphene [53].

3.2.7 Reinforced Materials

The outstanding mechanical and electrical conductivity properties of carbon nanotubes have sparked interest in employing them as reinforcing agents in elastomeric materials. Although there is no direct correlation between the electrical characteristics of black-loaded vulcanizates and reinforcement, it has been shown that changes in elastic modulus with amplitude at minor deformations are comparable to changes in conductivity with amplitude. Bokobza *et al.* studied rubber composites containing multiwall carbon nanotubes [54]. The formation of a percolating network may be seen in the rapid drop in resistivity between 2.9 and 3.8 wt% by many orders of magnitude. Those filled with MWNTs exhibit more conductivity than samples filled with carbon black at equal filler loadings. This demonstrates that in carbon black-filled systems, the filler-filler network starts to form at filler loadings greater than 9.1 wt%. Because of their superior mechanical qualities and high aspect ratio, carbon nanotubes (CNTs) have the potential to toughen and reinforce bioactive glass without compromising its bioactivity. As a result, the freeze casting approach has been used to successfully create scaffolds made of a composite of multiwall carbon nanotubes (MWCNT) and 45S5 Bioglass. MWCNTs elevate the 45S5 Bioglass scaffolds' compressive strength and elastic modulus from 2.08 to 4.56 MPa (a 119% increase) and 111.50 to 266.59 MPa (a 139% increase), respectively [55].

3.3 Conclusions and Future Directions

In terms of application and study in the field of biomedicine, carbon allotropes are very advantageous. A rising number of demonstrations have been published in the scientific literature, demonstrating that they are the best candidates for transporting and scaffolding biological molecules in a broad variety of effective implementations. In turn, this improves the productivity and use of the product itself. Carbon nanotubes are well renowned for their outstanding uses. Low-dimensional carbon nanoparticles, and their derivatives, have shown great promise in the field of environmental remediation, but there are still challenges and opportunities to be created and seized in practical application.

References

1. Buckley, J.D., *Carbon-Carbon-An Overview*, vol. 67, American Ceramic Society Bulletin, 1988.
2. Murakami, M. and Ishida, N., Cleavage of carbon–carbon σ-bonds of four-membered rings. *Chem. Rev.*, 121, 1, 264–299, 2020.
3. Zhang, Y., Zhang, Q., Chen, G., Carbon and carbon composites for thermoelectric application. *Carbon Energy*, 2, 3, 408–436, 2020.
4. Liu, W.-D., Yu, Y., Dargusch, M., Liu, Q., Chen, Z.-G., Carbon allotrope hybrids advance thermoelectric development and applications. *Renew. Sust. Energ. Rev.*, 141, 110800, 2021.
5. Fan, Q. *et al.*, Nanoribbons with nonalternant topology from fusion of polyazulene: Carbon allotropes beyond grapheme. *J. Am. Chem. Soc.*, 141, 44, 17713–17720, 2019.
6. Li, J. *et al.*, Two-dimensional carbon allotropes and nanoribbons based on 2, 6-polyazulene chains: Stacking stabilities and electronic properties. *J. Phys. Chem. Lett.*, 12, 2, 732–738, 2021.
7. Zhang, R.-S. and Jiang, J.-W., The art of designing carbon allotropes. *Front. Phys.*, 14, 1, 1–17, 2019.
8. Zhang, W., Chai, C., Fan, Q., Song, Y., Yang, Y., Two-dimensional carbon allotropes with tunable direct band gaps and high carrier mobility. *Appl. Surf. Sci.*, 537, 147885, 2021.
9. Chen, J., Liu, B., Gao, X., Xu, D., A review of the interfacial characteristics of polymer nanocomposites containing carbon nanotubes. *RSC Adv.*, 8, 49, 28048–28085, 2018.
10. Tripathi, A.C., Saraf, S.A., Saraf, S.K., Carbon nanotropes: A contemporary paradigm in drug delivery. *Materials*, 8, 6, 3068–3100, 2015.
11. Chung, D.D.L., Review graphite. *J. Mater. Sci.*, 37, 1475–1489, 2002.

12. Jara, A.D., Betemariam, A., Woldetinsae, G., Kim, J.Y., Purification, application and current market trend of natural graphite: A review. *Int. J. Min. Sci. Technol.*, 29, 5, 671–689, 2019.
13. Németh, P., McColl, K., Garvie, L.A., Salzmann, C.G., Murri, M., McMillan, P.F., Complex nanostructures in diamond. *Nat. Mater.*, 19, 11, 1126–1131, 2020.
14. Duan, X. *et al.*, Sp^2/sp^3 framework from diamond nanocrystals: A key bridge of carbonaceous structure to carbocatalysis. *ACS Catal.*, 9, 8, 7494–7519, 2019.
15. Kosowska, M., Majchrowicz, D., Sankaran, K.J., Ficek, M., Haenen, K., Szczerska, M., Doped nanocrystalline diamond films as reflective layers for fiber-optic sensors of refractive index of liquids. *Materials*, 12, 13, 2124, 2019.
16. Yu, W., Sisi, L., Haiyan, Y., Jie, L., Progress in the functional modification of graphene/graphene oxide: A review. *RSC Adv.*, 10, 26, 15328–15345, 2020.
17. Rao, C., Biswas, K., Subrahmanyam, K., Govindaraj, A., Graphene, the new nanocarbon. *J. Mater. Chem.*, 19, 17, 2457–2469, 2009.
18. Marsh, H. and Reinoso, F.R., *Activated Carbon*, Elsevier, Netherlands, 2006.
19. Bhatnagar, A., Hogland, W., Marques, M., Sillanpää, M., An overview of the modification methods of activated carbon for its water treatment applications. *Chem. Eng. J.*, 219, 499–511, 2013.
20. Shafeeyan, M.S., Daud, W.M.A.W., Houshmand, A., Shamiri, A., A review on surface modification of activated carbon for carbon dioxide adsorption. *J. Anal. Appl. Pyrolysis*, 89, 2, 143–151, 2010.
21. Ebbesen, T.W., Carbon nanotubes. *Annu. Rev. Mater. Sci.*, 24, 1, 235–264, 1994.
22. Dai, H., Carbon nanotubes: Opportunities and challenges. *Surf. Sci.*, 500, 1-3, 218–241, 2002.
23. Norizan, M.N. *et al.*, Carbon nanotubes: Functionalisation and their application in chemical sensors. *RSC Adv.*, 10, 71, 43704–43732, 2020.
24. Kinloch, I.A., Suhr, J., Lou, J., Young, R.J., Ajayan, P.M., Composites with carbon nanotubes and graphene: An outlook. *Science*, 362, 6414, 547–553, 2018.
25. Innocenzi, P. and Stagi, L., Carbon-based antiviral nanomaterials: Graphene, C-dots, and fullerenes. A perspective. *Chem. Sci.*, 11, 26, 6606–6622, 2020.
26. Cataldo, F., Fullerenes, in: *New Frontiers in Nanochemistry*, pp. 219–225, Apple Academic Press, Goldenrod Cir NE, Palm Bay, Florida, United States, 2020.
27. Azizi-Lalabadi, M., Hashemi, H., Feng, J., Jafari, S.M., Carbon nanomaterials against pathogens; The antimicrobial activity of carbon nanotubes, graphene/graphene oxide, fullerenes, and their nanocomposites. *Adv. Colloid Interface Sci.*, 284, 102250, 2020.
28. Dou, J. *et al.*, Functionalization of carbon nanotubes with chitosan based on MALI multicomponent reaction for Cu2+ removal. *Int. J. Biol. Macromol.*, 136, 476–485, 2019.

29. De Menezes, B. *et al.*, Effects of octadecylamine functionalization of carbon nanotubes on dispersion, polarity, and mechanical properties of CNT/HDPE nanocomposites. *J. Mater. Sci.*, 53, 20, 14311–14327, 2018.
30. Milowska, K.Z., Burda, M., Wolanicka, L., Bristowe, P.D., Koziol, K.K., Carbon nanotube functionalization as a route to enhancing the electrical and mechanical properties of Cu–CNT composites. *Nanoscale*, 11, 1, 145–157, 2019.
31. Ohadi, M., Rezaei, P., Mehrabani, M., Behnam, B., Ansari, M., Synthesis, characterization and toxicity assessment of the novel non covalent functionalized multi-walled carbon nanotubes with glycyrrhizin, curcumin and rutin. *J. Clust. Sci.*, 33, 3, 975–984, May 01, 2022.
32. Jang, J., Kang, Y.H., Lim, H.-W., Han, M., Highly conductive and mechanically robust multiwalled carbon nanotube-polyamide nanocomposites via noncovalent functionalization. *J. Mater. Sci.*, 57, 6, 4197–4209, Feb. 01, 2022.
33. Zhang, Y., Li, J., Shen, Y., Wang, M., Li, J., Poly-L-lysine functionalization of single-walled carbon nanotubes. *J. Phys. Chem. B*, 108, 39, 15343–15346, 2004.
34. Tan, S.H., Goak, J.C., Lee, N., Kim, J.Y., Hong, S.C., Functionalization of multi-walled carbon nanotubes with poly (2-ethyl-2-oxazoline), in: *Macromolecular Symposia*, vol. 249, Wiley Online Library, pp. 270–275, 2007.
35. O'Connell, M.J., *et al.*, Reversible water-solubilization of single-walled carbon nanotubes by polymer wrapping. *Chem. Phys. Lett.*, 342, 3-4, 265–271, 2001.
36. Bilalis, P., Katsigiannopoulos, D., Avgeropoulos, A., Sakellariou, G., Noncovalent functionalization of carbon nanotubes with polymers. *RSC Adv.*, 4, 6, 2911–2934, 2014.
37. Jiang, L., Gao, L., Sun, J., Production of aqueous colloidal dispersions of carbon nanotubes. *J. Colloid Interface Sci.*, 260, 1, 89–94, 2003.
38. Islam, M., Rojas, E., Bergey, D., Johnson, A., Yodh, A., High weight fraction surfactant solubilization of single-wall carbon nanotubes in water. *Nano Lett.*, 3, 2, 269–273, 2003.
39. Wenseleers, W., Vlasov, I., II, Goovaerts, E., Obraztsova, E.D., Lobach, A.S., Bouwen, A., Efficient isolation and solubilization of pristine single-walled nanotubes in bile salt micelles. *Adv. Funct. Mater.*, 14, 11, 1105–1112, 2004.
40. Fernando, K.S. *et al.*, Diminished band-gap transitions of single-walled carbon nanotubes in complexation with aromatic molecules. *J. Am. Chem. Soc.*, 126, 33, 10234–10235, 2004.
41. Niazvand, F., Wagh, P.R., Khazraei, E., Dastjerdi, M.B., Patil, C., Najar, I.A., Application of carbon allotropes composites for targeted cancer therapy drugs: A review. *J. Compos. Compd.*, 3, 7, 140–151, 2021.
42. Oliveira, S.F., Bisker, G., Bakh, N.A., Gibbs, S.L., Landry, M.P., Strano, M.S., Protein functionalized carbon nanomaterials for biomedical applications. *Carbon*, 95, 767–779, Dec. 01, 2015.

43. Kurwadkar, S., Hoang, T.V., Malwade, K., Kanel, S.R., Harper, W.F., Struckhoff, G., Application of carbon nanotubes for removal of emerging contaminants of concern in engineered water and wastewater treatment systems. *Nanotechnol. Environ. Eng.*, 4, 1, 12, Apr. 02, 2019.
44. Wang, S. *et al.*, Nitrogen-containing amino compounds functionalized graphene oxide: Synthesis, characterization and application for the removal of pollutants from wastewater: A review. *J. Hazard. Mater.*, 342, 177–191, 2018.
45. Chung, J.H., Hasyimah, N., Hussein, N., Application of carbon nanotubes (CNTs) for remediation of emerging pollutants-a review. *Trop. Aquat. Soil Pollution*, 2, 1, 13–26, 2022.
46. Verma, C., Quraishi, M., Ebenso, E.E., Hussain, C.M., Recent advancements in corrosion inhibitor systems through carbon allotropes: Past, present, and future. *Nano Select*, 2, 12, 2237–2255, 2021.
47. Thakur, A., Kumar, A., Sharma, P.K., Sharma, S., Ganjoo, R., Assad, H., Carbon allotropes as corrosion inhibitors, in: *Functionalized Nanomaterials for Corrosion Mitigation: Synthesis, Characterization, and Applications*, pp. 115–140, ACS Publications, Washington, DC, 2022.
48. Patil, A., Walunj, G., Ozdemir, F., Gupta, R.K., Borkar, T., Tribological behavior of carbon-based nanomaterial-reinforced nickel metal matrix composites. *Materials*, 14, 13, 3536, 2021.
49. Parfenov, A.S. *et al.*, Influence of various carbon allotropes on tribological and rheological characteristics of model lubricating systems. *J. Frict. Wear+*, 42, 3, 217–224, May 01, 2021.
50. Lee, W.J., Maiti, U.N., Lee, J.M., Lim, J., Han, T.H., Kim, S.O., Nitrogen-doped carbon nanotubes and graphene composite structures for energy and catalytic applications. *Chem. Commun.*, 50, 52, 6818–6830, 2014.
51. Melchionna, M., Marchesan, S., Prato, M., Fornasiero, P., Carbon nanotubes and catalysis: The many facets of a successful marriage. *Catal. Sci. Technol.*, 5, 8, 3859–3875, 2015.
52. Frank, B., Rinaldi, A., Blume, R., Schlögl, R., Su, D.S., Oxidation stability of multiwalled carbon nanotubes for catalytic applications. *Chem. Mater.*, 22, 15, 4462–4470, Aug. 10, 2010.
53. Shi, J., Wang, Y., Du, W., Hou, Z., Synthesis of graphene encapsulated Fe3C in carbon nanotubes from biomass and its catalysis application. *Carbon*, 99, 330–337, Apr. 01, 2016.
54. Bokobza, L. and Kolodziej, M., On the use of carbon nanotubes as reinforcing fillers for elastomeric materials. *Polym. Int.*, 55, 9, 1090–1098, 2006.
55. Touri, R., Moztarzadeh, F., Sadeghian, Z., Bizari, D., Tahriri, M., Mozafari, M., The use of carbon nanotubes to reinforce 45S5 bioglass-based scaffolds for tissue engineering applications. *BioMed. Res. Int.*, 2013, 465086, Nov. 04, 2013.

4

Carbon Allotropes in Lead Removal

Shippi Dewangan[1], Amarpreet K. Bhatia[2*] and Nishtha Vaidya[3]

[1]Department of Chemistry, SW Pukeshwar Singh Bhardiya Govt. College, Nikum, Durg, Chhattisgarh, India
[2]Department of Chemistry, Bhilai Mahila Mahavidyalaya, Bhilai Nagar, Chhattisgarh, India
[3]Department of Zoology, Bhilai Mahila Mahavidyalaya, Bhilai Nagar, Chhattisgarh, India

Abstract

By manipulating the size and structure of the materials at the nanoscale, nanotechnology is an advanced discipline of research that can address several environmental problems. Adsorbents made of carbon nanotubes have generated a lot of interest as prospective heavy metal removal sorbents. Due to their benign makeup, high area of surface, ease of biodegradation, and special utility in environmental cleanup, carbon nanoparticles are exceptional. Water contamination with lead is a serious issue that puts people's health at risk. Carbon nanomaterials (CNMs) are gaining attention as they posse's specific physicochemical qualities because of that; they are used to treat heavy metals present in some concentrations in wastewater. Because of their high surface area, nanoscale size, and presence of different functionalities as well as the ease with which they can be chemically modified and recycled, carbon nanomaterials (CNMs), specifically carbon nanotubes, fullerenes, graphene, and graphene oxide, have a specific property to treat lead-contaminated water. This chapter has examined recent developments in the use of these carbon nanomaterials for treating water contaminated with lead metal and has highlighted their use for environmental remediation.

Keywords: Water pollution, lead, carbon allotropes, adsorption, carbon nanotubes, fullerene

**Corresponding author*: amarpreet087@gmail.com

Chandrabhan Verma and Chaudhery Mustansar Hussain (eds.) Carbon Allotropes and Composites: Materials for Environment Protection and Remediation, (51–72) © 2023 Scrivener Publishing LLC

4.1 Introduction

Environmental contamination is now a major concern on a global scale. Environmental pollution has an impact on and has a negative impact on both emerging and developed countries. A significant expansion in industrialization over the past century has resulted in an increased need for energy and the unrestrained use of natural resources, the utilization of the above two has made the problem of environmental pollution worse and poses a harmful effect on the world's biodiversity and ecosystems. Demanding regulations are suggested each year to control pollutants connected to the environment especially water pollution [1]. Heavy metal water pollution has received a lot of attention lately. Lead is considered one of the most toxic heavy metals that are frequently released into the atmosphere by waste streams and are commonly used in various industries, which include industries used in the fabrication of vehicles, as an additive in brake pads, tyres, etc., agriculture (as an insecticide in the organic form of lead), paints, batteries, etc [2–4]. Due to their highly toxic character, the presence of these heavy metals in the environment poses a serious threat to the health and existence of people and other living things [5]. Through tainted food, water, and air, this heavy metal enters the human body and builds up in the bones. Animals and plants also get affected by air and soil which is heavily contaminated with lead [4]. Exposure to lead has elevated the danger to the kidneys, reproductive system, and liver, leading to renal disease, anaemia, cancer, and mental retardation [2, 6, 7].

Inorganic lead is ingested and inhaled by humans, where it is absorbed in amounts of 20% to 80% and 5% to 15%, respectively [4]. The lead pipes used for drinking water supply get corroded and leached and Pb/Sn soldered joints connected to the copper service lines frequently used in residential plumbing cause lead contamination in drinking water [8]. Lead poisoning can seriously harm a person's kidney, reproductive system, brain, neurological system, and liver, which can result in illness or even death. According to reports, severe lead exposure can result in sterility, abortion, stillbirths, and neonatal fatalities [9]. According to the Indian Standards Institution [11] and the United States Environmental Protection Agency (U.S. EPA) [10], the acceptable limit (mg/L) for Pb (II) in wastewater is 0.1 mg L^{-1}. However, the World Health Organization (WHO) states that the maximum permissible content of lead (II) in drinking water is 0.01 mg L^{-1} [12]. These effects of lead exposure are substantially more severe in youngsters and particularly hurt IQ levels. According to research conducted in the US, there are two main ways that children are exposed to lead: (i) by burning gasoline with lead-polluting air; (ii) by using paint in the home, which is made up of lead, respectively [13]. Presently, our aim is to reduce the utilization of lead-containing products globally. For reducing the harmful effects of lead in the environment, new methods have to introduce which will help in the reduction of lead concentration in the environment.

There are several different ways to remove Pb (II) from wastewater. The low-cost budget and level of heavy metal initial concentrations are the major determinants in choosing a certain treatment procedure. Heavy metals are often removed from water through chemical and physical treatment processes. Seawater, brackish groundwater, or surface water are all examples of this raw water. Precipitation, Bioremediation, electrochemical treatment, membrane filtration, ion exchange, lime softening, reverse osmosis, coagulation, and adsorption are some of the most commonly used methods to treat heavy metal contamination [14–24]. Table 4.1 summarizes the Benefits and Drawbacks of different methods adopted to treat lead contamination. Furthermore, many of the available techniques are not economically used in low-cost scale industries. Therefore a new technique shall be introduced. Presently, adsorption has been shown to be a practical and efficient way to remove a variety of contaminants from water. An appropriate desorption procedure also allows for the regeneration of the used adsorbent.

Recently, work has been done on replacing the current water treatment technique with one that uses nanotechnology. Numerous studies

Table 4.1 Benefits and drawbacks of the current methods for removing lead.

Method	**Benefits**	**Drawbacks**
Ion exchange method	• Capacity of treatment is high • Excessive removal efficiency. • Take less to complete the reaction.	• Use of synthetic resins increases the cost. • Regeneration of the resins release of secondary pollution.
Adsorption method	• Easy to operate in any condition. • metal binding capacities is High • Suitable at any pH range • Low-cost	• Production of waste products. • Selectivity is Low.

(*Continued*)

Table 4.1 Benefits and drawbacks of the current methods for removing lead. (*Continued*)

Method	Benefits	Drawbacks
Coagulation and Flocculation method	• Turbidly is removed easily. • Sludge produced has good settling and dewatering characteristics	• Generation of sludge gets Increased. • This technique requires an additional treatment process for the removal of heavy metal.
Chemical precipitation	• Easy to operate. • Economic cost	• Generate large volumes of Sludge which require an extra economic cost. • Not suitable for the treatment of water having minimum heavy metals concentrations.
Membrane filtration	• Selectivity toward the separation of heavy metal is high. • Need less space and Low-pressure conditions.	• Membrane fouling increases the operational cost. • Complexity in Process.
Photocatalysis	• Metals and other pollutants can be removed simultaneously. • By-products released are eco-friendly.	• Applications are limited. • Time-consuming process.
Flotation	• Show Excessive metal selectivity. • Excessive overflow rates • Economic cost • Production of more concentrated sludge • Requires less time. • Excessive treatment efficiency.	• Requires maintenance. • High operational costs

have demonstrated that nanotechnology, which was made possible by an effective and multifunctional method, is expected to provide an adequate water treatment solution that may not require any substantial and costly infrastructure. Several adsorbents, such as activated manganese oxides, carbon cloths, biomaterials, sugar beet pulp, kaolinite, activated carbon, recycled alum sludge; bagasse, etc. have been used to adsorb lead metal ions. These adsorbents, however, possess poor metal ion removal efficiencies or capacities. As a result, researchers are looking for new, promising adsorbents. Therefore, interest has been increased toward carbon-based material for the treatment of lead-contaminated water. In this chapter, we have discussed the extensive utilization of carbon-based nanomaterials for the treatment of lead-contaminated water.

4.2 Carbon Nanomaterials (CNMs)

Carbon can exist in various molecular forms, known as allotropes of carbon. These carbon allotropes consist of various structural changes and are made up of elements of carbon. CNMs are novel next-generation materials that have recently gained prominence for application in a variety of industries, including biomedicine, biosensors, superconductors, composite material reinforcing, delivery of drugs, and material having good electrical conductivity. Carbon Nanomaterials (CNMs) show an abnormally high level of electrical conductivity, as well as mechanical, thermal, and other physiochemical properties. Due to their excellent adsorption abilities and efficiencies, they have also attracted the attention of various scientists to use them as excellent adsorbents for the treatment of heavy metal ions pollutants from wastewater. They provide a promising alternative to current adsorbents for wastewater cure applications due to their advantageous structural characteristics, which include uniformly distributed pores, a high S/V ratio, and highly hollow and porous structures.

4.3 Dimension-Based Types of Carbon Nanomaterials

Zero-dimensional (0-D) nanomaterials are those with all three dimensions less than 100 nm; fullerene and quantum dots are two examples. One-dimensional (1-D) nanomaterials, such as carbon and titanium nanotubes, are defined as those that have just one dimension >100 nm and two dimensions< 100 nm. Two-dimensional nanomaterials are those with two dimensions >100 nm; graphene is a well-known example. Graphite and various nanomaterial composites are examples of three-dimensional (3-D) materials, which are defined as materials with all dimensions>100 nm. Figure 4.1 displays a few examples of well-known carbon compounds in various dimensions, including fullerene, single-wall carbon nanotube, graphene, and graphite.

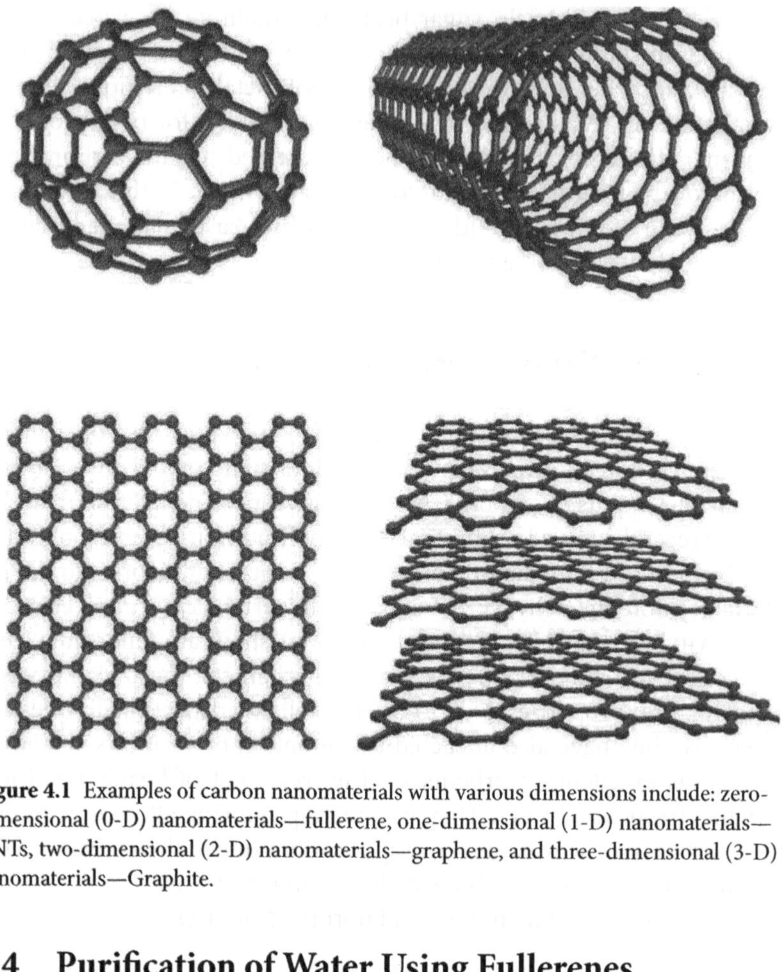

Figure 4.1 Examples of carbon nanomaterials with various dimensions include: zero-dimensional (0-D) nanomaterials—fullerene, one-dimensional (1-D) nanomaterials—CNTs, two-dimensional (2-D) nanomaterials—graphene, and three-dimensional (3-D) nanomaterials—Graphite.

4.4 Purification of Water Using Fullerenes

Fullerene is a kind of carbon sometimes referred to as buckminsterfullerene (C_{60}) or buckyballs. The molecules resemble convex closed polyhedral, hollow spheres, tubes, or ellipses and are made up of three-coordinated carbon atoms (carbon atoms are even in number). In 1985, Kroto and Heath used spectrometry to analyze interstellar dust and made the discovery. The most studied fullerenes, C_{60} and C_{70}, are made up of a variety of isomers and homologous series, as are the so-called higher fullerenes, such as C_{240}, C_{540}, and C_{720}. Arc-discharge vaporization of graphite, chemical vapor deposition techniques, and combustion processes are the traditional ways of producing fullerenes. Fullerenes are suitable to use as adsorbents for wastewater treatment due to their distinctive spherical structure, the low tendency of aggregation, high S/V ratio, and high electron affinity properties. The sorption capacity of these materials is typically linked to surface defects. The penetration of the species into the fullerene lattice is also thought to be the cause of their sorption on fullerenes. For instance, the fullerene C60 has micropores between its carbon nanoclusters where sorbate ions can pass through defects.

However, their direct use as a sorbent for the remediation of pollutants from wastewater is unprofitable due to the high price. To increase the sorption capacity of standard sorbents (activated carbons, zeolites, etc.), studies on the modification of these materials have been carried out. It was found that the introduction of fullerenes causes the sorbent surface to develop a porous structure, increases hydrophobicity, and enhances the material's ability to bind heavy metals from water. For example, activated carbons' sorption capacity for Pb (II) increased 1.5 to 2.5 times after the introduction of 0.001% to 0.004% of fullerenes [25]. Fullerene research, however, has diminished ground in recent years as research on CNTs and graphene-based materials has gained momentum. That might be because various reported literature works suggest that CNTs and graphene-based materials perform much better than fullerene-based materials.

4.5 Application of Graphene and Its Derivatives in Water Purification

The carbon allotrope known as graphene has a two-dimensional, hexagonal honeycomb structure. A layer of carbon atoms that are one atom thick, sp_2 hybridized, and connected to a two-dimensional, hexagonal, ccc lattice in the shape of a honeycomb by σ and π bonds make up graphene (the distance of the two nearest atoms is around 0.14 nm). Due to their high stabilities, S/V ratio, and also tunable functionalities, materials made up of graphene are receiving attention as excellent adsorbents for the treatment of wastewater. Additionally, through various bimolecular interactions, they get interact toward pollutants and these pollutants get adsorbed on their surface. Numerous studies have reported that the derivatives of grapheme posses excellent adsorption properties for the treatment of heavy metal-contaminated water.

There are numerous methods for producing graphene, including pyrolysis, chemical vapor deposition, chemical synthesis, and mechanical and chemical exfoliation of graphite. Because of the presence of various functional groups on its surface, the synthesized graphene composite exhibits better physiochemical properties than pristine graphene. By applying powerful oxidizing chemicals to oxidize graphite, Hummer's approach can create GO. Oxidation-treated GO surfaces can produce oxygen-containing functional groups like ketone, aldehyde, ethanolic, epoxy, carbonyl, etc. These functional groups have the ability to change the van der Waals interactions of GO, improving its solubility in organic and water-based solvents. Additionally, the exfoliated GO can be reduced to produce rGO. The above properties of rGO can be considered as a basis for obtaining the other form of carbon allotropes, i.e., graphite crystals. For lead removal from wastewater, functionalized graphene has been used by various researchers as a sorbent. In [26], the researcher has proved that Pb (II) show the maximum adsorption at 406.6 mg/g at the optimum pH of 5.0, and the equilibrium was achieved within 40 minutes.

In other work [27], the Pb (II) sorption capacities of three graphene-based composites: suspension of graphene nanoplatelets (SGN), a paste of multilayered oxidized graphene (PMOG), and a paste of few-layered oxidized grapheme (PFOG) was compared and was proved the adsorption equilibrium was achieved in less than 30 minutes under constant condition. The maximum sorption capacity of Lead (II) of a mixture of multilayered oxidised graphene (PMOG), a paste of few-layered oxidised grapheme (PFOG), and a suspension of graphene nanoplatelets (SGN) were 457, 103, and 38 mg/g, respectively and the pseudo-second-order model was found to fit the kinetic data well. Based on these findings, it was hypothesized that the oxygen-containing functional groups present on the surface of the adsorbent play a significant role in the remediation of Pb (II).

In [28], a few-layered GO from graphite was synthesized using the modified Hammer method and is used to treat lead (II)-contaminated water. After treatment, it was found that the sorption capacity of the sorbents increased up to 1850 mg g^{-1}. This increase in adsorption is due to the presence of oxygen-containing functional groups on the surface of the adsorbent. The pH range of the solution, just like in the other studies, was found to be a controlling factor in this one's sorption process.

4.6 Application of Carbon Nanotubes (CNTs) in Water Purification

Sumio Iijima of the NEC Laboratory in Tsukuba, Japan, first discovered CNTs in 1991 using the arc-discharge technique and was characterized using a high-resolution transmission electron microscope (HRTEM). One-dimensional hollow tubes make up CNTs. They resemble rolled-up carbon graphitic sheets that are hollow nanoscale cylinders. Since their discovery,

CNTs have been the subject of research in nanotechnology because of their distinct physicochemical characteristics. These carbon nanostructure allotropes with cylindrical shapes are used as filters for air and water, electronics, field emission, energy storage, semiconductors, and catalysis. With a high specific surface area (150–1500 m^2 g^{-1}), their diameter ranges from 1 to 100 nm. CNTs can have single or multiple walls (SWCNTs or MWCNTs). The first consists of a single layer of rolled graphene sheets that range in diameter from 0.8 to 2 nm, while the second is made up of many layers that have sizes between 5 and 20 nm. A tube-like structure made up of graphite sheets which are cylindrical in shape (an allotrope of carbon) makes up carbon nanotubes (CNTs). Single-walled carbon nanotubes are cylindrical structures made up of just one shell of graphene. Multi-walled carbon nanotubes, on the other hand, are composed of multiple layers of graphene sheets.

The orientation of the tube axis with respect to the hexagonal lattice determines the nanotubes' structure, which may be determined using the chiral vector, which is represented by chiral indices (n, m). The geometric configurations of carbon atoms at the seam of the cylinders are the basis for the armchair (n = m) and zigzag (n = 0) nanotube structures, while chiral nanotubes have two enantiomers with right and left side helicity (n-m). All of the chemical bonds in nanotubes are sp2 bonds, which are more powerful than the sp3 bonds present in alkanes and give nanotubes a special strength. They are the best choice for the adsorption-based removal of heavy metal ions due to their mesopores. Additionally, CNTs are easily functionalized with a variety of organic compounds, making them, particularly for the choice of adsorbates and potentially enhancing their adsorption capacity. Ion exchange, electrochemical potential, and surface characteristics all play a role in the sorption of heavy metals by CNTs. Because of properties like high surface area, highly porous and hollow structure, and average pore diameter, the adsorption percentage gets affected. CNTs have been widely used for the removal of different pollutants from aqueous solutions.

There are numerous methods used to remove Pb (II) from wastewater. However, the adsorption method utilizing carbon nanotubes exhibits remarkable lead from wastewater removal efficiency. The summary of numerous investigations on the use of various kinds of carbon nanotubes for the adsorption of lead metal ions is shown in Table 4.2. From the Table it can be seen that that the maximum adsorption capacity can be calculated using the Langmuir isotherm model. Additionally, practically all investigations show that the CNTs on treatment with acid show better adoption capacity than the raw CNTs. This increase in adsorption capacity of

Table 4.2 Lead metal ions showing maximum adsorption capacities with CNTs and their interaction mechanism.

Various adsorbents	Maximum sorption capacity, Q ($mg\ g^{-1}$)		Experimental conditions	Mechanism of interaction	Different model used	Refs.
	Calculated value from model	Experimental value				
Acidified multi-wall carbon nanotubes	17.44	11.2	The initial metal concentration was 10 mg L^{-1} while the pH range was 5 and the amount of acidified MWCNTS was 0.05 g	Ion exchange method	Langmuir	[29]
Acidified multi-wall carbon nanotubes (MWCNTs)	Not Applicable	30.32	The initial metal concentration was 30 mg L^{-1} while the pH range was 5 and the Temperature was fixed at 298 K.	Not Applicable	Freundlich	[35]
Oxidized carbon nanotubes (CNTs)	Not Applicable	82.6	The initial metal concentration was 10 mg L^{-1} while the pH range was 5	Not Applicable	Experimental	[36]
HNO_3-modified carbon nanotubes	97.10 ± 0.2	82 ± 0	The metal concentration was 10 mg L^{-1} while the pH range was 5. No temperature Conditions are required.	Ion exchange	Langmuir	[37]

(Continued)

Table 4.2 Lead metal ions showing maximum adsorption capacities with CNTs and their interaction mechanism. (*Continued*)

Various adsorbents	Maximum sorption capacity, Q ($mg\ g^{-1}$)		Experimental conditions	Mechanism of interaction	Different model used	Refs.
	Calculated value from model	Experimental value				
Carbon nanotubes	102.06 ± 0.02	37.60	The amount of Adsorbent was 10 mg works at pH 5 and contact time of 80 min	Chemisorption	Langmuir	[38]
Acidified multi-wall carbon nanotubes (MWCNTs)	Not Applicable	91	The Initial concentration of 50 $mg\ L^{-1}$ at temperature 298 K shows maximum adsorption.	Chemisorption	Experimental	[39]
Carbon nanotubes doped in nitrogen magnetically	Not Applicable	6.74 ±0.01 $mmol\ g^{-1}$	pH 8	Chemisorption	Experimental	[40]
Multi-wall carbon nanotubes mixed with acid	49.70± 0.01	85.00	The initial metal concentration was 50 $mg\ L^{-1}$ while the pH range was 5 and the Temperature was fixed at 298 K.	Physisorption	Langmuir	[41]
Functionalized multi-wall carbon nanotubes with iron	Not Applicable	Not Applicable	The initial metal concentration was 100 $mg\ L^{-1}$ while the pH range was 5	Not Applicable	Experimental	[42]

(*Continued*)

Table 4.2 Lead metal ions showing maximum adsorption capacities with CNTs and their interaction mechanism. (*Continued*)

Various adsorbents	Maximum sorption capacity, Q ($mg\ g^{-1}$)		Experimental conditions	Mechanism of interaction	Different model used	Refs.
	Calculated value from model	Experimental value				
MnO_2/ carbon nanotubes (CNTs)	Not Applicable	Not Applicable	The initial metal concentration was 30 mg L^{-1} while the pH range was 5 and the Temperature was fixed at 298 K.	Electrostatic interaction, surface complexation	Langmuir	[43]
3-Mercaptopropyltriethoxy-silane (MPTS) grafted multi-wall carbon nanotubes	65.40 ± 0.01	42.12± .02	The initial metal concentration was 50 mg L^{-1} while the pH range was 6.5 and the Temperature was fixed at 25 ± 0.2 °C.	Interactions due to acid-base which shows Lewis's character	Langmuir	[44]
Multi-wall carbon nanotubes (MWCNTs) treated with oxygen	Not Applicable	17	pH 7	Not Applicable	Experimental	[45]
Oxidized multi-wall carbon nanotubes (MWCNTs)	0.0211 mmol/g	Not Applicable	The initial surfactant concentration was 0.83 mmol L^{-1} while pH range was 4.1 ± 0.1 and adsorbent dosage = 0.75 g L^{-1}	Chemical, hydrophobic, pi–pi Electrostatic interaction	Langmuir	[46]

(*Continued*)

Table 4.2 Lead metal ions showing maximum adsorption capacities with CNTs and their interaction mechanism. (*Continued*)

Various adsorbents	Maximum sorption capacity, Q (mg g^{-1})		Experimental conditions	Mechanism of interaction	Different model used	Refs.
	Calculated value from model	Experimental value				
Titanium oxide/multi-wall carbon nanotubes (MWCNTs) composites	137.0	4.63	The initial metal concentration was 10 mg L^{-1} while pH was fixed 6	Not Applicable	Langmuir	[47]
Multi-wall carbon nanotubes (MWCNTs)	33	4.86	The initial metal concentration was 10 mg L^{-1} while pH was fixed 6	Not Applicable	Langmuir	[47]
Multi-wall carbon nanotubes grafted with 2-vinyl pyridine	37.00	Not Applicable	pH 6	Electrostatic and Ion exchange Interaction	Langmuir	[48]
Multi-wall carbon nanotubes	15.9 ±.00	Not Applicable	pH 6	Electrostatic and Ion exchange Interaction	Langmuir	[48]
Multi-wall carbon nanotubes with ferric oxide	22.04 ± .02	13.04 ± .02	The initial metal concentration was 30 mg L^{-1} while pH was fixed at 5.3	Electrostatic, hydrophobic, p–p interaction	Langmuir	[49]

(*Continued*)

Table 4.2 Lead metal ions showing maximum adsorption capacities with CNTs and their interaction mechanism. (*Continued*)

Various adsorbents	Maximum sorption capacity, Q (mg g^{-1})		Experimental conditions	Mechanism of interaction	Different model used	Refs.
	Calculated value from model	Experimental value				
MWCNTs/Fe_3O_4 modified with 3-aminopropyltriethoxy-silane (APTS) multi wall carbon nanotubes	75.02± .02	37.64± .01	The initial metal concentration was 30 mg L^{-1} while pH was fixed at 5.3	Electrostatic, hydrophobic, p–p interaction	Langmuir	[49]
Oxidized carbon nanotubes (CNTs)	117.65 ± .02	101.05 ± .02	At pH 7, when the initial metal concentration was 1200 mg L^{-1} at temperature 298 K shows maximum adsorption	Chemisorption	Langmuir	[50]
Single-walled carbon nanotubes	33.55	Not Applicable	At pH 5, when the initial metal concentration was 20 mg L^{-1} and the adsorbent amount of 50 mg L^{-1} at temperature 298 K shows maximum adsorption in 120 minutes.	Physisorption	Langmuir	[51]

(*Continued*)

Table 4.2 Lead metal ions showing maximum adsorption capacities with CNTs and their interaction mechanism. (*Continued*)

Various adsorbents	Maximum sorption capacity, Q (mg g^{-1})		Experimental conditions	Mechanism of interaction	Different model used	Refs.
	Calculated value from model	Experimental value				
Single-walled carbon nanotubes with fictionalized acid	96.02 ± .02	Not Applicable	At pH 5, when the initial metal concentration was 20 mg L^{-1} and the adsorbent amount of 50 mg L^{-1} at temperature 298 K shows maximum adsorption in 120 minutes.	Chemisorption	Langmuir	[51]
Oxidized multi-wall carbon nanotubes (MWCNTs)	Not Applicable	2.96	At pH 9, when the initial metal concentration varies from 1 to 20 mg L^{-1} and the adsorbent amount of 50 mg L^{-1} at temperature 298 K shows maximum adsorption.	Electrostatic interactions	Freundlich	[52]

acidified CNTs is because of bimolecular interaction that takes place between Pb (II) and the negatively charged CNT surface. Additionally, the efficiency of removal of heavy metal using CNTs can be increased by adding other nanoparticles. Improved CNTs have demonstrated great removal effectiveness. Table 4.2 also lists the mechanisms by which lead metals and CNTs interact during each process.

The removal efficiency of Lead (II) using CNTs produced by catalytic pyrolysis of a hydrogen-propylene mixture demonstrates that CNTs have low Pb^{2+} adsorption capacity and can be increased by varying pH from 3–7. When concentrated nitric acid was refluxed over CNTs, the adsorption capacity increased significantly. From the experimental data, it was proved that the maximum adsorption capacity is 49.95 mg g^{-1} at a pH 7 and was proved using the Freundlich isotherm [29].

Using modified CNTs, lead is removed with a proportion of nearly 100% at pH 7, compared to 17% and 25% at pH 4 and pH 5, respectively. This is due to the fact that altering the surface of CNTs can increase their negative charge, hydrophilicity, and ability to generate different functional groups which facilitate in adsorption [30]. When the pH is more than 7, $Pb(OH)_2$ precipitates, removing Pb^{2+}. Adsorption is therefore preferable in acidic conditions; otherwise, adsorption and precipitation will work together to remove Pb^{2+} [29].

In another work [31], it was reported that modification on the surface of MWCNTs by oxidation using plasma causes increases in the adsorption efficiency of lead. Oxidized CNTs increase the oxygen-containing functional groups (such as COOH, OH, etc.) on their surface, which increases their specific surface area and surface defects. Additionally, some amorphous carbon and MWCNT caps were removed, and their dispersion in water was enhanced.

According to additional studies [32, 33], the elimination at pH 7 was primarily related to the adsorption of Pb^{2+} and $Pb(OH)^{+}$. At a pH range of 7 to 9, the lead removal maintained a constant maximum value. This results from the simultaneous adsorption of Pb^{2+} and Pb $(OH)^{+}$ ions and precipitation of Pb $(OH)_2$ that occurred when the pH was in the range of 7 to 9 [32, 34].

4.7 Conclusion

In this chapter, we have discussed about the application of fullerene, carbon nanotubes, graphene, and graphene oxide to the removal of lead metal from heavy metal pollutants. With tremendous effectiveness, these carbon nanoparticles are utilized to treat heavy metal-contaminated water. Due to their intriguing qualities, such as high surface area, simplicity in recycling, and ease in desorbing the adsorbed metal ions, only acidic conditions, e.g., nitric acid, sulphuric acid, etc. may be utilized to regenerate the material and reuse it while retaining its adsorption capabilities. In addition to these characteristics, the carbon nanoparticles are simple to produce in combination with some other technique which produce some different nanoparticles which can be functionalize easily. The nanoparticle synthesized by above technique shows multifunctional properties and can be utilized in better way to treat lead (II)-contaminated water. Materials made of carbon are very biocompatible with both the environment and biological organisms. This lead (II) adsorption method is greatly impacted by a number of other variables, including pH, kind of sorbents, and experimental time. It is clear from this chapter of the literature that carbon nanostructures have exciting physicochemical characteristics and significant potential for use in water purification.

References

1. Dewangan, S., Bhatia, A.K., Singh, A.K., Carabineiro, S.A.C., Removal of hydrophobic contaminants from the soil by adsorption onto carbon materials and microbial degradation. *C*, 7, 83, 2021.
2. Eqani, S.A.M.A.S., Khalid, R., Bostan, N., Saqib, Z., Mohmand, J., Rehan, M., Ali, N., Katsoyiannis, I.A., Shen, H., Human lead (Pb) exposure via dust from different land use settings of Pakistan: A case study from two urban mountainous cities. *Chemosphere*, 155, 259–265, 2016.
3. Ghorbani, M., Aghamohammadhasan, M., Shams, A., Tajfirooz, F., Pourhassan, R., Khosravi, S.R.B., Karimi, E., Jampour, A., Ultrasonic-assisted magnetic dispersive solid phase microextraction for preconcentration of two nonsteroidal anti-inflammatory drugs in real water, biological and milk samples employing an experimental design. *Microchem. J.*, 145, 1026–1035, 2019.
4. Matta, G. and Gjyli, L., Mercury, lead and arsenic: Impact on environment and human health. *J. Chem. Pharm. Sci.*, 9, 718–725, 2016.
5. Abadin, H., Ashizawa, A., Stevens, Y.W., *ATSDR, Division of Toxicology and Environmental Medicine*, Toxicology Profile for Lead, Atlanta, 2007.
6. Abbas, A., Al-Amer, A.M., Laoui, T., Al-Marri, M.J., Nasser, M.S., Khraisheh, M., Atieh, M.A., Heavy metal removal from aqueous solution by advanced carbon nanotubes: Critical review of adsorption applications. *Separ. Purif. Technol.*, 157, 141–161, 2016.

7. Mason, L.H., Harp, J.P., Han, D.Y., Pb neurotoxicity: Neuropsychological effects of lead toxicity. *BioMed. Res. Int.*, 2014, 1–8, 2014. https://doi.org/10.1155/2014/840547.
8. Shanmugavalli, R., SyadShabudeen, P.S., Vencktesh, R., Kadirvelu, K., Madhavakrishnan, S., Pattbhi, R.E., Uptake of Pb (II) ion from aqueous solution using silk cotton hull carbon: An agricultural waste biomass. *J. Chem.*, 3, 13, 218–229, 2006.
9. El-Said, A.G., Biosorption of Pb (II) ions from aqueous solution onto rice husk and its ash. *J. Am. Sci.*, 6, 10, 143–150, 2010.
10. *General Standard for Discharge of Environmental Pollutants, Effluent as per Schedule VI for Environment (Protection) Rules*, 1986. https://www.cpcb.nic.in/GeneralStandards.pdf
11. BIS, *Tolerance Limits for Industrial Effluents Prescribed by Indian Standard Institution. IS: 2490(Part I)*, 1981. http://www.worldenviro.com/effstd.htm
12. World Health Organization, *International Standards for Drinking Water*, WHO, Geneva, 2008.
13. O'Connor, D., Hou, D., Ye, J., Zhang, Y., Ok, Y.S., Song, Y., Coulon, F., Peng, T., Tian, L., Lead-based paint remains a major public health concern: A critical review of global production, trade, use, exposure, health risk, and implications. *Environ. Int.*, 121, 85–101, 2018.
14. Fu, F. and Wang, Q., Removal of heavy metal ions from wastewaters: A review. *J. Environ. Manage.*, 92, 407–418, 2011.
15. Awual, M.R., Ismael, M., Yaita, T., Safty, S.A.E., Shiwaku, H., Okamoto, Y., Suzuki, S., Trace copper (II) ions detection and removal from water using novel ligand modified composite adsorbent. *Chem. Eng. J.*, 222, 67–76, 2013.
16. Al-Rashdi, B.A.M., Johnson, D.J., Hilal, N., Removal of heavy metal ions by nanofiltration. *Desalination*, 315, 2–17, 2013.
17. Liu, C., Bai, R., Ly, Q.S., Selective removal of copper and lead ions by diethylenetriamine-functionalized adsorbent: Behaviors and mechanisms. *Water Res.*, 42, 1511–1522, 2008.
18. Peric, J., Trgo, M., Medvidovic, N.V., Removal of zinc, copper and lead by natural zeolite-a comparison of adsorption isotherms. *Water Res.*, 38, 1893–1899, 2004.
19. Khosravi, J. and Alamdari, A., Copper removal from oil-field brine by coprecipitation. *J. Hazard. Mater.*, 166, 695–700, 2009.
20. Babel, S. and Kurniawan, T.A., Low-cost adsorbents for heavy metals uptake from contaminated water: A review. *J. Hazard. Mater.*, 45, 219–243, 2003.
21. Rio, S., Brasquet, C.F., Coq, L.L., Courcoux, P., Cloirec, P.L., Experimental design methodology for the preparation of carbonaceous sorbents from sewage sludge by chemical activation—Application to air and water treatments. *Chemosphere*, 58, 423–437, 2005.
22. Rahman, M.S. and Islam, M.R., Effects of pH on isotherms modeling for Cu(II) ions adsorption using maple wood sawdust. *Chem. Eng. J.*, 149, 273–280, 2009.

23. Bailey, S.E., Olin, T.J., Bricka, R.M., Adrian, D.D., A review of potentially low-cost sorbents for heavy metals. *Water Res.*, 33, 2469–2479, 1999.
24. Al-Rashdi, B., Somerfield, C., Hilal, N., Heavy metals removal using adsorption and nanofiltration techniques. *Sep. Purif. Rev.*, 40, 209–259, 2011.
25. Samonin, V.V., Nikonova, V.Y., Podvyaznikov, M.L., Carbon adsorbents on the basis of the hydrolytic lignin modified with fullerenes in producing. *Russ. J. Appl. Chem.*, 87, 2, 190–193, 2014.
26. Deng, L.P., Su, Y.Y., Su, H., Wang, X.T., Zhu, X.B., Sorption and desorption of lead (II) from wastewater by green algae cladophora fascicularis. *J. Hazard. Mater.*, 143, 220–225, 2007.
27. Kucherova, A.E., Burakova, I.V., Burakov, A.E., Krasnyansky, M.N., Memetov, N.R., Graphene based nano-composites for enhanced Pb^{2+} adsorption. *Nano Hybrids Compos.*, 13, 323–329, 2016.
28. Zhao, G., Ren, X., Gao, X., Tan, X., Li, J., Chen, C., Huang, Y., Wang, X., Removal of Pb(II) ions from aqueous solutions on few-layered graphene oxide nanosheets. *Dalton Trans.*, 40, 10945–10952, 2011.
29. Li, Y.H., Wang, S.G., Wei, J.Q., Zhang, X.F., Xu, C.L., Luan, Z.K., Wu, D.H., Wei, B.Q., Lead adsorption on carbon nanotubes. *Chem. Phys. Lett.*, 357, 263–266, 2002.
30. Atieh, A., Bakather, O.Y., Al-Tawbini, B., Bukhari, A.A., Abuilaiwi, F.A., Fettouhi, M.B., Effect of carboxylic functional group functionalized on carbon nanotubes surface on the removal of lead from water. *Bioinorg. Chem. Appl.*, 2010, 603978, 2010. https://doi.org/10.1155/2010/603978.
31. Yu, X., Luo, T., Zhang, Y., Jia, Y., Zhu, B., Fu, X., Huang, X., Adsorption of lead(II) on O2-plasma-oxidized multi-walled carbon nanotubes. *ACS Appl. Mater. Interfaces*, 3, 7, 2585–2593, 2011.
32. Gupta, V.K., Agarwal, S., Saleh, T.A., Synthesis and characterization of alumina-coated carbon nanotubes and their application for lead removal. *J. Hazard. Mater.*, 185, 1, 17–23, 2011.
33. Wang, H., Zhou, A., Peng, F., Yu, H., Yang, J., Mechanism study on adsorption of acidified multiwalled carbon nanotubes to Pb (II). *J. Colloid Interface Sci.*, 316, 2, 277–283, 2007.
34. Xu, D., Tan, X., Chen, C., Wang, X., Removal of Pb(II) from aqueous solution by oxidized multi-walled carbon nanotubes. *J. Hazard Mater.*, 154, 1–3, 407–416, 2008.
35. Li, Y.H., Di, Z., Ding, J., Wu, D., Luan, Z., Zhu, Y., Adsorption thermodynamics, kinetic and desorption studies of Pb2+ on carbon nanotubes. *Water Res.*, 39, 605–609, 2005.
36. Li, Y.H., Zhu, Y., Zhao, Y., Wu, D., Luan, Z., Different morphologies of carbon nanotubes effect on the lead removal from aqueous solution. *Diam. Relat. Mater.*, 15, 90–94, 2006.
37. Li, Y.H., Ding, J., Luan, Z., Di, Z., Zhu, Y., Xu, C., Wu, D., Wei, B., Competitive adsorption of Pb2+, Cu2+, and Cd2+ ions from aqueous solutions by multi-walled carbon nanotubes. *Carbon*, 41, 2787–2792, 2003.

38. Kabbashi, N.A., Atieh, M.A., Mamun, A.A., Mirghami, M.E.S., Alam, M.D.Z., Yahya, N., Kinetic adsorption of application of carbon nanotubes for Pb(II) removal from aqueous solution. *J. Environ. Sci.*, 21, 539–544, 2009.
39. Wang, H.J., Zhou, A.L., Peng, F., Yu, H., Yang, J., Mechanism study on adsorption of acidified multiwalled carbon nanotubes to Pb (II). *J. Colloid Interface Sci.*, 316, 277–283, 2007.
40. Shin, K., Hong, J., Jang, J., Heavy metal ion adsorption behavior in nitrogendoped magnetic carbon nanoparticles: Isotherms and kinetic study. *J. Hazard. Mater.*, 190, 36–44, 2011.
41. Wang, H.J., Zhou, A.L., Peng, F., Yu, H., Chen, L.F., Adsorption characteristic of acidified carbon nanotubes for heavy metal Pb(II) in aqueous solution. *Mater. Sci. Eng. A*, 466, 201–206, 2007.
42. Wang, H., Yan, N., Li, Y., Zhou, X.H., Chen, J., Yu, B.X., Gong, M., Chen, Q.W., Fe nanoparticle-functionalized multi-walled carbon nanotubes: One-pot synthesis and their applications in magnetic removal of heavy metal ions. *J. Mater. Chem.*, 22, 9230–9236, 2012.
43. Wang, S.G., Gong, W.X., Liu, X.W., Yao, Y.W., Gao, B.Y., Yue, Q.Y., Removal of lead (II) from aqueous solution by adsorption onto manganese oxide-coated carbon nanotubes. *Sep. Purif. Technol.*, 58, 17–23, 2007.
44. Zhang, C., Sui, J., Li, J., Tang, Y., Cai, W., Efficient removal of heavy metal ions by thiol-functionalized superparamagnetic carbon nanotubes. *Chem. Eng. J.*, 210, 45–52, 2012.
45. Li, C., Zhang, Y., Wang, X., Zhao, J., Chen, W., Removal and recovery of lead(II)ions from contaminated licorice extracts using oxidized multi-walled carbon nanotubes. *J. Nanosci. Nanotechnol.*, 11, 9731–9736, 2011.
46. Li, J.X., Chen, S.Y., Sheng, G.D., Hu, J., Tan, X.L., Wang, X.K., Effect of surfactants on Pb(II) adsorption from aqueous solutions using oxidized multiwall carbon nanotubes. *Chem. Eng. J.*, 166, 551–558, 2011.
47. Zhao, X.W., Jia, Q., Song, N.Z., Zhou, W.H., Li, Y.S., Adsorption of Pb(II) from an aqueous solution by titanium dioxide/carbon nanotube nanocomposites: Kinetics, thermodynamics, and isotherms. *J. Chem. Eng. Data*, 55, 4428, 2010.
48. Ren, X.M., Shao, D.D., Zhao, G.X., Sheng, G.D., Hu, J., Yang, S.T., Wang, X.K., Plasma induced multiwalled carbon nanotube grafted with 2-vinylpyridine for preconcentration of Pb(II) from aqueous solutions. *Plasma Process. Polym.*, 8, 589–598, 2011.
49. Ji, L.Q., Zhou, L.C., Bai, X., Shao, Y.M., Zhao, G.H., Qu, Y.Z., Wang, C., Li, Y.F., Facile synthesis of multiwall carbon nanotubes/iron oxides for removal of tetrabromobisphenol A and Pb(II). *J. Mater. Chem.*, 22, 15853–15862, 2012.
50. Tofighy, M.A. and Mohammadi, T., Adsorption of divalent heavy metal ions from water using carbon nanotube sheets. *J. Hazard. Mater.*, 185, 140–147, 2011.

51. Moradi, O., The removal of ions by functionalized carbon nanotube: Equilibrium, isotherms and thermodynamic studies. *Chem. Biochem. Eng. Q.*, 25, 229–240, 2011.
52. Stafiej, A. and Pyrzynska, K., Adsorption of heavy metal ions with carbon nanotubes. *Sep. Purif. Technol.*, 58, 49–52, 2007.

5

Carbon Allotropes in Nickel Removal

Amarpreet K. Bhatia[1], Nishtha Vaidya[2] and Shippi Dewangan[3]*

[1]Department of Chemistry, Bhilai Mahila Mahavidyalaya, Bhilai Nagar, Chhattisgarh, India
[2]Department of Zoology, Bhilai Mahila Mahavidyalaya, Bhilai Nagar, Chhattisgarh, India
[3]Department of Chemistry, SW Pukeshwar Singh Bhardiya Govt. College, Nikum, Durg, Chhattisgarh, India

Abstract
The stock of secure drinking and uncontaminated water is getting to be progressively perplexing recommendations all through the world. The arrangement of naturally economical nanomaterials with one-of-a-kind preferences specifically selectivity, elevated effectiveness, great quantity, eco-friendly, low-charge of generation forms, and steadiness, has been a need even though a few critical tasks and imperatives quite remained uncertain. Carbon nanomaterials, to be specific, single- and multiwalled nanotubes of carbon, fullerene have been created and connected for instance adsorbents for the management of contaminated water and refinement; nanomaterials based on graphene and graphene oxide have appeared as noteworthy guarantees for the management of contaminated water and water filtration, particularly, for mechanical- and pharmacological-loaded squanders. The expulsion of nickel from the sea-going environment could be a genuine natural issue in see of open well-being. Consequently, this chapter includes progressed carbonaceous nanomaterials and techniques that are used for the disposal of nickel particles in aquatic environments, and instance innovative nanosorbents for drinking groundwater and contaminated water management. Moreover, later patterns and tasks relating to the maintainable graphene and carbon-inferred nanomaterials and their apparatuses for considering and decontaminating polluted water are emphasized.

Keywords: Nanotubes of carbon, graphene, fullerene, nickel ion, contaminated water, adsorption

**Corresponding author*: shippidewangan2729@gmail.com

Chandrabhan Verma and Chaudhery Mustansar Hussain (eds.) Carbon Allotropes and Composites: Materials for Environment Protection and Remediation, (73–90) © 2023 Scrivener Publishing LLC

5.1 Introduction

Quick suburbanization and industrial development have troubled the natural steadiness of the environment by ruining nearly all the natural components with a run of toxins [1]. It has triggered numerous water figures to get masses of harmful metals which influence the feature of water [2]. For instance, as a consequence, the accessibility of elevated quality clean water has been reduced definitely. Natural, social, and financial variables have driven a negligible quantity of uncontaminated water, particularly to the destitute. Almost 200 million individuals pass on each year due to nonfatal and water-borne contaminations. More than one billion individuals do not have to get to clean water sources which have come about in major well-being issues around the world. Dairy ranches, rural, and poultry, as well as mechanical, family, and civil squander eras are contaminating the global water assets definitely and diminishing the accessibility of uncontaminated water exponentially [1].

Heavy metal defilement in water physiques has ended up with most of the main worldwide problems nowadays due to their carcinogenicity and harmfulness. The dishonorable release of mechanical sewages from different areas such as metallurgical, chemical, mining, tannery, shipping, and horticulture are the major source of heavy metal pollution (Figure 5.1) [3].

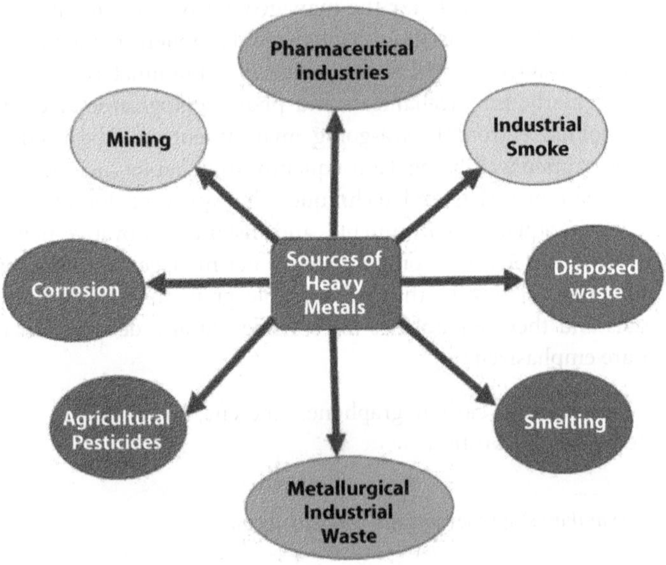

Figure 5.1 Various causes of heavy metal pollution in the environment [3].

The unsafe heavy metal ions which are recognized to be the major dangers to the surroundings are lead (Pb), nickel (Ni), copper (Cu), mercury (Hg), arsenic (As) zinc (Zn), iron (Fe), cobalt (Co), cadmium (Cd), and [3]. These heavy metal ions are nonbiodegradable and posture unfavorable impacts on human well-being, living beings, and biological systems [3].

Ni is recognized as a poisonous component of human well-being. Ni is discharged into the watery bodies via diverse sources. The important cause of nickel in water is filtering from metals in interaction through water, for instance, fittings and channels. In any case, it might also stay in a little underground water as a result of disintegration from nickel ore-bearing rocks [2, 3]. In different floras and faunas, Ni is found in a small amount which is nearly 0.1 to 0.6 mg L^{-1}. There are many industrial sources counting galvanization, mining, electroplating, metallurgy, catalyst, batteries, printing, amalgams, shades, electronic gear, stainless steel industry, and phosphate fertilizers are related to Ni pollution [2]. All these discharge Ni into the environment. Long-standing introduction to Ni has been originating to reason well-being issues, such as hypersensitivity, gastrointestinal aggravation, central anxious framework, the mental framework, cardiovascular, respiratory framework, and kidney and lung illnesses because it can effortlessly permit via the brain-blood obstruction [2]. Ni causes kidney issues, aspiratory fibrosis and skin dermatitis [4]. Consequently, it is of most extreme significance to expel Ni from polluted water successfully in arranging to avoid their hurtful influences on the environment and human wellbeing. The current advances utilized for the management of heavy metal refluent comprise Ni incorporate electrochemical management, extraction, precipitation by use of chemicals, adsorptions, ion exchange, filtrations, and invert osmosis [1–3].

Parenthetically, most of the present innovations are constrained by elevated working prices and small competencies for handling huge dimensions of water forms of weakened concentrations of heavy metal. Among entirely these current procedures, adsorptions have been documented using foremost broadly utilized strategies due to their being sustainable, the comfort of the handle, and elevated fetched adequacy. The foremost frequently utilized adsorbents for the expulsion of heavy metal counting Ni from polluted water are zeolites, biomass, soils, polymers, silica, and metal oxides. In later a long time, nanomaterials based on carbon (CNMs) have strained colossal consideration using capable adsorbents aimed at the expulsion of heavy metal counting Ni from polluted water on account of their particular belongings, for instance, expansive exterior to volume proportion, exceedingly permeable and empty assembly, in addition to abundant adsorption capabilities and efficacies. Despite these focal points, perfect CNMs have a few deficiencies, such as small dissolvability and dispersibility in many solvents, in addition to the inclination to make packages and masses. These restrictions might be overwhelmed by exterior adjustment of CNMs. This chapter gives an inclusive study of the evacuation of Ni from polluted water utilizing exterior adjusted CNMs, especially nanotubes of carbon (CNTs) and materials based on graphene (GR) [1–4].

5.2 Carbon and Its Allotropes: As Remediation Technology for Ni

The elemental carbon has been utilized for instance a basic of vitality for the previous few hundred long times and presented in this period of innovation, carbon has occupied a critical and exceptionally noticeable part in nearly all areas of science and innovation [5]. The innovation of carbon and its different physical forms have changed electrical and optical-based electrical manufacturing attributable to their empowering assets in an expansive range of uses. The curiously specific of carbon is its capacity to create numerous different physical forms owing to its valency. In later eras, different physical forms and characters of carbon have been designed, counting CNTs, GR, and fullerenes (Figure 5.2) [6]. Meanwhile, at the beginning of nanotechnology, nanocomposite-based carbon allotropes have gotten to be a driving segment of investigation and progression owing to their special holding characteristics. Polymer nanocomposites based on CNTs and fullerenes have pulled in critical inquire about intrigued due to their tremendous applications in each circle of science and innovation. The present investigative driving force uncovers that carbon and its different physical

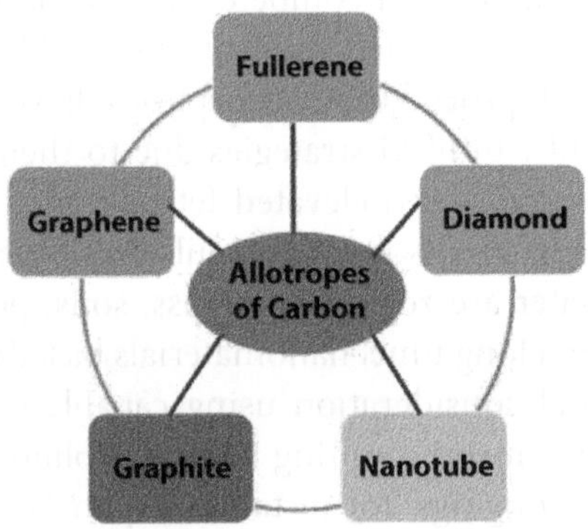

Figure 5.2 Allotropes of carbon [6].

forms have transformed the industry and the scholarly world owing to their captivated features [6].

5.2.1 Nanotubes Based on Carbon

5.2.1.1 Overview

Adsorbents based on carbon nanotubes have pulled in significant intrigued as probable adsorbents for heavy metals expulsion counting Ni [7]. The CNTs have arisen as of late as effective adsorbents for liquid filtration since of their momentous polluted water management competencies and their reasonableness for the evacuation of natural, mineral and carbon-based liquid toxins. Tunable chemical, electrical, physical, and basic properties, huge particular surface zone in addition to little, empty and coated assembly sort the CNTs as an exceptionally effective adsorbent for an extent of heavy metals. The CNTs are inferred as graphite and made of numerous layers of a solo particle in a hexagonal assembly in a honeycomb gem grid named using graphene. These graphene assemblies may collapse addicted to round and hollow form to allow make-to-solo-enclosed CNTs (SWCNTs), two-fold enclosed CNTs (DWCNTs), and/or multienclosed CNTs (MWCNTs). The SWCNTs are made of tube-like molded graphene layers which are composed and see comparative to the carbon atoms of benzene rings. It as a rule comprises ten molecules nearby their boundary taking the width of approximately one molecule. The DWCNTs and MWCNTs are bigger in measure and comprise two and a few SWCNTs, individually, loaded interior one another. All type of CNTs comprises empty assemblies comprising a distance across 0.4 to 3.0 nm. The MWCNTs may advance be characterized into two fundamental bunches i.e., (i) parchments like assembly and (ii) Russian doll show, whereas SWCNTs may be categorized encourage into (i) crisscross design, (ii) helical, (iii) rocker, and (iv) chiral assemblies [1].

5.2.1.2 Features of CNTs

The CNTs have elite electrical and basic features which sort them, probable applicants, for many uses. A few noticeable belongings of CNTs are displayed in Table 5.1 [1]. For instance in SP^2 hybridization, the elemental carbon marks an assortment of forms, subsequently separated from graphite, carbon can also sort sealed and vulnerable enclosures. Therefore, owing to SP^2 hybridization, the CNTs are rigid and have advanced axial power. After numerous enclosures of carbon, Iijima [8] detected nanoscale tube-shaped

Table 5.1 A number of noticeable features of nanotubes of carbon [1].

Type	Modification	Micropore volume (g/cm^3)	Mesopore volume (g/cm^3)	Surface area (g/m^2)
CNT	Methane	-	-	122
CNTs	H_2O_2 oxidized	-	-	130
	$KMnO_4$ oxidized	-	-	128
	HNO_3 oxidized	-	-	154
	Pristine	0.0025	0.3404	64.8
	NaOCl	0.0382	1.11	194.08
SWCNTs	Catalytic chemical vapor deposition	0.0249	0.093	576.67
	Pristine	-	-	625
	Hydroxylated	-	-	526
	Carboxylated	-	-	552
	Purified	-	-	228
DWCNTs	Catalytic chemical vapor deposition	0.149	0.098	382.24
MWCNTs	Pristine	0.029	0.58	258.6
	Oxidized	0.041	0.89	427.5
	Pristine	0.01	0.08	156
	Co/SiO at 700 °C	0.76	0.003	411
	Co/SiO at 900 °C	0.66		396
	Co/NaY at 600 °C	0.43	0.004	128
	NaOCl	-	-	88.56
	Graphitized	-	-	117
	Hydroxylated	-	-	228

carbon assembly earliest in 1991. These nanoscale tube-shaped assemblies were fabricated of numerous tens of graphitic ammunitions and consequently were designated for instance MWCNTs. The CNTs are physical forms of carbon and are elucidated employing graphene layers revolved into a cylinder. Physical features of CNTs for instance advanced mechanical quality, small electrical resistivity and advanced warm conductivity are owing to the special assembly of CNTs [8]. Then comes about of hypothetical calculations and exploratory showings clarified that CNTs are greatly adaptable through an elevated youthful modulus. The auxiliary features of CNTs uncovered that CNTs are either metallic or semiconductors [1].

The CNTs show greater electrical conductivity, auxiliary keenness, chemical steadiness and lesser edge electric arena, totalled holes (shaped by CNTs owing to van der Waals strengths of fascination), expansive outside exterior zone, empty contribute of nanotubes, interstitial hole seats concerning the cylinders packages, and forest at the edge. Utilitarian features of CNTs incorporate form recuperation, desiccated grip, elevated checking, tetra hertz division, huge hit activation, close perfect dark body assimilation and thermoacoustic sound emanation. The chemical well-suited with DNA and proteins prepared the CNTs as constituents of biosensors and restorative gadgets. The CNTs networks in convenient channels and decontaminate the sullied liquid by expelling microbes, infections and electrochemically corroded natural pollutants. A few of the CNTs have hydrophilic behaviour; hence, owing to their greater solubility in water, their adsorption capability for heavy metals from polluted water makes strides altogether. The execution of the CNTs is influenced by their features, which can be adjusted amid its creation through actuation and adjustments to attain greater effectiveness. A few actuation methodologies have been utilized to adjust the exterior and auxiliary features of CNTs for specific use. Be that as it may, for the most part, the CNTs may be functionalized by either covalent or noncovalent methodologies.

The covalent functionalization of the CNTs may be achieved through corroding with corrosive to make –COOH or OH groups. The fullerene-alike tips were extra receptive than the dividers, subsequently, the useful functional groups (–COOH, –OH and –ClO) can be produced on CNTs through conclusion functionalization. The exterior of CNTs can also be functionalized with amino groups to upgrade its competence to detect organophosphorus pesticides. Within the noncovalent functionalization, different bio-chemically dynamic particles may be connected to the CNTs via noncovalent sidewalls using H–bonding, van der Waals strengths, π-π stacking, hydrophobic bonding or electrostatic strengths. The noncovalent functionalization is measured as more appealing because it keeps up the auxiliary characteristics of flawless CNTs. The CNTs can also be functionalized with ultra-sonication, weaken corrosive action, corrosive oxidation, radiation and microwave corrosive absorption to deliver unadulterated CNTs without the pollutions of graphite units, shapeless carbon and metal catalysts [1].

5.2.2 Fullerenes

Fullerenes were founded and union sometime recently the generation of CNTs. In reality, fullerenes are recognized as the point of reference generation which clue to the generation of different nanomaterials based on carbon. Other than the different physical forms of carbon like graphite and diamond, fullerenes or moreover recognized for instance buckyballs the 3rd frame of the presence of carbon. The basic distinction concerning CNTs and fullerenes is the form of the carbons in which CNTs display within the frame of nanotubes whereas fullerenes are made up of enclosure like an assembly which included twelve 5-member rings and an indefinite number of six members. Fullerenes for the most part display within the form of hexagonal rings in which the carbon particles are put.

By and large, it is soundly recognized that structures with fewer hexagons show more noteworthy sp^3 holding character, greater strain energies and accessibility of extra responsive carbon destinations. Moreover, isomers with adjoining pentagons also show lesser solidness and comparative plenitude associated with the isomers comprises of separated pentagons in which reverberation assemblies delocalize π bonds over the fullerene assemblies. On that note, C-60 is an icosahedral symmetric fullerene which is steadied through reverberation assemblies associated with an identical electric form and holding shape for separate carbon atoms. This comparing steadiness has carried up C-60 as the beginning fabric for chemical responses within the claim right.

Additionally, covalent, supramolecular and Endohedral changes empower atomic control and polymeric fabric improvement for particular natural uses. Fullerene's features were fine misused to extend the use within the natural segment [9].

5.2.3 Graphene

5.2.3.1 Overview

Graphene (GR) is a recently found physical form of carbon. It is enchanted: the basis of entirely graphitic carbon, one particle dense, two-dimensional carbon fabric. GR has risen as a modern trust for the complete logical society. A molecularly dense sole coating of GR is compared to only sheet of the scratchpad. And carbon particles are fortified and composed in a hexagonal honeycomb cross-section through sp^2 hybridization. This grid comprises two comparable subgrids of carbon molecules reinforced to more additional. In this fashion, graphite is fabricated of numerous coatings of GR and each sheet is consistent with each other by Van Der Waal strengths. Utilizing the nearby orbital inhabitance approach and second-order perturbation hypothesis, it was detailed that the binding energy concerning every layer of graphite is roughly 60 to 72 meV [6].

Since the previous era, there has been a vast development within the utilization of GR and GR-based materials for natural remediation, owing to their one-of-a-kind possessions which make a difference to modern conceivable outcomes to advance the execution of various natural forms. There is the option of whether to apply GR by way of a nanocomposite based on carbon will be decided by the taken toll, prepare capacity and natural suggestions of each fabric. In this respect, natural uses based on graphene oxide (GO) proposed additional practical conceivable outcomes related to perfect graphene due to GO's lower generation prices. In expansion to financial contemplations, natural suggestions of materials based on GR will express a critical calculation within the advancement of innovations based on GR. GR may be an auxiliary for CNT and a perfect fabric for water management. Associated with CNTs, the application of GR-based materials as adsorbents may propose a few points of interest. To begin with, solo enclosure GR materials have two basal planes accessible for poison adsorption. In differentiation, the internal layers in CNTs are not opened by the adsorbates. Moment, GO and decreased graphene oxide (rGO) can be effortlessly synthesized through the chemical shedding of graphite, without utilizing complex devices or metallic catalysts. The coming about GR fabric is free of catalyst buildups, and no assist filtration steps are required. Within the particular case of GO, the as-arranged fabric now has an expansive number of oxygen-containing functional groups and no extra corrosive medicines are required to confer a hydrophilic character and reactivity to GO. This can be a critical advantage, subsequently, those functional assemblies are possibly capable of the adsorption of metal particles by GO sheets [10].

GR-based materials help as active adsorbents, owing to their huge particular surface region and electron-wealthy environment. Owing to the solid functional assemblies on the GO exterior, GO will be a possible adsorbent for metal particle complexation via equally electrostatic and facilitate methodologies. An assortment of things about has depicted the application of materials based on graphene as adsorbents for the evacuation of inorganic species from fluid arrangements [1]. Much of this research about have employed GO as a demonstrated adsorbent for the removal of metal particles in water. GO is best to perfect graphene for metal particle adsorption due to GO's raised substance of oxygen groups available to associated with metal particles. The consequence of these oxygen-having functional assemblies was illustrated by comparing the Pb (II) adsorption implementation of impeccable and corroded graphene layers. GO and graphene nanosheets (GNs) can moreover be joined with metal oxides to make composite materials [10].

5.2.3.2 Properties

GR nanomaterials have been connected for handling wastewater for instance they appear critical important physicochemical features that incorporate elevated surface region, hydrophobicity and, basic functionalization. Nanocomposites based on graphene may be applied for instance nanosorbents for killing dangerous and poisonous blends of numerous poisons, and for their ensuing catalytic corruption utilizing catalytic, photocatalytic, electrocatalytic and photoelectrocatalytic corrosion, and lessening forms using their profitable generation, and apply is quiet an imposing task. Constant nanoporous GR layers may be connected for purifying and filtrating liquid via shifting their aperture estimate and thus the weight, the most disadvantage being their mechanical steadiness subsequently expanding the aperture number. Yan *et al.* [12], delivered N-incapacitated roll-like GR-composites and connected them for instance capacitive deionization anodes through an association strategy. This roll complex advertised considerably lower electronic resistivity and open surface region and showed significant salt adsorption competence with prevalent alteration capacity attached to reusing processes. It has appeared that the GO nanolayers might be connected as reasonable specific boundaries for liquid penetrability; superoleophobic and super-hydrophobic permeable GR has been created as permeable for different materials, illustrating extraordinary selectivity, appropriate recyclability, and critical retention capacities of more than 90%.

Also, GO-based nanofiltration film with a massively permeable polyacrylonitrile nanofibrous tangle has been manufactured for handling water; graphene oxide might produce a blockade on a polyacrylonitrile nanofibrous tangle through well-measured width. Not at all like CNT-based nanomaterials, the watery expulsion of harmful colors by GR-based nanoadsorbents have changed points of interest wherein solo sheet GR or GO nanolayers have two basal planes open for pollutant adsorption; CNTs internal dividers are not accessible to adsorbates. Additionally, GR and its equivalent can be essentially created through mechanical chemical/ peeling of graphite deprived of utilizing metallic catalysts (e.g. Ni) and composite devices. Eminently, the coming about nanomaterials based GR are destitute of catalytic buildups, hence not requiring extra refinement steps. Concerning GO, the nanosorbents comprise an endless amount of O_2-bearing functionalities that blocks extra corrosive medications to improve reactivity and hydrophilicity feature to GO. It may be emphatically associated through differences of carbon-based/mineral pollutants in ionic and/or atomic shapes via the instruments, usually recognized for instance the hydrophobic and electrostatic π-π intuitive, amongst others. Hence, GO has strained impressive consideration such as a compelling nano sorbent for sullied water/wastewater [13].

5.3 Removal of Ni in Wastewater by Use of Carbon Allotropes

5.3.1 Carbon Nanotubes for Ni Adsorption From Aqueous Solutions

Ni is viewed as unique of the foremost damaging heavy metals that are considered as a nondecomposable harmful heavy metal ion affecting many disputes to the human physique. Ni is broadly utilized in industry and is frequently found as an emanating; more than 40% of Ni is utilized in steel generation, Ni batteries and a few amalgams. Numerous division forms have broadly been connected for Ni evacuation. In any case, the adsorption procedure is found to be additionally compelling for Ni expulsion since of the tall productivity and low price. Abdel-Ghani *et al.* [14] studied the FTIR spectra of Ni (II)-encumbered MWCNTs and crude MWCNTs. The extending band of C–C at 1420 per cm on crude MWCNTs was part into 1419 and 1461 per cm after the adsorption of Ni ions. The presences of crests at about 400 per cm allude to the solid holding amid the Ni (II) particles and the nanotube via oxygen-having functional assemblies. These oxygen-having functional assemblies increment the adsorption ability of Ni ions as they give various adsorption destinations. Acid-corroded MWCNTs were connected within the expulsion of Ni; the evacuation productivity expanded from 10%–80% on expanding the pH range from 3.5–8.

The adsorption information was well fixed utilizing Langmuir isotherm with an intended adsorption capability of 9.8 mg/g at the temperature of 60°C and pH 6.55. Wilson *et al.* [15] detailed an adsorption ability of 49.26 mg/g by corroded CNTs. Additionally, in an alternative test utilizing tarnished MWCNTs, adsorption of Ni (II) expanded steadily from almost 0% to 99% at a pH range of 2 to 9. The increment in evacuation rate by pH is owing to the reality that underneath pH 9 the leading type of nickel is Ni (II). Hence, assist expanding the pH sorts the exterior of the CNTs additional negative, which indorses adsorption. The MWCNTs were enhanced with polyacrylic corrosive which in turn made the exterior additional negative and improved the adsorption over the corrosion of the nanotubes only [7]. Additional research about utilizing corroded MWCNTs and SWCNTs with a NaClO arrangement was conveyed for Ni (II) adsorption from watery arrangements [16] counting HNO_3 oxidized MWCNTs [17].

5.3.2 Ni Adsorption From Aqueous Solutions on Composite Material of MWCNTs

The adsorption of heavy metal ions from polluted water has been carried out by using the complex fabric of MWCN. The MWCNTs-Fe_3O_4, have been effectively connected for the expulsion of the Ni (II) from polluted water. In a heterogeneity adsorbent surface, locales pooled double are fixed within the Langmuir-Freundlich isotherms models equation which was employed to distinguish between dual sorts of adsorption destinations with more prominent and reduced energy empathies for the Ni (II). It is accepted that Ni (II) adsorption happens at the locales of energy with altered nanocomposites of MWCNTs and the nanomodification mains to a 20% increment within the adsorption ability at little up to 0.1 per mole equilibrium adsorbate amounts [18].

5.3.3 GR and GO-Based Adsorbents for Removal of Ni

GR is a single atom profuse coated hexagonal grid of carbon molecules and is recognized using the most slender fabric with a quality 200 times than that of steel. 2-D GR is being utilized broadly in nearly every area for example in semiconductors, touch screens, computer chips, mobiles, LCDs, energy production, batteries, water channels, sun-oriented cells, supercapacitors, pharmaceutical, and natural disciplines. These materials based on 2-D GR are receiving increasing consideration in water management owing to their one-of-a-kind physicochemical features to be specifically elevated surface region, elevated mechanical quality, electronic properties, thermal portability, and tunable surface chemistry. Tabish *et al.* [19] outlined permeable GR and connected it as an adsorbent for the expulsion of heavy metal ions in addition to extra toxins from water. GO was functionalized with 4-sulfophenylazo and connected it for the expulsion of a Ni ion from the fluid arrangement. The planned fabric appeared the most extreme adsorption ability of 66 to 191 mg g^{-1} for the Ni (II). Vilela D *et al.* [20] outlined a GR-based self-driven microdot framework whose organization was fabricated of nanosized multidimensional comprising GO, Ni, and Pt. Each coating accomplished a distinctive work, e.g., GO incarcerations the heavy metal Pb(II) ions, the center layer of Ni empowers the switch of microdots with the assistance of an outside attractive arena, and the internal layer of Pt makes a difference in the motor in self-propelling [18].

Hoan *et al.* [21] detailed that GO/Fe_3O_4 appeared amazing competencies in evacuating Ni ions from wastewater with an assimilation ability of 76.34 mg/g for Ni(II). Using GO incapacitated through Fe_3O_4, the nanoscale round units gave extra adsorption locales to the fabric, expanding its compelling exterior region [21, 22]. The adsorption capacity of Ni (II) on GO and GO-G surfaces from watery arrangements was well explored and illustrated by Najafi *et al.* [23]. The consequence appeared that the improved values of introductory concentration of Ni (II), pH contact time and adsorbent measurements were found to be 15 mg/L, 6, 50 min, and 20 mg separately; progressing through the effect of temperature it was seen that, when the temperature was expanded from the range of 283 to 308 K, the expulsion of Ni (II) by GO surface diminished but the expulsion of Ni (II) by functionalized GO-G surface diminished in anticipation of 298 K and later that it was expanded. Four sorts of the Langmuir and Freundlich isotherm models were utilized, and they come about to clearly portray that the adsorption of Ni (II) on GO and GO-G surfaces was well fixed and found to be in great understanding with sort (II) of the Langmuir isotherm demonstrate and Freundlich isotherm demonstrate separately [23]. Table 5.2 [3, 7, 16, 24] shows the adsorption of Ni (II) on distinctive carbon nanomaterials as adsorbents.

Table 5.2 Different carbon nanomaterials for adsorption of Ni (II) [3, 7, 16, 24].

Adsorbents	Adsorption capacity (mg/g)	pH	Temperature	References
MWCNTs	7.53	7.0	-	[16]
SWCNTs	9.22	7.0	-	
SWCNTs (NaOCl)	47.85	7.0	298 K	[7]
MWCNTs (NaOCl)	38.4	7.0	298 K	
O-MWCNT	9.43	6.55±0.02	300 K	
O-MWCNTs	$6.346*10^{-5}$ mol/g	5.4±0.1	293±2 K	
Polyacrylic acid functionalized O-MWCNTs	$6.615*10^{-5}$ mol/g	5.4±0.1	293±2 K	
MWCNTs	6.09	7.0	298 K	
O-MWCNT	17.86	6.5	298 K	
Oxidized CNTs	49.261	6.0	298 K	
Iron oxide-MWCNTs composite	9.18	6.4±0.2	298±2 K	
O-CNTs	50.09	7.0	293 K	

(Continued)

Table 5.2 Different carbon nanomaterials for adsorption of Ni (II) [3, 7, 16, 24]. *(Continued)*

Adsorbents	Adsorption c apacity (mg/g)	pH	Temperature	References
Oxidized CNTs	38 (18.08)	6.0	298 K	[3]
PAA-modified MWCNTs	2.71	5.4	293 K	
Sodium hypochlorite (NaClO)-oxidized MWCNT	13.05	7.0	298 K	
Triphenylphosphine (Tpp)-linked MWCNTs	48.84	6.0	298 K	
Graphene oxide (GO)	122	-	-	[24]
Graphitic carbon nitride (G-C_3N_4)	168	-	-	

5.4 Conclusion

In this chapter, the natural and distinct decontamination of Ni (II) from heavy metal toxins with the help of nanomaterials based on carbon, specifically CNTs, GR, fullerene, and GO are discussed. These nanomaterials based on carbon have been applied within the filtration of Ni (II)-polluted water with remarkable accomplishment. The purpose arrears the fruitful use is owing to their interesting kinds of stuff comparable elevated exterior region, easiness of reusing, and ease to desorb the adsorbed metal ions; as it was utilizing the mineral corrosive solution and recovered fabric can be used again with the maintenance of adsorption competence. In expansion to this stuff, the nanomaterials based on carbon can effortlessly be created with further nanomaterials and are simple to be functionalized coming about in multifunctional nanoadsorbent. C nanomaterials based on carbon are exceedingly eco-friendly with existing beings and the atmosphere. There is also a monstrous impact of distinctive factors for instance interaction period, pH, and sort of adsorbents on the method of metal ion adsorption. Established on this works survey, it can be determined that nanomaterials based on carbon have intriguing physicochemical characteristics and have incredible latent to be oppressed within the natural treatment and water decontamination.

References

1. Ahmad, J., Naeem, S., Ahmad, M., Usmand, A.R.A., Al-Wabel, M. I., A critical review on organic micropollutants contamination in wastewater and removal through carbon nanotubes. *J. Environ. Manage.*, 246, 214–228, 2019.
2. Abd El-Magied, M.O., Hassan, A.M.A., Gad, H.M.H., Mohammaden, T.F., Youssef, M.A.M., Removal of nickel (II) ions from aqueous solutions using modified activated carbon: A kinetic and equilibrium study. *J. Dispers. Sci. Technol.*, 39, 862–873, 2017.
3. Lau, Y.J., Khan, F.S.A., Mubarak, N.M., Lau, S.Y., Chua, H.B., Khalid, M., Abdullah, E.C., Functionalized carbon nanomaterials for wastewater treatment, in: *Industrial Applications of Nanomaterials*, pp. 283–311, 2019, https://doi.org/10.1016/B978-0-12-815749-7.00010-4.
4. Aji, M.P., Wiguna, P.A., Karunawan, J., Wati, A.L., Sulhadi, Removal of heavy metal nickel-ions from wastewaters using carbon nanodots from frying oil. *Proc. Eng.*, 170, 36–40, 2017.
5. Karthik, P.S., Himaja, A.L., Singh, S.P., Carbon-allotropes: Synthesis methods, applications and future perspectives. *Carbon Lett.*, 15, 4, 219–237, 2014.

6. Tiwari, S.K., Kumar, V., Huczko, A., Oraon, R., De Adhikari, A., Nayak, G.C., Magical allotropes of carbon: Prospects and applications. *Crit. Rev. Solid State Mater. Sci.*, 41, 1–61, 2016.
7. Bassyouni, M., Mansi, A.E., Elgabry, A., Ibrahim, B.A., Kassem, O.A., Alhebeshy, R., Utilization of carbon nanotubes in removal of heavy metals from wastewater: A review of the CNTs' potential and current challenges. *Appl. Phys. A*, 126, 38, 2020.
8. Iijima, S., Helical microtubules of graphitic carbon. *Nature*, 354, 56, 1991.
9. Thines, R.K., Mubarak, N.M., Nizamuddin, S., Sahu, J.N., Abdullah, E.C., Ganesan, P., Application potential of carbon nanomaterials in water and wastewater treatment: A review. *J. Taiwan Inst. Chem. Eng.*, 72, 116–133, 2017.
10. Santhosh, C., Velmurugan, V., Jacob, G., Jeong, S.K., Grace, A.N., Bhatnagar, A., Role of nanomaterials in water treatment applications: A review. *Chem. Eng. J.*, 306, 1116–1137, 2016.
11. Sabzehmeidani, M.M., Mahnaee, S., Ghaedi, M., Heidari, H., Roy, V.A.L., Carbon based materials: A review of adsorbents for inorganic and organic compounds. *Mater. Adv.*, 2, 598—627, 2021.
12. Yan, T., Liu, J., Lei, H., Shi, L., An, Z., Park, H.S., Zhang, D., Capacitive deionization of saline water using sandwich-like nitrogen-doped graphene composites via a self-assembling strategy. *Environ. Sci. J. Integr. Environ. Res. Nano*, 5, 2722–2730, 2018.
13. Nasrollahzadeh, M., Sajjadi, M., Iravani, S., Varma, R.S., Carbon-based sustainable nanomaterials for water treatment: State-of-art and future perspectives. *Chemosphere*, 263, 128005, January 2021.
14. Abdel-Ghani, N.T., El-Chaghaby, G.A., Helal, F.S., Individual and competitive adsorption of phenol and nickel onto multiwalled carbon nanotubes. *J. Adv. Res.*, 6, 3, 405–415, 2015.
15. Wilson, K., Yang, H., Seo, C.W., Marshall, W.E., Select metal adsorption by activated carbon made from peanut shells. *Biosour. Technol.*, 97, 18, 2266–2270, 2007.
16. Lu, C., Liu, C., Rao, G.P., Comparisons of sorbent cost for the removal of Ni2+ from aqueous solution by carbon nanotubes and granular activated carbon. *J. Hazard. Mater.*, 151, 1, 239–246, 2008.
17. Chen, C. and Wang, X., Adsorption of Ni (II) from aqueous solution using oxidized multiwall carbon nanotubes. *Ind. Eng. Chem. Res.*, 45, 26, 9144–9149, 2006.
18. Baby, R., Saifullah, B., Hussein, M.Z., Carbon nanomaterials for the treatment of heavy metal-contaminated water and environmental remediation. *Nanoscale Res. Lett.*, 14, 341, 2019.
19. Tabish, T.A. *et al.*, A facile synthesis of porous graphene for efficient water and wastewater treatment. *Sci. Rep.*, 8, 1, 1817, 2018.
20. Vilela, D. *et al.*, Graphene-based microbots for toxic heavy metal removal and recovery from water. *Nano Lett.*, 16, 4, 2860–2866, 2016.

21. Hoan, V., Thi, N., Thu, A., Thi, N., Duc, H.V., Cuong, N.D. *et al.*, Fe3O4/reduced graphene oxide nanocomposite: Synthesis and its application for toxic metal ion removal. *J. Chem.*, 2016, 1–10, 2016.
22. Piumie Rajapaksha, P., Power, A., Chandra, S., Chapman, J., Graphene, electrospun membranes and granular activated carbon for eliminating heavy metals, pesticides and bacteria in water and wastewater treatment processes. *Analyst*, 143, 5629, 2018.
23. Najafi, F., Moradi, O., Rajabi, M., Asif, M., Tyagi, I., Agarwal, S., Gupta, V.K., Thermodynamics of the adsorption of nickel ions from aqueous phase using graphene oxide and glycine functionalized graphene oxide. *J. Mol. Liq.*, 208, 106–113, 2015.
24. Chai, W.S., Cheun, J.Y., Kumar, P.S., Mubashir, M., Majeed, Z., Banat, F., Ho, S.-H., Show, P.L., A review on conventional and novel materials towards heavy metal adsorption in wastewater treatment application. *J. Clean. Prod.*, 296, 126589, 2021.

6

Molybdenum-Modified Carbon Allotropes in Wastewater Treatment

Madhur Babu Singh[1,2], Anirudh Pratap Singh Raman[1,2], Prashant Singh[1*], Pallavi Jain[1,2] and Kamlesh Kumari[3†]

[1]*Department of Chemistry, Atma Ram Sanatan Dharma College, University of Delhi, New Delhi, India*
[2]*Department of Chemistry, Faculty of Engineering and Technology, SRM Institute of Science and Technology, NCR Campus, Ghaziabad, India*
[3]*Department of Zoology, University of Delhi, Delhi, India*

Abstract
Molybdenum disulfide is a pioneering example of a two-dimensional material, and due to its extraordinary properties, it has lately attracted a significant amount of study attention. These characteristics include a large surface area and porous character, the ability to be tuned by intercalation, excellent chemical and thermal stability, and wide availability in their natural state. MoS_2 is a strong candidate for combining with carbon nanomaterials to develop unique hybrid nanostructures with extraordinary properties, such as increased catalytic activity. This would be accomplished by mixing MoS_2 with the carbon nanoparticles. In this chapter, we have made an effort to review and consolidate the characteristics that are related to the synthesis of MoS_2-carbon allotropes. Further, we have provided a brief description of the synthesis of these allotropes, as well as their applications. Researchers are focusing their attention on it for a variety of reasons, including the fact that it is simple to use, cost-effective, energy-efficient, and environmentally friendly.

Keywords: Molybdenum disulphide, carbon allotropes, waste treatment, nanocomposite

**Corresponding author*: psingh@arsd.du.ac.in
†*Corresponding author*: kkumari@zoology.du.ac.in

Chandrabhan Verma and Chaudhery Mustansar Hussain (eds.) Carbon Allotropes and Composites: Materials for Environment Protection and Remediation, (91–112) © 2023 Scrivener Publishing LLC

6.1 Introduction

Carbon is one of the most interesting element, it displays a wide range of chemically motivated allotropic forms, and it may range from one to three-dimensional structures [1]. Examples include a vast difference in mechanical strength, conductivity, and appearance, and these are caused by the difference in specific bonding orientation between graphite and diamond. The various carbon allotropes including fullerene, carbon nanotubes (CNT), graphene, and their composites have drawn significant attention in energy storage and conversion over the last few years due to their exceptional mechanical, electrical, and optical capabilities caused by structural variety. According to this perspective, fuel cells and supercapacitors have attracted interest as workable options for dependable power-generating technology that is clean, safe, economical, efficient, and assures minimal emissions of pollutants [2, 3].

The fundamental structure and physicochemical features of the material used have a significant influence on their success. Most of these developments may be attributed to the creation of new materials that possess the appropriate attributes to improve the efficiency and stability of the device being manufactured. On comparing the organized carbon allotropes-based nanomaterials to amorphous or disordered carbon materials, one can see that the ordered carbon allotropes-based nanomaterials provide better activity, stability, and oxidation resistance. Researchers in academia and business are continuously working on a variety of carbon-based materials to understand the links between the physical and chemical characteristics of materials. New advance technologies are towards development to improve their performance and stability. Examples of carbon allotropes include graphene oxide, carbon dots, graphene carbon nanotubes, and fullerenes. Due to the extraordinary scope of possibilities, they are also referred to as "wonder materials" [4]. Some of the properties that make it feasible to employ materials such as sensors, electrodes, supercapacitors, batteries, and other electronic components include conductivity, electrochemical activity, surface area, and ease of functionalization. Carbon allotrope blending or mixing has the potential to enhance the mechanical properties of materials, which might lead to a wider variety of applications for these materials. Utilizing carbon-based materials as bridges presents significant potential to more efficiently use of these materials. It is feasible to transform rigid thermoelectric materials into flexible thermoelectric materials. Some possible future research directions are in the areas of carbon-hybrid thermoelectric materials [5, 6].

6.2 Carbon-Based Allotropes

Diamond and graphite are the two natural allotropes of carbon that include long networks of sp^3- and sp^2-hybridized carbon atoms, respectively. The physical characteristics of these forms are distinct, including hardness, lubrication behaviour, thermal conductivity and electrical conductivity. Modifying the periodic binding in networks made up of sp^3-, sp^2-, and sp-hybridized carbon atoms theoretically opens up a wide range of additional possibilities to create carbon allotropes. There are various forms of carbon allotropes and their properties are discussed below in Table 6.1 and shown in Figure 6.1.

6.2.1 Graphene

Carbon has the extraordinary ability to be able to generate the chemically stable, two-dimensional (2D), one-atom-thick membrane known as graphene. Graphene has found various applications in a various sectors due to its favourable features. The crucial use was discovered in biomedical engineering like drug delivery. Covalent bonds are formed between each carbon atom in graphene and three additional pi-bond through a process known as sp^2 hybridization. Graphene is presently known to be the thinnest as well as the strongest substance ever. The strength of graphene is about six folds of copper. It has exceptionally high conductivity, as well as high levels of stiffness and strength, and it does not let gases pass through it [7, 8].

6.2.2 Graphite

This form of carbon allotrope is a superb thermal and electrical conductor, making it an excellent material for electrodes in a lamp.The electrical conductivity of the graphene is due to the delocalization of the pi electrons of the carbon atoms, a phenomenon which is not possible in diamonds. Graphite displays the lowest energy state at ambient pressure and temperature. A graphene layer or sheet is a single carbon layer that is part of the crystalline honeycomb graphite lattice. The graphite crystal lattice is made up of stacks of parallel 2D-graphene sheets that have tightly linked sp^2 hybridised carbon atoms. The graphene sheet contains less energy when it is perfectly flat. This is because the 2pz orbitals of the carbon atoms get overlapped with each other most effectively. Graphite is characterized by its anisotropic nature because the carbon atoms may form bonds both

Table 6.1 A comparison of physical and chemical properties of carbon allotropes.

Allotropes	Thermal conductivity	Electrical conductivity	Tenacity	Density (g/cm^3)	Hardness	References
Graphite	Anisotropic 1500-2000	Anisotropic $2\text{-}3 \times 10^4$	Flexible, Non-elastic	2.25	High	[17, 18]
Graphene	4840-5300	2000	elastic	2.27	Uppermost layer	[18]
CNT	3500	Structure-dependent	elastic	1.74	High	[10, 19]
Fullerene	0.4	10^{-10}	elastic	1.51	High	[20]

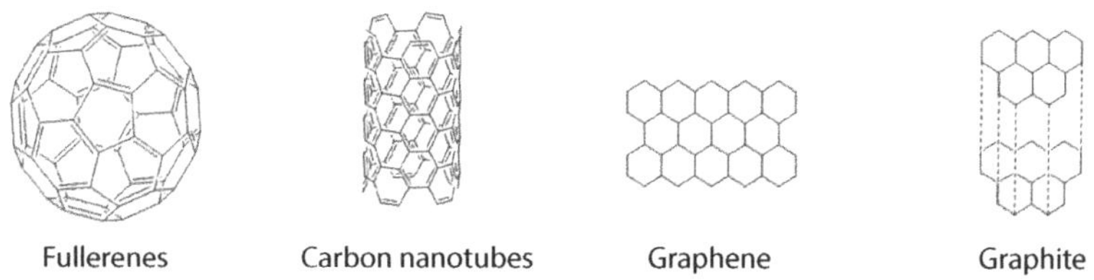

Figure 6.1 Diagrammatic presentation of carbon allotropes.

in and out of the plane of the crystal. Graphite has a lower density and high elasticity than diamond [9].

6.2.3 Carbon Nanotubes

Nanomaterials have been considered to use in a vast applications, most notably in electronics, medicine, and the clean-up of the environment. Carbon nanotubes (CNTs) have surpassed the performance of other carbon-based and metal-based nanomaterials because of their remarkable properties, such as great mechanical strength, and low weight. A single-walled carbon nanotube (SWNT) is a graphene sheet that has been flipped over and has an average diameter of 1.4 nm. A multiwall carbon nanotube (MWNT) is made up of concentric cylinders that have an interlayer spacing of 3.4 and a typical diameter of 10–20 nm. The lengths of the two distinct kinds of tubes may range anywhere from tens to hundreds of microns, and sometimes even centimetres. Two primary structural aspects define an SWNT, which is short for single-walled carbon nanotube. In terms of their ability to withstand mechanical stress, nanotubes are up there with some of the most solid and hardy things found in nature. A nanotube has a high tensile strength (about 100 times that of steel), and its Young's modulus is 1.2 TPa, thus it can sustain considerable pressures before breaking mechanically [10]. This is because Young's modulus is so high. The electrical properties of carbon nanotubes are significantly influenced by a variety of factors, including their diameter and chirality, indices etc [11, 12].

6.2.4 Glassy Carbon (GC)

Materials based on carbon have tuneable properties because the arrangement of carbon atoms may vary depending on the processing technique and the precursors that are utilized. Glassy carbon (GC) is an isotropic and brittle material. The presence of numerous structural pores may be responsible for their characteristically low density. In addition, GC has excellent chemical stability as well as high thermal and electrical conductivities. Further, it demonstrates a high level of permeability to gases as well as liquids. Graphene represents an intriguing alternative and has the potential to be explored in a variety of applications, like energy storage devices, purification of wastewater, sensors etc. GC electrodes were also used to test for the electrochemical formation of H_2O_2 for disinfecting drinking water. Anodic polarisation was used by Jin *et al.* to functionalize the RVC electrode and could be used as the cathode in an electrochemical device for the formation of H_2O_2 [13]. The corrosive functionalization led to the development of holes in this GC electrode as well as an increase in the surface roughness [14–16].

6.3 Molybdenum Disulfide

Molybdenum disulfide (MoS_2) nanostructures allotropes exhibit a wide range of morphologies like nanoplatelets, nanostrips, and nanorods that can be understood as portions of the mass improvements, to significantly bigger fullerenes and nanotubes with a polymorphic atomic structure that distinguishes from the stable bulk state. The minute particles include a great number of atoms that are chemically active yet have loose connections between them [21].

In both of its forms, the naturally occurring layered solid MoS_2 is useful in a variety of industrial applications. The individual S-Mo-S is sandwiched by van der Waals interactions in a form of hexagonally packing. This is quite similar to the way graphite's layers of graphene are arranged. MoS_2 is a good choice for use as a dry lubricant. Because of the substantial excitation of the metal-centred d-d transition, leading to distinctive electronic properties [2]. It is used in a wide variety of applications, like hydrogen generation, catalysis, and coating materials. Both theoretical and experimental work is being performed by different research groups to discover the fundamental principles that dictate the stability, reactivity, and electrical characteristics of these materials.

6.3.1 Synthesis of MoS_2

There have been several ways used to create MoS_2 nanostructures, which may be split into two categories: (1) physical methods, and (2) chemical methods. Physical approaches often require quick, high-energy processes such as pulsed laser deposition, laser ablation, microwave plasma or arc discharge. The MoS_2 nanostructures that are produced, however, are sparse and have a propensity to aggregate, which might limit their surface areas and prevent further functionalization or dispersion [22].

6.3.2 Physical Methods

In the plasma reaction process, a quartz tube serves as a single-mode microwave cavity. $Mo(CO)_6$ was vaporised and then added to the reaction just ahead of the plasma zone to create nanoparticles. The H2S gas was heated first to prevent the compounds from precipitating. using hexacarbonyl, a precursor called $Mo(CO)_6$, and H_2S in argo, Vollath *et al.* were able to successfully synthesize MoS_2 by using a microwave plasma with a frequency of either 0.915 or 2.45 GHz under argon [23, 24]. Further, Liu and Wang have successfully generated nanoparticles of MoS_2 supported on graphene sheets (GS) by applying a hybrid catalyst (M–MoS_2/GS) on microwave irradiation [25]. Figure 6.2 shows the schematic representation of the preparation of M-MoS_2/GS catalyst.

Closed-field imbalanced magnetron sputtering is used to deposit solid lubricants on end mills, drills, etc. An array of magnetrons is used to regulate both the chemical makeup and the order of its application. Instruments treated with a "soft," solid lubricating coating are effective. Minimal contact and propensity for aluminium, and molybdenum disulfide make it possible to produce at high rates. For solid lubrication of moving parts, Spalvins announced physical DC sputtering of MoS_2 in 1969. Varying the temperature of the substrate changes the shape of the MoS_2 films produced by the sputtering method. Scanning electron microscopy reveals a

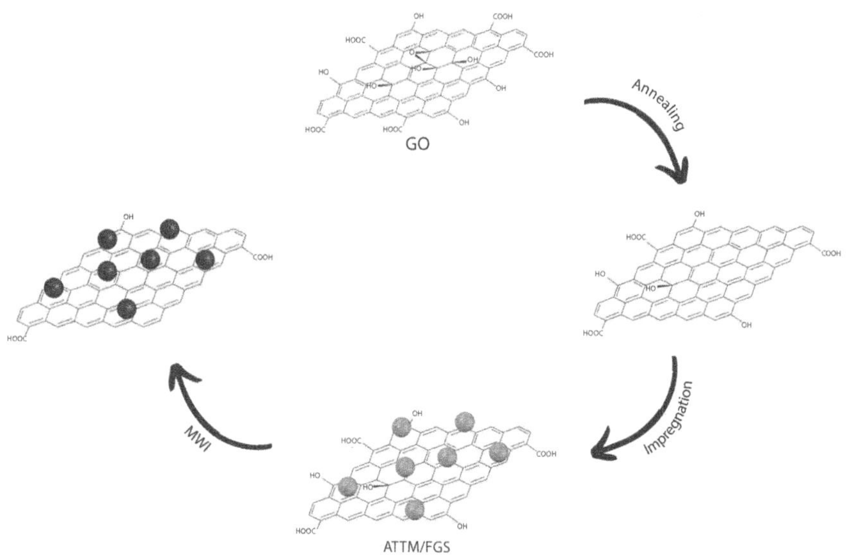

Figure 6.2 Schematic representation of the preparation of M-MoS_2/GS catalyst [25].

microstructure characteristic of MoS_2 crystals that are thick and lamellar in shape but have no discernible orientation. MoS_2 may also be dipped to create thin coatings.

6.3.3 Chemical Methods

The MoS_2 has been recognised for its exceptional/unique qualities, which prompted the introduction of cutting-edge methods to create the material. The ion intercalation technique, commonly known as the Morrison approach or methodology, was first presented in 1986 to manufacture the single-layered MoS_2. Utilizing 2H-MoS_2, single-layered MoS_2 was synthesized using two-step synthesis procedures. Small ions were introduced into the layer in the first phase to weaken and widen it. These ions react with water in the second stage, releasing hydrogen gas that separates the layers and creates single-layered MoS_2 as a result. Although this process proved effective for producing single-layered MoS_2, it has numerous drawbacks, including slow response times and inadequate intercalation procedure control [26].

The prospective uses of MoS_2 in a variety of optoelectronic devices make the synthesis of this material with crystallinity, defects, and shape very important. Historically, MoS_2 was created via chemical or mechanical exfoliation methods, but in current history, chemical vapour deposition (CVD) technology has drawn interest from researchers due to its advantages over other methods. 2D materials of outstanding quality and purity can be created by applying the CVD technique [27]. The CVD method uses a chemical interaction between the substrate and vapour to create thin layers. The vapor solid growth approach, commonly known as the direct evaporation method, produces excellent quality irregular 2D MoS_2 thin films. For large-scale synthesis, the CVD method uses three distinct types of molybdenum precursors: molybdenum trioxide (MoO_3), ammonium thiomolybdate $(NH_4)_2MoS_4$, and elemental molybdenum [28, 29].

At low temperatures, the CVD approach has also been used as a template to create the MoS_2 on rGO sheets. Ammonium thiomolybdate was utilized as the source of both molybdenum and sulphur [30]. They used the CVD method for the formation of MoS_2 on rGO sheets, and as a result, they got extremely crystalline hexagonal flakes of MoS_2 on rGO sheets. The synthesis of MoS_2 was also achieved by Lin *et al.* [31] using the CVD technique, which was applied to rGO. The development of a flower-like MoS_2/graphene composite was accomplished by Liu *et al.* [32] by the hydrolysis of $LiMoS_2$ (lithiated-MoS_2). On graphene sheets, flowers that looked like MoS_2 were produced. The researchers found that the aggregation of the

MoS_2 flowers is effectively suppressed by the graphene sheets' development of a 2D conductive network, which results in the formation of homogenous MoS_2/graphene heterostructures. Not only is the manufacturing of single-layer graphene very important, but so are its uses in electrochemical processes. He *et al.* [33] have created ultrathin sheets of vertically aligned MoS_2 on single layers of graphene using the CVD method for use in lithium ion battery applications.

The developed MoS_2/rGO showed outstanding performance in terms of the current density for the use of sodium-ion batteries. The existence of hetero-interfaces increased the electrochemical activity of MoS_2/rGO [34]. Additionally, freestanding metallic 1T MoS_2/graphene composites have been produced for use in battery production, and these composites have shown promising performance [35]. Li *et al.* [36] have published their findings on the rational design and synthesis of MoS_2/rGO sponges. The procedure is used to produce the MoS_2/rGO sponges involved in freezing ammonium tetrathiomolybdate and GO, followed by a thermal treatment in an atmosphere with N_2 and H_2. The higher charge transport activity seen in the MoS_2/rGO sponges was found to be caused by their porous and conductive properties. Xiong *et al.* [37] described an easy formation of mesoporous MoS_2 foam (mPF-MoS_2) modified graphene (mPF- MoS_2/graphene) hetero-structure.

6.3.4 Properties

MoS_2 has a 2-D structure that tends to switch between an indirect band gap (1.2 eV) and a direct band gap (1.9 eV). At room temperature, the monolayer form of MoS_2 has outstanding mobility of 200 cm^2 $(V\text{-}s)^{-1}$ and a reasonable on/off current ratio of 10^8 [38]. The characteristics of MoS_2 are reliant on its thickness, which is one reason why this material is an appealing choice for a wide range of optoelectronic applications. A new class of nanomaterials for use in optoelectronic applications has been discovered as hybrid composites of graphene and MoS_2 [39]. In the past, hybrid MoS_2/rGO composites with a high surface area have been manufactured utilizing several synthetic processes. MoS_2 and rGO are similar in that they both have a layered structure, are held together by van der Waals forces, and have excellent properties. Because of its high electron mobility, rGO has the potential to be used as an electron acceptor as well as a transporter to achieve improved charge separation. In addition to this, rGO prevents MoS_2 from clumping together and acts as a conductive matrix, both of which speed up the movement of electrons. Similarly, rGO shields MoS_2 from the damaging effects of harsh conditions. Enhancing electron transfer from MoS_2 to rGO and maintaining the stability of the MoS_2/rGO composite have both been attributed to the formation of C-O-Mo bonds between MoS_2 and rGO. Because of its great qualities, the MoS_2/rGO composite has been the subject of a lot of research as a possible electrode material for electrochemical sensing, wastewater treatment and dye-sensitized solar cell applications [27].

To larger nanotubes and fullerenes with a versatile atomic structure that diverges from the stable bulk state, MoS_2 can be deduced as fragments of the bulk modifications. In addition, these MoS_2 nanoscale allotropes differ from the second species, which is coordinatively saturated and more inert, in that they contain a greater proportion of chemically active atoms with dangling bonds. The efficiency of bulk MoS_2 in its typical industrial applications, such as tribology, catalysis, electrochemistry, and electronics could be improved by both of these things. This is because both types of objects are made up of MoS_2. A lot of research, both theoretical and practical, is now looking at the basic patterns that control the stability, reactivity, and electrical properties of these nanoobjects [21].

6.4 Application of MoS_2

MoS_2 is a two-dimensional nanomaterial that has several outstanding properties and is widely employed. MoS_2 nanoparticles may be used for a variety of purposes, such as a dry lubricant, a catalyst for the hydrogen evolution process (HER), a catalyst for hydrodeoxygenation (HDO), a catalyst for hydrodesulfurization (HDS), and energy storage, amongst other applications. When carbon nanomaterials (CNMs) and nanoscale MoS_2 are combined, it is possible to create unique hybrid nanostructures with extraordinary qualities. These attributes may include improved electrocatalytic and photocatalytic activity, as well as enhanced rheological and tribological capabilities.

In the field of desalination and membrane technology, CNMs (CNTs and graphene/GO) are beneficial materials. Because of their huge surface area, porosity, stability, and availability of functionalities, several innovative functional CNMs have concurrently shown exceptional adsorption effectiveness for the treatment of wastewater. In general, CNMs have shown their potential to be effective in the laboratory for the filtration of water. While the investigation is still underway, it is predicted that the CNMs may lead to more fascinating findings and perhaps low-cost commercial adsorbents that can be used for the filtration of water. They may be transformed into new goods and nanotechnologies shortly [5].

6.4.1 Dye-Sensitized Solar Cells (DSSCs)

DSSCs, which are photovoltaic devices that can change solar energy into electrical energy, have garnered a lot of attention recently because they are relatively inexpensive, simple to produce, and have a high power conversion efficiency (PCE). The first distributed solar thermal energy converters DSSCs were developed by Gratzel *et al.* in 1991. These early DSSCs had a respectable PCE of 6% [40]. A few of the components that are used in the construction of DSSCs include the light sensitizer, the counter electrode, and the electron transport layer. According to the research that had been analysed, the photovoltaic performance of DSSCs may be greatly enhanced by using cutting-edge counter-electrode materials or hybrid composites. The remarkable performance of the DSSCs is directly attributable to the greater surface area of the counter electrode. As a direct consequence of this, it has been shown that the catalytic capabilities of the counter electrode may be significantly improved by combining rGO with MoS_2. In this particular scenario, Yuan *et al.* [7] made use of a hydrothermal process to produce MoS_2 nanosheets that were layered vertically atop rGO sheets.

6.4.2 Catalyst

MoS_2 nanoparticles as the catalyst for hydrogen evolution processes have garnered considerable attention as a consequence of the continued look for new renewable energy sources. HER refers to the cathodic reaction that occurs during water splitting [6]. The process of water splitting may provide a clean and sustainable source of hydrogen, which may then be stored, utilized in fuel cells, or put to several industrial purposes [41]. MoS_2-based catalysts have a low price relative to other kinds of catalysts, high chemical stability, and powerful electrocatalytic properties. As a consequence, nano-MoS_2 has attracted a great deal of scientific interest as an electrocatalyst for the hydrogen evolution process. In addition, MoS_2/CNM hybrid nanomaterials demonstrate higher catalytic activity [42]. This is due to the synergistic effect produced by the interaction of carbon nanomaterials with MoS_2 nanoparticles. This effect has many advantages, including a spike in the volume of photocatalytic reaction centres and the number of exposed edges per catalyst volume [43].

6.4.3 Desalination

The process of eliminating salt and minerals from saltwater to get pure, fresh water is referred to as desalination. Researchers have been paying a lot of attention to MoS_2 in recent years since it is a novel sort of nanomaterial that consists of two dimensions. It had a lower potential for harm and required fewer resources to produce. In addition to this, it has a high resistance to swelling as a result of its stability in the water as well as the straight and smooth channels that are present inside the membrane. This is a positive development for the movement of water molecules. These characteristics are much superior to those found in carbon-based materials, which have been the subject of substantial research in the past. It is important to point out that electrons that are shared between molybdenum and sulfur atoms are what charge the nanopores [44]. This strengthens the connection between the membrane and the ions, which in turn improves ion selectivity. Sapkota *et al.* [45] found that the existence of holes in a membrane-based desalination system hindered both the sheet length and the ability of nanodisks (NDs) to operate as spacers [46]. They were responsible for an increase in membrane porosity, which led to the formation of an interface that was extremely porous and charged. This improved both the water penetration rate and the ion selectivity. As a consequence of this, they were able to manufacture a multilayer MoS2 film by successively stacking one to two layers of porous NSs and NDs of varying sizes. The membrane consists of a multimodal porous network that varies in surface charge, pore size, and interlayer spacing throughout its construction. In the investigations conducted with forward osmosis and reverse osmosis, the salt rejection rate may reach 99% at high salinity, which indicates that the rejection of small molecule dyes is possible [47].

6.4.4 Lubrication

Lubrication is one of MoS_2's major applications. It is a well-known fact that lubrication has a significant impact on a variety of operations, including bearings, pistons, and gears. An approach for saving time and labour is to use lubricant additives to increase the efficacy of lubrication. MoS_2 has outstanding tribological behaviour in the reduction of friction and wear because of its high chemical inertness, enormous strength, and ease of shearing [48]. This is a result of its extraordinary strength. If the 2D layered structure is aligned with the direction in which the fluid is travelling, it may also be capable to alter the mass and heat transfer during friction [8]. There have been several investigations and successful applications of MoS_2 in space technology, as well as numerous dispersions in grease, oil, and lacquers. These applications include, among others: Graphene and MoS_2 nanocomposites make excellent solid lubricants and lubricant additives. A GO/nano-MoS_2 nanohybrid was produced by Xin *et al.* [49] by chemically fusing the two components separately. The multidimensional assembly of the nanohybrid results in a significant increase in the material's mechanical, thermodynamic, surface, and tribological characteristics when mixed with polyimide. Hou *et al.* in their recent work [50] produced rGO/MoS_2 heterostructures using a simple hydrothermal one-pot technique.

6.4.5 Sensor

To make a Hg^{2+} ion sensor, MoS_2 nanosheets and carbon dots are both necessary components. Samira *et al.* [51] exhibit a very sensitive field-effect transistor sensor for identifying Hg ions in water samples. The sensor depended on the newly discovered MoS_2 material. Because of its semiconductivity and layered structure in two dimensions, MoS_2 is a suitable material for measuring applications. MoS_2 nanosheets with just a few layers might be easily manufactured by employing a simple solvent exchange method. A significant number of carbon dots that contained DNA probes were able to be maintained by MoS_2 nanosheets that had a considerable surface area. According to the results, the DNA-CD/ MoS_2 FET aptasensor can identify Hg ions with a low identification limit of 0.65 AM and remarkable sensitivity in the linear range of 1:00 AM to 10:00 PM. The detection limit was determined by the amount of mercury that could be detected with the sensor. In addition to this, the aptasensor exhibited a higher degree of selectivity for mercury ions in comparison to other ions found in water. After the practical application, Hg ions in samples of tap water and mineral water were successfully measured using the aptasensor. Because of its quickness, sensitivity, and selectivity throughout the detection process, the DNA-CD/MoS_2 hybrid sensor was shown to be a successful mercury ion detector [51].

6.4.6 Electroanalytical

The production of chemically produced MoS_2-based electrodes is a field of research that has a vast amount of potential applications and is of significant interest in the field of electroanalytical chemistry [52]. Researchers in materials science have been prompted by layered MoS_2 to design and produce unique, fascinating 2D or 3D hybrid components that are built of MoS_2 and other 1D/2D blocks. These components may be either layered or unlayered. Despite the substantial progress that has been made in the manufacturing of 3D-MoS_2, the material is still constrained in its use by factors such as the difficulty of the preparation operations and its inherent activity [53]. As a result, it is still a question that has to be answered as to how large amounts of high-quality MoS_2 can be consistently prepared for use in technological applications. The amazing electrocatalytic properties of extremely sensitive and selective MoS_2 nanohybrids have been highlighted [54]. These nanohybrids include functional components such as metal NPs, metal oxide NPs, polymeric NPs, and carbon nanomaterials. In addition to being produced through hydrothermal synthesis, ultrasonic exfoliation, electrochemical deposition, and physical procedures, MoS_2 nanohybrids may also be made utilizing these methods. To generate nanoscale materials that have a compelling form, high conductivity, and stable functional qualities is the overarching goal of each of these techniques; however, the specific means by which this goal is accomplished differ. However, to comprehend their function in enhancing the kinetics, charge transport, sensitivity, and selectivity of electrochemical sensors, it is necessary to overcome obstacles in the management of size, crystallinity, and defects (edges) in 2D MoS_2. This is because 2D MoS_2 presents several challenges in this regard [55].

MoS_2 and rGO are similar as they both have a layered structure, and are held together by Van der Waals forces, and have excellent properties. Because of its high electron mobility, rGO has the potential to be used as an electron acceptor as well as a transporter to achieve improved charge separation. In addition to this, rGO prevents MoS_2 from clumping together and acts as a conductive matrix, both of which speed up the movement of electrons. Similarly, rGO shields MoS_2 from the damaging effects of poor conditions. Enhancing electron transfer from MoS_2 to rGO and maintaining the stability of the MoS_2/rGO composite have both been attributed to the formation of C-O-Mo bonds between MoS_2 and rGO. As a consequence of its outstanding qualities it has, the MoS_2/rGO composite has been the subject of a significant amount of research as a potential electrode material for dye-sensitized solar cells and electrochemical sensing applications [27]. In the construction of Pt-free counter electrodes for dye-sensitized solar cells and as an electrode modification for electrochemical sensors, the MoS_2/rGO composite has shown exceptional performance. Dye-sensitized solar cells made with the MoS_2/rGO composite were comparable to those made with Pt due to their remarkable open circuit voltage and acceptable power conversion percentages. MoS_2/rGO composite-modified electrodes showed exemplary electrochemical performance for the identification of several analytes, including hydrogen peroxide, glucose, uric acid, dopamine, catechol, and ascorbic acid, among others. As a result, one may deduce that composites made of MoS_2/rGO may have applications in the fields of sensing and photovoltaics.

6.4.7 Biomedical

MoS_2/GO has shown a great affinity to specifically target lung tissue. In other words, MoS_2/GO demonstrated the same level of lung localization as GO, which enabled it to act as a "guided missile" when directing its effects towards the lung. In addition to this, the drug loading capacity of the MoS_2/GO composites increased, and the tumour-killing effectiveness against lung-metastasizing cancer cells was significantly increased. Considerably, MoS_2/GO composites prevented B16 murine melanoma cancer cells from metastasizing in the lungs of mice. This is an important finding. The MoS_2/GO nano-bio interface exhibited impaired reactivity toward macrophages when compared to GO, which led to the contact and absorption of nanosheets by macrophages as a result of phagocytosis and activation of macrophages. This was the cause of the contact and absorption of nanosheets by macrophages [56].

6.5 Molybdenum-Modified Carbon Allotropes in Wastewater Treatment

Heavy metal contamination of water presents a significant concern to both the environment and humans since heavy metals cannot biodegrade and are hazardous [57]. Adsorption is an effective strategy to cope with contaminated water. An excellent adsorbent, which has a large capacity for adsorption, rapid adsorption kinetics, and great selectivity, is crucial to this procedure. Activated carbon has been utilised extensively as a conventional adsorbent, however, due to its poor selectivity, delayed adsorption kinetics, and low adsorption capacity, many different novel adsorbents have been explored. Due to its superior adsorption capability, MoS_2-based material is one of the most widely used adsorbents among them. MoS_2 in two dimensions has grown to be one of the most prominent although disseminated in an aqueous solution, MoS_2 nanosheets did not perform to their fullest in the adsorption. To get around this, MoS_2 nanosheets are made into 3D gels with macroporous network structures or the interlayer gap is widened by intercalating polymers to expose additional surface and S sites to heavy metals. For instance, the extraordinarily high mercury absorption capacity of MoS_2 nanosheets with expanded interlayer spacing (2563 mg/g) surpasses the theoretically expected adsorption capacity of monolayer MoS_2 under the assumption of a stoichiometric S/Hg ratio of 1:1 [58]. Zhang *et al.* studied the elimination of heavy metals from water using thin film membranes and two-dimensional MoS_2 nanosheets suspended in aqueous solutions, respectively. Through these investigations, a novel heavy metal ion removal method was revealed, one that uses MoS_2 nanosheets and a reduction-oxidation (redox) process between heavy metal ions and the latter. MoS_2 nanosheets were created by chemically exfoliating bulk powder using Ag^+ as a model species [59]. The synthesis of 3D aerogel for the adsorption of water-soluble organic contaminants is accomplished by the use of a one-step hydrothermal technique to bind graphene and MoS_2 sheets to polydopamine (PDA). For there to be no aggregation on the surface of PDA-modified graphene throughout the process of developing an aerogel with equally dispersed MoS_2 nanocrystallites, the biomolecule DA is necessary. When compared to a scenario in which additional bio-inspired adhesion interactions between graphene and MoS_2 are present, the accumulation of MoS_2 nanocrystallites in the absence of interfacial adhesion leads to a substantial reduction in both specific surface area and total pore volume, as well as a considerable increase in the average pore size. This is the case even though the average pore size is significantly increased [60].

Because of human activity, bio-resistant and potentially hazardous organic chemicals, most often dyes and antibiotics, are being discharged into wastewater systems all over the globe. Extensive study has been conducted on MoS_2 and other materials in a similar vein for the adsorptive removal of these persistent molecules [61]. Erythromycin, 17-beta-estradiol, and triclosan are examples of pharmaceutical and personal care items that have the potential to be effectively adsorbed on molybdenum disulphide. Natural organic matter (humic acid) limits adsorption capacity by competing for adsorption sites [38], whereas MoS_2 is 40 to 800 times more efficient per unit surface area than activated carbon. Formaldehyde adsorption on monolayer MoS_2 was calculated using DFT. The adsorption energy of monolayer MoS_2 was 0.11 eV, whereas it was 1.59 eV for Ti-doped MoS_2. The potential of g-MoS_2–coated biochar composites for ciprofloxacin absorption from soil revealed a large pH dependence. Zhao *et al.* [39] studied the adsorption capacities of a core-shell magnetic nanocomposite constructed of Fe_3O_4@MoS_2 to remove sulfonamide antibiotics (SA) from water, milk, hog meat, and fish flesh [40].

Self-assembly of modified MoS_2 nanosheets onto an HPAN ultrafiltration membrane led to the production of MoS_2@PDA nanofibrous membranes, which were then used for the treatment of wastewater. Zaho *et al.* conducted research on the permeability of MoS_2@PDA nanofiber membranes to various substances by varying the amount of MoS_2 loadings and the amount of time that PDA was modified. PDA variation led to an increase in the hydrophilicity of MoS_2@PDA NFMs. At the same time, PDA nanoaggregates prevented the stacking of MoS_2 nanosheets, which resulted in the formation of an unusually loose membrane structure. In addition to having an impermeable nanosheet surface, this structure also has the potential to provide transmembrane channels for the diffusion of molecules, which led to a rise in membrane permeance while maintaining high salt permeance [44].

Extensive research is now being conducted on MoS_2 and nanomaterials based on MoS_2 as potential photocatalysts for the destruction of organic and inorganic pollutants as well as the killing and/or inactivation of microorganisms. As a result of the excellent physicochemical properties of MoS_2, a wide range of synthetic methods have been developed for the production of MoS_2 and materials based on molybdenum disulfide, and uses have been found in a variety. Taking a look at some of the most recent advancements in the modification of MoS_2 such as doping it with metals and nonmetals, coupling it with other semiconductors or metals, and using carbon-based supports [62]. Water contamination caused by several industrial contaminants poses a significant danger to the ecosystem. Various challenges impede the removal of these contaminants using conventional methods. Due to its favourable properties, photocatalysis has been investigated by several industries as a method for removing harmful contaminants. As photocatalysts, semiconducting transition metal dichalcogenides with distinct attributes are receiving a great deal of attention. Due to the presence of a straight band gap, MoS_2, a member of the metal dichalcogenide family, is a potential material for photocatalysis. However, it is limited by a lack of emission, high recombination, and stacking defects. These restrictions are solved by MoS_2-based nanocomposites, which have a high degradation rate for all main contaminants and dyes [57].

6.6 Conclusion

MoS_2 nanoparticles possess an extremely high capture capacity, rapid adsorption kinetics, notable affinity, and excellent selectivity, they can preferentially adsorb softer heavy metals in different ions in water. This is made possible by the fact that MoS_2 nanoparticles have a remarkable affinity. Not only have MoS_2-based heterostructures demonstrated application promise in water remediation technologies, but they have also shown application potential in water splitting, supercapacitors, electrical devices, batteries, photodetectors, and other fields. Many different properties of MoS_2, such as the architectures of its optical and electric bandgaps have been investigated. MoS_2 has a low coefficient of friction, excellent mechanical strength, a large surface area an adaptable band gap, features of visible light absorption, and potential electrical transport capabilities, among other desirable qualities. MoS_2 has several opportunities; nevertheless, there is still a significant obstacle in the way of producing large quantities of homogeneous, single-layer MoS_2 suitable for use in industry.

References

1. Malhotra, M., Puglia, M., Kalluri, A., Chowdhury, D., Kumar, C.V., Adsorption of metal ions on graphene sheet for applications in environmental sensing and wastewater treatment. *Sensors Actuators Rep.*, 4, 100077, 2022. https://doi.org/10.1016/j.snr.2022.100077.
2. Hirsch, A., The era of carbon allotropes. *Nat. Mater.*, 9, 11, 2–5, 2010. https://doi.org/10.1038/nmat2885.
3. Yadav, S., Singh Raman, A.P., Meena, H., Goswami, A.G., Bhawna, V., Jain, P., Kumar, G., Sagar, M., Rana, D.K., Bahadur, I., Singh, P., An update on graphene oxide: Applications and toxicity. *ACS Omega*, 7, 35387–35445, 2022. https://doi.org/10.1021/acsomega.2c03171.
4. Zaytseva, O., Differential impact of multi-walled carbon nanotubes on germination and seedling development of glycine max, phaseolus vulgaris. *Eur. Chem. Bull.*, 5, 202–210, 2016.
5. Gusain, R., Kumar, N., Ray, S.S., Recent advances in carbon nanomaterial-based adsorbents for water purification. *Coord. Chem. Rev.*, 405, 213111, 2020. https://doi.org/10.1016/j.ccr.2019.213111.
6. Bae, D., Seger, B., Vesborg, P.C.K., Hansen, O., Chorkendorff, I., Strategies for stable water splitting via protected photoelectrodes. *Chem. Soc. Rev.*, 46, 1933–1954, 2017. https://doi.org/10.1039/C6CS00918B.
7. Yuan, X., Li, X., Zhang, X., Li, Y., Liu, L., MoS_2 vertically grown on graphene with efficient electrocatalytic activity in Pt-free dye-sensitized solar cells. *J. Alloys Compd.*, 731, 685–692, 2018. https://doi.org/10.1016/j.jallcom.2017.08.208
8. Berman, D., Erdemir, A., Sumant, A.V., Graphene: A new emerging lubricant. *Mater. Today*, 17, 31–42, 2014. https://doi.org/10.1016/j.mattod.2013.12.003
9. Sengupta, R., Bhattacharya, M., Bandyopadhyay, S., Bhowmick, A.K., A review on the mechanical and electrical properties of graphite and modified graphite reinforced polymer composites. *Prog. Polym. Sci.*, 36, 638–670, 2011. https://doi.org/10.1016/J.PROGPOLYMSCI.2010.11.003.
10. Liu, W., Yu, Y., Dargusch, M., Liu, Q., Chen, Z., Carbon allotrope hybrids advance thermoelectric development and applications. *Renew. Sust. Energ. Rev.*, 141, 110800, 2021. https://doi.org/10.1016/j.rser.2021.110800.
11. Gacem, A., Modi, S., Yadav, V.K., Islam, S., Patel, A., Dawane, V., Jameel, M., Inwati, G.K., Piplode, S., Solanki, V.S., Basnet, A., Recent advances in methods for synthesis of carbon nanotubes and carbon nanocomposite and their emerging applications: A descriptive review. *J. Nanomater.*, 16, 2022. https://doi.org/10.1155/2022/7238602
12. Zaytseva, O. and Neumann, G., Carbon nanomaterials: Production, impact on plant development, agricultural and environmental applications. *Chem. Biol. Technol. Agric.*, 3, 1–26, 2016. https://doi.org/10.1186/s40538-016-0070-8.
13. Jin, Y., Shi, Y., Chen, Z., Chen, R., Chen, X., Zheng, X., Liu, Y., Combination of sunlight with hydrogen peroxide generated at a modified reticulated

vitreous carbon for drinking water disinfection. *J. Clean. Prod.*, 252, 119794, 2020. https://doi.org/10.1016/j.jclepro.2019.119794.
14. Vieira, L.D.S., A review on the use of glassy carbon in advanced technological applications. *Carbon N. Y.*, 186, 282–302, 2021. https://doi.org/10.1016/j.carbon.2021.10.022.
15. Rani, U.A., Ng, L.Y., Ng, C.Y., Mahmoudi, E., A review of carbon quantum dots and their applications in wastewater treatment. *Adv. Colloid Interface Sci.*, 278, 102124, 2020. https://doi.org/10.1016/j.cis.2020.102124.
16. Ahmad Zawawi, N., Abdul Majid, Z., Aini Abdul Rashid, N., Effect of acid oxidation methods on multi-walled carbon nanotubes (MWCNT) for drug delivery application. *Int. J. Adv. Sci. Res. Manage.*, 1, 14–22, 2016.
17. Raccichini, R., Varzi, A., Passerini, S., Scrosati, B., The role of graphene for electrochemical energy storage. *Nat. Mater.*, 14, 271–279, 2015. https://doi.org/10.1038/nmat4170.
18. Park, B., Current and future applications of nanotechnology. *Issues Environ. Sci. Technol.*, 24, 1–18, 2007. https://doi.org/10.1039/9781847557766-00001.
19. Kim, S.H., Mulholland, G.W., Zachariah, M.R., Density measurement of size selected multiwalled carbon nanotubes by mobility-mass characterization. *Carbon N. Y.*, 47, 1297–1302, 2009. https://doi.org/10.1016/j.carbon.2009.01.011.
20. Ruoff, R.S., Thornton, T., Smith, D., Density of fullerene containing soot as determined by helium pycnometry. *Chem. Phys. Lett.*, 186, 456–458, 1991. https://doi.org/10.1016/0009-2614(91)90208-Q.
21. He, Z. and Que, W., Molybdenum disulfide nanomaterials: Structures, properties, synthesis and recent progress on hydrogen evolution reaction. *Appl. Mater. Today*, 3, 23–56, 2016. https://doi.org/10.1016/j.apmt.2016.02.001.
22. Yu, H., Liu, Y., Brock, S.L., Synthesis of discrete and dispersible MoS_2 nanocrystals. *Inorg. Chem.*, 47, 1428–1434, 2008. https://doi.org/10.1021/ic701020s.
23. Vollath, D. and Szabó, D.V., Synthesis of nanocrystalline MoS_2 and WS_2 in a microwave plasma. *Mater. Lett.*, 35, 236–244, 1998. https://doi.org/10.1016/S0167-577X(97)00247-4.
24. Vollath, D. and Szabó, D.V., Nanoparticles from compounds with layered structures. *Acta Mater.*, 48, 953–967, 2000. https://doi.org/10.1016/S1359-6454(99)00395-X
25. Liu, N., Wang, X., Xu, W., Hu, H., Liang, J., Qiu, J., Microwave-assisted synthesis of MoS_2/graphene nanocomposites for efficient hydrodesulfurization. *Fuel*, 119, 163–169, 2014. https://doi.org/10.1016/j.fuel.2013.11.045.
26. Novoselov, K.S., Jiang, D., Schedin, F., Booth, T.J., Khotkevich, V.V., Morozov, S.V., Geim, A.K., Two-dimensional atomic crystals. *Proc. Natl. Acad. Sci. U. S. A.*, 102, 10451–10453, 2005. https://doi.org/10.1073/pnas.0502848102.
27. Ahmad, K., Shinde, M.A., Kim, H., Molybdenum disulfide/reduced graphene oxide: Progress in synthesis and electro-catalytic properties for electrochemical sensing and dye sensitized solar cells. *Microchem. J.*, 169, 106583, 2021. https://doi.org/10.1016/j.microc.2021.106583.

28. Liu, K.-K., Zhang, W., Lee, Y.-H., Lin, Y.-C., Chang, M.-T., Su, C.-Y., Chang, C.-S., Li, H., Shi, Y., Zhang, H., Lai, C.-S., Li, L.-J., Growth of large-area and highly crystalline MoS_2 thin layers on insulating substrates. *Nano Lett.*, 12, 1538–1544, 2012. https://doi.org/10.1021/nl2043612.
29. Pondick, J.V., Woods, J.M., Xing, J., Zhou, Y., Cha, J.J., Stepwise sulfurization from MoO_3 to MoS_2 via chemical vapor deposition. *ACS Appl. Nano Mater.*, 1, 5655–5661, 2018. https://doi.org/10.1021/acsanm.8b01266.
30. Wang, D., Pan, Z., Wu, Z., Wang, Z., Liu, Z., Hydrothermal synthesis of MoS_2 nanoflowers as highly efficient hydrogen evolution reaction catalysts. *J. Power Sources*, 264, 229–234, 2014. https://doi.org/10.1016/j.jpowsour.2014.04.066
31. Lin, M.-Y., Chang, C.-E., Wang, C.-H., Su, C.-F., Chen, C., Lee, S.-C., Lin, S.-Y., Toward epitaxially grown two-dimensional crystal hetero-structures: Single and double MoS_2/graphene hetero-structures by chemical vapor depositions. *Appl. Phys. Lett.*, 105, 73501, 2014. http://dx.doi.org/10.1063/1.4893448
32. Liu, Y., Zhao, Y., Jiao, L., Chen, J., A graphene-like MoS_2/graphene nanocomposite as a highperformance anode for lithium ion batteries. *J. Mater. Chem. A*, 2, 13109–13115, 2014.
33. He, J., Zhang, C., Du, H., Zhang, S., Hu, P., Zhang, Z., Ma, Y., Huang, C., Cui, G., Engineering vertical aligned MoS_2 on graphene sheet towards thin film lithium ion battery. *Electrochim. Acta*, 178, 476–483, 2015. https://doi.org/10.1016/j.electacta.2015.08.025
34. Sun, D., Ye, D., Liu, P., Tang, Y., Guo, J., Wang, L., Wang, H., MoS_2/graphene nanosheets from commercial bulky MoS_2 and graphite as anode materials for high rate sodium-ion batteries. *Adv. Energy Mater.*, 8, 1702383, 2018. https://doi.org/10.1002/aenm.201702383
35. Geng, X., Jiao, Y., Han, Y., Mukhopadhyay, A., Yang, L., Zhu, H., Freestanding metallic 1T MoS_2 with dual ion diffusion paths as high rate anode for sodium-ion batteries. *Adv. Funct. Mater.*, 27, 1702998, 2017. https://doi.org/10.1002/adfm.201702998
36. Li, J., Qin, W., Xie, J., Lin, R., Wang, Z., Pan, L., Mai, W., Rational design of MoS_2-reduced graphene oxide sponges as free-standing anodes for sodium-ion batteries. *Chem. Eng. J.*, 332, 260–266, 2018. https://doi.org/10.1016/j.cej.2017.09.088
37. Xiong, P., Ma, R., Sakai, N., Nurdiwijayanto, L., Sasaki, T., Unilamellar metallic MoS_2/graphene superlattice for efficient sodium storage and hydrogen evolution. *ACS Energy Lett.*, 3, 997–1005, 2018. https://doi.org/10.1021/acsenergylett.8b00110
38. Pumera, M., Sofer, Z., Ambrosi, A., Layered transition metal dichalcogenides for electrochemical energy generation and storage. *J. Mater. Chem. A*, 2, 8981–8987, 2014. https://doi.org/10.1039/C4TA00652F
39. Forsberg, V., Zhang, R., Bäckström, J., Dahlström, C., Andres, B., Norgren, M., Andersson, M., Hummelgård, M., Olin, H., Exfoliated MoS_2 in water without additives. *PloS One*, 11, e0154522, 2016. https://doi.org/10.1371/journal.pone.0154522

40. Gratzel, M., O'Regan, B., A low cost high-efficiency solar cell based on dye-sensitized colloidal TiO_2 films. *Nature*, 353, 737–740, 1991. https://doi.org/10.1038/353737a0
41. Benck, J.D., Hellstern, T.R., Kibsgaard, J., Chakthranont, P., Jaramillo, T.F., Catalyzing the hydrogen evolution reaction (HER) with molybdenum sulfide nanomaterials. *ACS Catal.*, 4, 3957–3971, 2014. https://doi.org/10.1021/cs500923c.
42. Bojarska, Z., Mazurkiewicz-Pawlicka, M., Gierlotka, S., Makowski, Ł., Production and properties of molybdenum disulfide/graphene oxide hybrid nanostructures for catalytic applications. *Nanomaterials*, 10, 1865, 2020. https://doi.org/10.3390/nano10091865.
43. Zhou, H., Liu, Y., Zhang, L., Li, H., Liu, H., Li, W., Transition metal-doped amorphous molybdenum sulfide/graphene ternary cocatalysts for excellent photocatalytic hydrogen evolution: Synergistic effect of transition metal and graphene. *J. Colloid Interface Sci.*, 533, 287–296, 2019. https://doi.org/10.1016/j.jcis.2018.07.084
44. Zhao, H., Liu, G., Zhang, M., Liu, H., Zhang, M., Zhou, L., Gao, J., Jiang, Y., Bioinspired modification of molybdenum disulfide nanosheets to prepare a loose nanofiltration membrane for wastewater treatment. *J. Water Process Eng.*, 40, 101759, 2021. https://doi.org/10.1016/j.jwpe.2020.101759
45. Sapkota, B., Liang, W., VahidMohammadi, A., Karnik, R., Noy, A., Wanunu, M., High permeability sub-nanometre sieve composite MoS_2 membranes. *Nat. Commun.*, 11, 1–9, 2020. https://doi.org/10.1038/s41467-020-16577-y
46. Yadav, S., Singh, P., Jain, P., Polymeric membranes as sustainable material for water desalination: A review. *Desal. Water Treat.*, 269, 1–20, 2022. https://doi.org/10.5004/dwt.2022.28753.
47. Wei, H., Zhao, S., Zhang, X., Wen, B., Su, Z., The future of freshwater access: Functional material-based nano-membranes for desalination. *Mater. Today Energy*, 22, 100856, 2021. https://doi.org/10.1016/j.mtener.2021.100856.
48. Wu, J., Mu, L., Zhu, J., Feng, X., Lu, X., Larsson, R., Shi, Y., Synthesis of hollow fullerene-like molybdenum disulfide/reduced graphene oxide nanocomposites with excellent lubricating properties. *Carbon N. Y.*, 134, 423–430, 2018. https://doi.org/10.1016/j.carbon.2018.04.021.
49. Li, Y., Wang, Q., Wang, T., Pan, G., Preparation and tribological properties of graphene oxide/nitrile rubber nanocomposites. *J. Mater. Sci.*, 47, 730–738, 2012.
50. Hou, K., Wang, J., Yang, Z., Ma, L., Wang, Z., Yang, S., One-pot synthesis of reduced graphene oxide/molybdenum disulfide heterostructures with intrinsic incommensurateness for enhanced lubricating properties. *Carbon N. Y.*, 115, 83–94, 2017. https://doi.org/10.1016/j.carbon.2016.12.089
51. Mansouri Majd, S., Ghasemi, F., Salimi, A., Sham, T.-K., Transport properties of a molybdenum disulfide and carbon dot nanohybrid transistor and its applications as a Hg2+ aptasensor. *ACS Appl. Electron. Mater.*, 2, 635–645, 2020. https://doi.org/10.1021/acsaelm.9b00632.

52. Youn, D.H., Han, S., Kim, J.Y., Kim, J.Y., Park, H., Choi, S.H., Lee, J.S., Highly active and stable hydrogen evolution electrocatalysts based on molybdenum compounds on carbon nanotube a graphene hybrid support. *ACS Nano*, 8, 5164–5173, 2014. https://doi.org/10.1021/nn5012144.
53. Youn, D.H., Han, S., Kim, J.Y., Kim, J.Y., Park, H., Choi, S.H., Lee, J.S., Highly active and stable hydrogen evolution electrocatalysts based on molybdenum compounds on carbon nanotube-graphene hybrid support. *ACS Nano*, 8, 5164–5173, 2014. https://doi.org/10.1021/NN5012144/SUPPL_FILE/NN5012144_SI_001.PDF.
54. Chhowalla, M., Shin, H.S., Eda, G., Li, L.-J., Loh, K.P., Zhang, H., The chemistry of two-dimensional layered transition metal dichalcogenide nanosheets. *Nat. Chem.*, 5, 263–275, 2013. https://doi.org/10.1038/nchem.1589
55. Vilian, A.T.E., Dinesh, B., Kang, S.-M., Krishnan, U.M., Huh, Y.S., Han, Y.-K., Recent advances in molybdenum disulfide-based electrode materials for electroanalytical applications. *Microchim. Acta*, 186, 203, 2019.
56. Liu, Y., Peng, J., Wang, S., Xu, M., Gao, M., Xia, T., Weng, J., Xu, A., Liu, S., Molybdenum disulfide/graphene oxide nanocomposites show favorable lung targeting and enhanced drug loading/tumor-killing efficacy with improved biocompatibility. *NPG Asia Mater.*, 10, e458–e458, 2018. https://doi.org/10.1038/am.2017.225.
57. Kaur, M., Krishnan, U., Kumar, A., MoS_2 Based nanocomposites for treatment of industrial effluents, in: *Metal, Metal-Oxides and Metal-Organic Frameworks for Environmental Remediation*, S. Rajendran, M. Naushad, L. Cornejo Ponce, E. Lichtfouse (Eds.), pp. 97–115, Springer International Publishing, Cham, 2021, https://doi.org/10.1007/978-3-030-68976-6_4.
58. Liu, C., Wang, Q., Jia, F., Song, S., Adsorption of heavy metals on molybdenum disul fi de in water: A critical review. *J. Mol. Liq.*, 292, 111390, 2019. https://doi.org/10.1016/j.molliq.2019.111390
59. Zhang, W., Wang, S., Ji, J., Li, Y., Zhang, G., Zhang, F., Fan, X., Primary and tertiary amines bifunctional graphene oxide for cooperative catalysis. *Nanoscale*, 5, 6030–6033, 2013. https://doi.org/10.1039/c3nr01323e.
60. Jin, Y., Shi, Y., Chen, Z., Chen, R., Chen, X., Zheng, X., Liu, Y., Combination of sunlight with hydrogen peroxide generated at a modified reticulated vitreous carbon for drinking water disinfection. *J. Clean. Prod.*, 252, 119794, 2020. https://doi.org/10.1016/J.JCLEPRO.2019.119794.
61. Zhang, M., Gao, J., Liu, G., Zhang, M., Liu, H., Zhou, L., Liu, Y., Zheng, X., Jiang, Y., High-throughput zwitterion-modified MoS_2 membranes: Preparation and application in dye desalination. *Langmuir*, 37, 417–427, 2020. https://doi.org/10.1021/acs.langmuir.0c03068
62. Rahman, A., Jennings, J.R., Tan, A.L., Khan, M.M., Molybdenum disulfide-based nanomaterials for visible-light-induced photocatalysis. *ACS Omega*, 7, 22089–22110, 2022. https://doi.org/10.1021/acsomega.2c01314.

7

Carbon Allotropes in Other Metals (Cu, Zn, Fe etc.) Removal

Manoj Kumar Banjare[1]*, Kamalakanta Behera[2†] and Ramesh Kumar Banjare[3]

[1]*MATS School of Sciences, MATS University, Pagaria Complex, Pandri, Raipur (C.G.), India*
[2]*Department of Chemistry, University of Allahabad, Prayagraj, Uttar Pradesh, India*
[3]*Department of Chemistry (MSEIT), Mats University, Gullua Arang Raipur, Chhattisgarh, India*

Abstract

Heavy metal contamination in water is a serious issue that puts people's health at risk. Carbon nanomaterials are gaining popularity as a result of their superior physicochemical features, which can be used to remediate heavy metal-contaminated water more effectively. Because of their large surface area, nanoscale size, and availability of various functionalities, carbon nanomaterials, such as carbon nanotubes, fullerenes, graphene, graphene oxide, and activated carbon, have great potential for heavy metal removal from water. Toxic metal pollution (heavy metals, radioactive metals, etc.) is one of the most serious global problems; hence, removing toxic metals from contaminated water appears to be very vital. Nanotechnology plays a critical role in environmental monitoring and pollution management to address these concerns. Carbon nanotubes (CNTs) and their composites have gained a lot of attention because of their high adsorption capability in removing metals from contaminated water or enriching metals from wastewater. By selectively functionalizing CNTs with organic ligands, the removal efficiency for metal ions by CNTs was found to be around 10% to 80%, which may be enhanced to nearly 100%. We examine the use of carbon nanotubes (CNTs) in the treatment of toxic metal-containing wastewater for environmental monitoring and metal recovery in this paper. The newest study advancement of employing CNT composites for metal treatment is also mentioned, due to their increased

**Corresponding author*: manojbanjare7@gmail.com; manojbanjarechem111@gmail.com
†*Corresponding author*: kamala.iitd@gmail.com; kkbehera@allduniv.ac.in

Chandrabhan Verma and Chaudhery Mustansar Hussain (eds.) Carbon Allotropes and Composites: Materials for Environment Protection and Remediation, (113–154)

sensitivity and selectivity towards metal enrichment or detection of harmful metal contamination of the environment.

Keywords: Carbon allotropes, heavy metals, techniques

7.1 Introduction

"Pollution" term refers to the existence of an unfavorable chemical entity that interferes with natural processes or has detrimental impacts on the environment and living things. Modern heavy metals have atomic weights (at wt.) among 63 and a density of more than 5 g per cubic centimeter [1]. The concern is widespread due to the pollution of water caused by the release of heavy metals into the environment. The effluent from contemporary chemical industries, i.e., those producing steel plating, batteries, fertilizer, mined materials, pesticides, metallurgical products, fossil fuels, tanneries, and various polymers, including polyvinyl chloride, are the main sources of heavy metal reasserts. Heavy metals have been released into the environment in large part as a result of recent rapid industrialization. The recent fast industrialization is largely to blame for the leakage of heavy metals into the environment. Heavy metals often tend to accumulate in living organisms, in contrast to natural contaminants, because they can no longer be biodegraded [2]. Toxic heavy metals of special concern for treating commercial wastewater include lead, chromium, cadmium, mercury, arsenic, nickel, copper, and zinc. There has been extensive research on the harmful effects of heavy metals like arsenic, cadmium, chromium, mercury, zinc, and lead on human health. High blood pressure, difficulty speaking, fatigue, sleep issues, aggressive behaviour, lack of attention, irritability, temper swings, depression, improved allergic reactions, autoimmune diseases, vascular occlusion, and memory loss are all potential toxic metal symptoms [3]. Heavy metals can also interfere with human cell enzymes, which are dependent on dietary components like selenium, magnesium, and zinc. Some of the hazardous heavy metals that may be damaging to the human body are mercury, cadmium, lead, and arsenic. Even though certain heavy metals, such as manganese, iron, chromium, copper, and zinc, are essential for human health, their presence in high amounts can be exceedingly dangerous [4]. This study tests the effectiveness of employing carbon nanotubes to remove specific steel ions from water, including lead, chromium, cadmium, arsenic, copper, zinc, and nickel, summarizes the permissible concentration limits for the chosen heavy metals as stated in Table 7.1 by the World Health Organization (WHO) and the US Environmental Protection Agency (US EPA).

Table 7.1 The allowable concentrations limits for the selected heavy metals, as reported by U.S. Environmental Agency (U.S. EPA) and World Health Organization (WHO) [5, 6].

Heavy metal	Symbol	level (MCL) (mg/L)	goal (MCLG) (mg/L)	(mg/L)
Copper	Cu	1.3	1.3	2
Zinc	Zn	5	–	3
Cadmium	Cd	0.005	0.005	0.003
Arsenic	As	0.010	0	0.01
Mercury	Hg	0.002	0.002	0.006

7.2 Carbon-Allotropes: Synthesis Methods, Applications and Future Perspectives

The letter "C" stands for carbon, the sixth element on the periodic chart. One of the most prevalent materials on earth, carbon is present practically everywhere. It is the fifteenth most frequent element in the crust of the planet and the fourth most frequent element overall. The Latin term "carbo" for coal and charcoal is the source of the name carbon, which is also derived from the French word "charbon" for charcoal [7].

The distribution of carbon on and within the globe is depicted in the accompanying picture (Figure 7.1), and it can be summed up as follows:

1. The crust of the Earth contains 2700 ppm of carbon.
2. Carbon content in oceanic crusts is 200 ppm.
3. There is 400 ppm of carbon in our atmosphere.
4. Carbon content in the hydrosphere is 30 ppm.

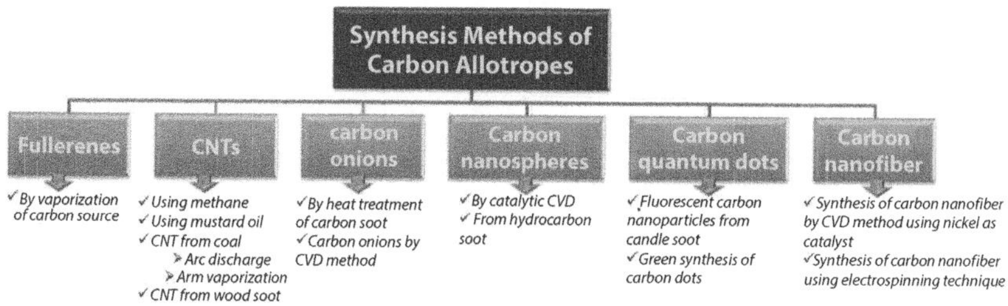

Figure 7.1 Synthesis method of carbon allotropes.

7.3 Reaffirmations of Heavy Metal Contaminations in Water and Their Toxic Effects

The major heavy metals in water as pollutants are shown in Figure 7.2 and Table 7.2.

7.3.1 Copper

Copper offers some benefits in industrial and agricultural processes. Many reassert have the potential to release copper into the surrounding area. Severe copper exposure may require a large supply of drinking water. Mercury, which poses a serious threat to the safety of drinking water, is the only metal more poisonous than copper [8]. Even though copper is necessary for animal metabolism. However, consuming too much copper can have dangerous adverse effects, including rapid breathing and blood pressure increases, liver and kidney damage, convulsions, cramping, vomiting, and even death.

7.3.2 Zinc

A multitude of metabolic activities and the physiological features of resident tissue is tightly regulated by zinc. But the presence of zinc also adds to other major medical conditions like anemia, stomachaches, rashes, cramps, and skin rashes. Among the commercial reassertions of zinc is the manufacture of brass plating, wood pulp, floor and newspaper paper, metallic works with

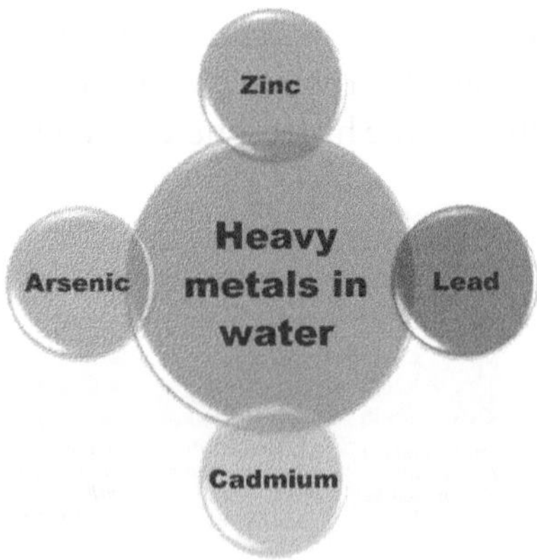

Figure 7.2 Heavy metals in water.

Table 7.2 Sources of contaminations and potential toxic effects of some heavy metals.

Heavy metal	Sources of contamination	Potential toxic effects	Available treatment techniques	Refs.
Cadmium	i. Steel and plastics industries ii. Cooling tower blow down iii. Electroplating, metal plating and coating operations, etc. iv. Nickel-cadmium batteries	i. Damage to kidney ii. Cancers iii. Bronchiolitis, COPD, emphysema, fibrosis iv. Skeletal damage	i. Coagulation ii. Ion-exchange iii. Precipitation iv. Softening v. Membrane separation	[13–17]
Chromium	i. Industrial wastewater discharge to environment. ii. Cooling tower blowdown iii. Electroplanting iv. Metal planting and coating operation	i. Severe diarrhea ii. Vomiting iii. Pulmonary congestions liver and kidney damage	i. Membrane technologies ii. Ion exchange iii. Coangulation iv. Floatation v. Solvent extraction vi. Cyanide treatment vii. Adsorption	[18–20]
Lead	i. Paint ii. Pesticides iii. Smoking iv. Automobile emissions v. Burning of coal vi. Mining	i. Anaemia ii. Cancer iii. Renal kidney disease iv. Nervous system damage v. Metal retardation vi. Impaired intellectual ability and behavioural problems in children	i. Reverse osmosis ii. Ion exchange iii. Membrane separation iv. Filtration v. Adsorption vi. Cementation vii. Chemical precipitation	[21–30]

(Continued)

Table 7.2 Sources of contaminations and potential toxic effects of some heavy metals. (*Continued*)

Heavy metal	Sources of contamination	Potential toxic effects	Available treatment techniques	Refs.
Zn	i. Brass platting ii. Wood pulp production, iii. Ground and newsprint paper production iv. Steel works with galvanizing lines v. Zinc and brass metal works vi. Refineries vii. Plumbing	i. Stomach nausea ii. Skin irritations iii. Cramps iv. Vomiting, and anemia	i. Precipitation ii. Membrane processes iii. Ion exchange resins iv. Adsorption	[31]
Cu	i. Pesticides industry ii. Mining iii. Metal piping iv. Chemical industry	i. Increased ii. Blood pressure and respiratory rates iii. Damaged in kidney and liver iv. Convulsions, cramps, vomiting, or even death	i. Ion exchange ii. Reverse osmosis iii. Membrane technologies iv. Chemical precipitation v. Electrochemical treatment	[32–42]

galvanizing lines, and zinc and brass metallic works. The literature has found zinc concentrations in a variety of waste streams ranging from less than 1 to more than 48,000 mg L^{-1}. A combination of agricultural activities, groundwater intrusion, sediment remobilization or entrainment, or any one of these alone can release zinc into the environment [9].

7.3.3 Lead

Lead is a dangerous heavy metal that may readily accumulate in the human body and cannot be broken down. Drinking water, which also includes a sizable quantity of lead, is the main source of lead in a person's body. Initially, may enter the body through the respiratory and digestive systems. Blood is then used to distribute throughout the body. Massive lead levels in drinking water have been associated with cerebral impairment, anemia, cancer, kidney illness, fear-related device damage, and cancer [10].

7.3.4 Cadmium

Herbal deposits that also include other elements can contain the heavy metal cadmium. This particular substance is regarded to be one of the main contaminants in drinking water and is quite harmful. It was discovered that cadmium frequently builds up inside the kidneys and has a biological half-life of 10-35 years in humans [11]. Because cadmium is a water contaminant, the kidney is the primary target organ for poisoning.

7.3.5 Arsenic

The source of arsenic present in the water is both human activity and herbal sources. The geological formation that takes place in sedimentary rocks, geothermal water, and weathered volcanic rocks aids in the formation, they are propelled into the floor water. Arsenic is also introduced into the water in our bodies by human activities such as mining, manufacturing, metallurgy, timber maintenance, and insecticides [12]. Heavy metals treatment techniques are shown in Figures 7.3 and 7.4.

7.4 Technology is Used to Treat Heavy Ions of Metal

7.4.1 Chemical Precipitation

This approach is the most successful and widely used in businesses due to its straightforward operating principles and low operational costs. Chemical substances are combined with toxic metal ions to form precipitates during

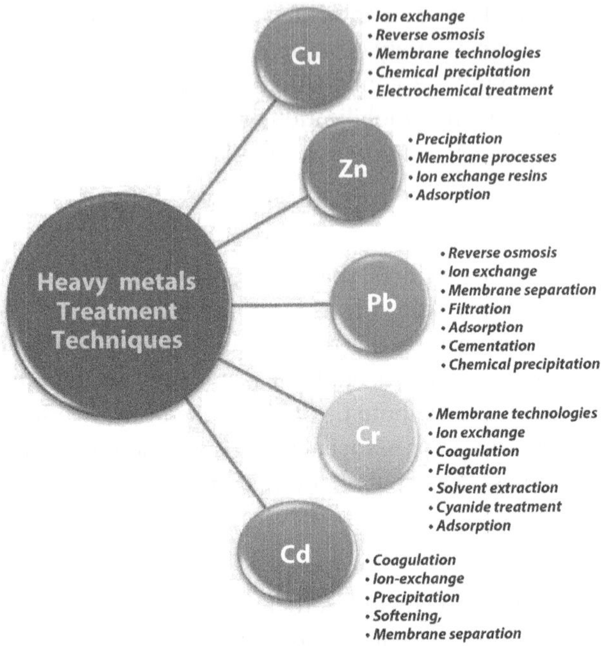

Figure 7.3 Heavy metals treatment techniques.

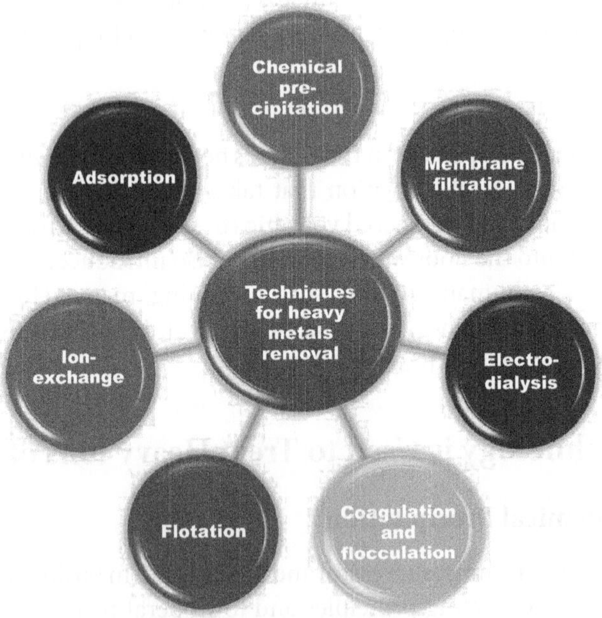

Figure 7.4 Techniques for heavy metals removal.

chemical precipitation. The insoluble precipitates are separated using two methods, the first of which is filtration and the second of which is sedimentation. There are two methods for conventional precipitation: chemical precipitation of sulphide and hydrogen oxide [43].

a) Hydroxide chemical precipitation
The hydroxide precipitation method is the most frequently used in chemical precipitation because it can manage pH costs and offers significant cost savings. Lime is an excellent alternative for hydroxide chemical precipitation in enterprises due to its low cost and ease of handling. It has been shown that a remarkable number of hydroxides are used to detect the precipitation of harmful ions from contaminated water. The hydroxide chemical precipitation method employs $Ca(OH)_2$ and NaOH to remove Cr(VI) and Cu(II) ions from polluted water. To increase the amount of lime precipitation, we used fly ash-based seed cloth [44]. This carbonation therapy enhances the particle's size and provides better ion-elimination efficiency for dangerous metals. Even though this approach is widely used, it also has a few main drawbacks The first is that HCP offers significant volumes of low-density muck, which presents disposal and dewatering issues. The second is that some hydroxide metals are amphoteric, which affects compound steel.

b) Sulfide precipitation
Sulphide chemical precipitation, the most popular technique, is especially efficient for treating heavy steel ions. The fact that this system is less soluble than the HCP system and isn't always amphoteric is its greatest benefit. As a result, heavy steel ions have a substantially higher removal efficiency than hydroxide chemical precipitation. Ozverdi and Erdem (2006) claim that Cd2+, Pb2+, and Cu2+ are removed using pyrite sulphide iron. With the aid of a jar tester utilized as a precipitator and hydroxide precipitation, Brbootl *et al.* were able to remove Iron (III), Chromium (III), Copper (II), Lead (II), Cadmium (II), and Nickel (II) from the water [45]. The pH range for the MgO precipitant is 9.6 to 10 [46].

7.4.2 Ion-Exchange

The second method is ion change, which takes the benefits into excessive consideration. This method is widely employed to treat commercial wastewater effluent. These advantages include its high kinetics, high cure potential, and high elimination performance. Ion-change resin can be either synthetic or stable herbal resin. It can transform its cations into the steel ions present in the contaminated water thanks to a few unique potentials.

Because it is so good at eliminating dangerous steel ions from effluent, synthetic resin is the substance that is used in the ion-change process the most commonly. The carboxyl (-COOH) and sulfonic (-S03H) acidic organizations with weak acidic resin and strong acidic organization, respectively, make up the cation exchanger [47]. Heavy steel-ions are exchanged for H+ at the resins in the cations col, where heavy metals that are administered during the response process go through:

$$nR\text{- }SO3H + M^{a} \longrightarrow (R\text{- }SO3^{-})n\ M^{a} + nH+ \ (a)$$
$$nR\text{- }COOH + M^{a} \longrightarrow (R\text{- }COOH^{-})n\ M^{a} + nH+ \ (b)$$

Numerous researchers have found that, in variable conditions, zeolite exhibits greater cation update capability for hazardous ions. According to several researchers working on current paintings, clinoptilolite's surface is wrapped by amorphous iron oxide for better changing functionality. Zeolites and montmorillonites are used as ion-changing resins to remove steel ions, according to the study. In contrast to synthetic resins, their stock is tiny. This is an ion-change strategy with improved performance, and Alyuz and Veli were the first to adequately explain maximum green strategies for the removal of undesired steel ions from the contaminated water [48]. Zeolite is an inorganic ion-exchanger fabric that is used in Zagorodni's ion-change method for the removal of heavy steel ions [49]. Steel sulphide is described by Rathor *et al.* as an ion exchanger that performs better at removing heavy steel ions from wastewater effluent [50]. For the excessive capacity of putting off heavy steel ions from the wastewater effluent, Fathima and Pandith *et al.* describe hybrid strategies wherein natural and inorganic are hybrid [51].

7.4.3 Adsorption

When the most economical way for removing heavy steel ions is taken into account, adsorption is a much more effective method for treating wastewater effluent. For the removal of heavy steel ions, a variety of adsorbents are available, and adsorption can occasionally be reversed. The adsorbent for the desorption method can be renewed [52].

a) Activated carbon adsorbents—The most prevalent kind of adsorbent used to remove heavy metallic ions is activated carbon. The huge amount of mesopore and micropore this adsorbent includes, which gives a significant amount of floor area, is its key advantage. Many researchers have found that activated carbon works well as an adsorbent to remove harmful metallic ions. Due to the depletion of coal, tannic acid, surfactants, alginate components, magnesium, and most crucially activated carbon composite may be effective absorbents for heavy metallic contaminants. As a result, we can employ activated carbon composite and its parts in the area of ACA without having an impact on this aspect. The abundant supply of AC adsorbent makes it possible [53].

b) Carbon nanotube adsorbents—Iijima created CNT in 1991. (carbon nanotube). He offers a basic understanding of its applicability in addition to describing the properties of CNTs. The utilization of carbon nanotubes facilitates the removal of harmful steel ions, such as cadmium, copper, lead, chromium, and nickel from contaminated water. And the remarkable result of the heavy metals' removal is made possible by carbon nanotube adsorption. Single-walled carbon nanotubes (SWCNT) and multiwalled carbon nanotubes are two different types of carbon nanotubes (MWCNT). The mechanism between the surface of the functional organization and the heavy steel ions involves chemical interaction, electrostatic attraction, and sorption precipitation due to the complexity of the carbon nanotube adsorbent system [54].

c) Low-cost adsorbents—The findings show that activated carbon is a cheap and frequently used adsorbent for removing steel ions. Therefore, the researcher continued to research to find an inexpensive adsorbent. For the treatment of harmful steel ions, hundreds of articles have been provided [55].

d) Bio-adsorbents—Bioadsorption is the process of removing toxic metals from water using a brand-new adsorbent known as bio-adsorbent. The advantages of this adsorbent are that it is less expensive than other adsorbents and that it is incredibly effective at removing steel ions [56].

7.4.4 Membrane Filtration

Different types of membranes are used in membrane filtration methods to treat wastewater effluent. It is simple to carry out saving space and also has excellent metal removal performance. The membrane employed in opposite osmosis, electrodialysis, ultra-filtration, and nano-filtration is to remove ions from polluted water [57]. Types of membrane filtration are shown in Figure 7.5 and membrane filtration for contaminant water filtration is shown in Figure 7.6.

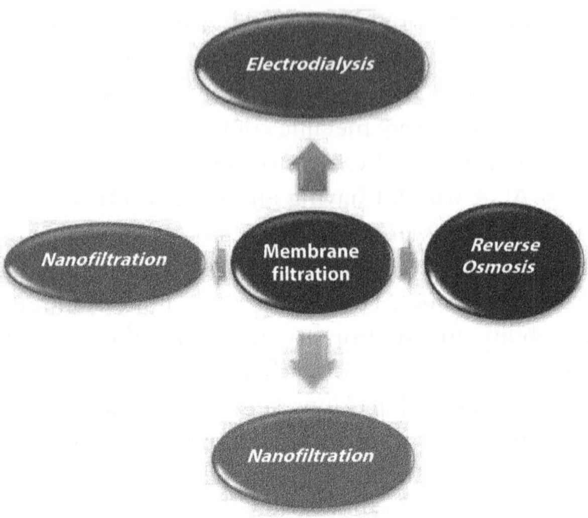

Figure 7.5 Types of membrane filtration.

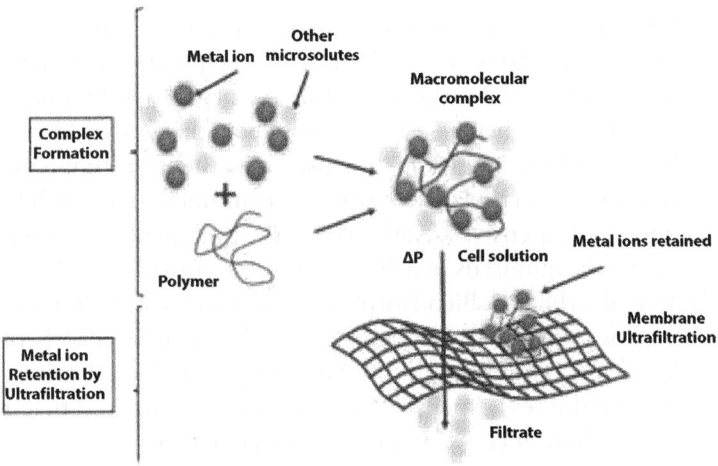

Figure 7.6 Membrane filtration for contaminant water filtration.

7.4.5 Electrodialysis

An electrical discharge through the solution is used in ED membrane filtration to separate ions from solutions. In the electrodialysis procedure, an ion-change membrane is used. One of the exchangers, an anion exchanger, and the other, a cation exchanger, are currently in operation electrodialysis is frequently used to remove water from seawater and salty water to create process water. Electrodialysis is a method for effectively removing advantageous metallic ions from sewage effluent and seawater [58].

a) Reverse Osmosis—This will allow the solution to be cleansed to pass through the semipermeable barrier while keeping out the undesirable elements. Reverse osmosis is a useful technique for eliminating a sizable amount of dissolved contaminants from water. 20% of the world's desalination potential is accomplished via opposite osmosis techniques [59].
b) Nanofiltration—Between reverse osmosis and ultra-filtration, a process known as NF (Nano-filtration) takes place. For the removal of toxic metallic ions from contaminated water, such as Cr, As, Ni, and Co, nano-filtration is a worthwhile strategy. There are numerous advantages to using nanofiltration, including the fact that it uses less energy, removes harmful metallic ions more effectively than other filtration methods, and is more reliable in nature [60].
c) Ultrafiltration—A membrane technique known as UF uses low trans-membrane pressure to reject colloidal and dissolved material from aqua solution (ultra-filtration). MEUF (micellar enhanced ultra-filtration) and PEUF (polymer enhanced ultra-filtration) were developed to achieve high performance of metal removal [61]. Hani Abu-Qudais uses reverse osmosis and nano-filtration as a component of membrane filtration to remove cadmium and copper from wastewater effluent [62]. The hybrid strategies for removing heavy metal ions from contaminated water are described by Dorda J *et al.* in their study [63].

7.4.6 Flotation

One often used technique for the removal of wastewater treatment is flotation. With the aid of an attachment of bubble, mineral-derived method, we use flotation techniques to extract hazardous ions from the aqua solution. The three primary flotation processes used to remediate wastewater effluent are dissolved air flotation, ion flotation, and precipitation. A thin, low-density film that holds the harmful metallic ions that are discharged as a sludge formation is created by dissolved air flotation, which enables air bubbles to connect to suspended particles in the aqua solution [64]. To remove cadmium and copper hazardous metal ions from wastewater effluent, Mavrov *et al.* develop a hybrid flotation process [65]. To remove zinc, copper, and nickel from the wastewater effluent, Blocher *et al.* 2003 also provided a hybrid theory for the removal procedure [66] that included flotation and membrane filtering [91]. For the removal of cadmium in 2002, Rubio *et al.* employed collectors like SOS and MIBC, and the results ranged from 89.3% to 97.6% [67].

7.4.7 Electrochemical Treatment

The steel ions are spread out in this method on a cathode floor, and with this system, advantageous metals are improving in the elemental condition. However, it has inherent risks, such as its extremely expensive techniques due to high electric-powered consumption. Therefore, this method isn't more effective, but because of its controlled fundamental qualities, it has been widely employed for the past two years. And here, we looked at how this technology's distinctive Electroflotation (EF), electrocoagulation (EC), and electrodeposition processes work (ED) [68].

Hasson *et al.* present the methods for phosphate removal using electrochemical treatment, as well as the hardness of calcium and magnesium. Hard metal can be successfully removed from industrial effluent using electrochemical techniques [69]. It is the best method for removing heavy metal ions from industrial wastewater effluent, according to Petsriprasit *et al.*'s description of the electrochemical process [70].

7.4.8 Electroflotation

In the electro-flotation (EF) method of separating liquids from solids, the contaminant settles on the water's surface in the form of oxygen- and hydrogen-containing fuel bubbles. The coolest method to eliminate damaging steel ions is electro-flotation. The researcher examines electroflotation methods for the treatment of wastewater. Advantages and disadvantages of the current techniques for heavy metals removal were shown in Table 7.3. This technique makes it simple to remove steel ions from wastewater, including Ni, Zn, Fe, Co, and Pb. It delivers a 99.9% eradication rate, which is unquestionably a great outcome of elimination. Checking the rising performance of Ni without or with micro-, filter-paper, and ultrafiltration in the electro-flotation method with hybrid Fe electrode [71]. Maximum adsorption capacities and interaction mechanisms of selected heavy metal ions with CNTs were shown in Table 7.4.

Table 7.3 Advantages and disadvantages of the current techniques for heavy metals removal.

Methods	Advantages	Disadvantages	Ref.
Ion exchange	i. High treatment capacity ii. High removal efficiency iii. Fast kinetics	i. High cost due to synthetic resins ii. Regeneration of the resins cause serious secondary pollution	[73]
Adsorption	i. Easy operating conditions ii. High metal binding capacities iii. Having wide pH range iv. Low-cost	i. Production of waste products ii. Low selectivity	[74–76]
Chemical precipitation	i. Simple operation ii. Low capital cost	i. Sludge generation ii. Extra operational cost for sludge disposal iii. Ineffective for treatment of water with low concentration of heavy metals	[77, 78]
Membrane filtration	i. High separation selectivity ii. Small space requirement iii. Low pressure requirement	i. High operational cost due to membrane fouling ii. Process complexity iii. Low permeate flux	[78–80]

(Continued)

Table 7.3 Advantages and disadvantages of the current techniques for heavy metals removal. (*Continued*)

Methods	Advantages	Disadvantages	Ref.
Electrodialysis	i. High separation selectivity. ii. High operational cost due to energy Consumption	i. Limited applications ii. Long duration time	[74, 79]
Coagulation and flocculation	i. Remove the turbidly in addition to heavy metal removal ii. Produced sludge with good sludge settling and dewatering characteristics	i. Increased sludge volume generation ii. Coagulation flocculation can't treat the heavy metal wastewater and must be followed by other treatment techniques	[73, 78, 80]
Flotation	i. High metal selectivity ii. High removal efficiency iii. High overflow rates iv. Low detention periods v. Low operating cost vi. Production of more	i. High initial capital cost High maintenance and operation costs	[73, 78, 81]

Table 7.4 Maximum adsorption capacities and interaction mechanisms of selected heavy metal ions with CNTs.

				Q (mg/g)		Adsorption capacity,	
Adsorbates	**Adsorbents**	**Conditions**	**Model used**	**Experimental**	**Interaction mechanism**	**Calculate from model**	**Ref.**
Cu(II)	HNO_3-modified CNTs	pH 5, initial concentration = 10 mg/L, room temperatiure	Langmuir	29	Ion exchange	28.49	[82]
	CNTs	pH 5, copper equilibrium concentration = 5 mg/L	Langmuir	N/A	Surface complexation/ ion exchange	26.41	[83]
	CNTs/calcium alginate composites	pH 5, copper equilibrium concentration = 5 mg/L	Langmuir	N/A	Surface complexation/ ion exchange	84.88	[83]
	Purified multi-walled carbon nanotubes (p-MWCNTs)	pH 5, initial concentration = 20 mg/L	D–R model	N/A	Physical adsorption	36.82	[84]

(*Continued*)

Table 7.4 Maximum adsorption capacities and interaction mechanisms of selected heavy metal ions with CNTs. (*Continued*)

Adsorbates	Adsorbents	Conditions	Model used	Q (mg/g)	Interaction mechanism	Adsorption capacity,	Ref.
				Experimental		Calculate from model	
	Sulfonated multi-walled carbon	pH 5, initial concentration = 20 mg/L	D–R model	N/A	Electrostatic interactions	43.16	[84]
	Nanotubes (s-MWCNTS) MWCNTs	pH 5, initial concentration = 2.36 mg/L × 10^{-4} mol/L, T = 303 K	Langmuir	N/A	Electrostatic interaction, surface complexation, surface precipitation	3.19 × 10^{-5} mol/g	[85]
	MWCNTs impregnated with di-(2-ethyl hexyl phosphoric acid) (D2EHPA) and tri-n-octyl Phosphine oxide (TOPO)	pH 5, initial concentration = 500 lg/L, adsorbent dosage = 500 mg	Experimental	4.90	Electrostatic interaction	N/A	[86]

(*Continued*)

Table 7.4 Maximum adsorption capacities and interaction mechanisms of selected heavy metal ions with CNTs. (*Continued*)

Adsorbates	Adsorbents	Conditions	Model used	Q (mg/g)	Interaction mechanism	Adsorption capacity,	Ref.
				Experimental		Calculate from model	
	Magnetic-functionalized MWCNTs	pH 6, initial concentration = 100 mg/L	Experimental	N/A	N/A	N/A	[87]
	Oxidized CNTs sheets	pH 7, initial concentration = 1200 mg/L, T = 298 K	Langmuir	50.37	Chemical interactions	64.93	[88]
	Single walled carbon nanotubes	pH 5, initial concentration = 20 mg/L; contact time,	Langmuir	N/A	Physisorption	24.29	[89]
	(SWCNTs)	120 min, adsorbent dosage, 50 mg/L, T = 298 ± 1 K					

(*Continued*)

Table 7.4 Maximum adsorption capacities and interaction mechanisms of selected heavy metal ions with CNTs. (*Continued*)

Adsorbates	Adsorbents	Conditions	Model used	Q (mg/g)	Interaction mechanism	Adsorption capacity,	Ref.
				Experimental		Calculate from model	
	SWCNTs-COOH	pH 5, initial concentration = 20 mg/L; contact time, 120 min, adsorbent dosage, 50 mg/L, *T* = 298 ± 1 K	Langmuir	N/A	Chemisorption	77	[89]
	Oxidized MWCNTs	pH 9, initial concentration = 1–20 mg/L, adsorbent dosage = 50 mg	Freundlich	3.49	Electrostatic interactions	N/A	[90]
	As-produced CNTS	pH 6, T = 300 K	Langmuir	N/A	Electrostatic interactions	8.25	[91]
	NaOCI-modified CNTs	pH 6, T = 300 K	Langmuir	N/A	Electrostatic interactions	47.39	[91]
	HNO_3-modified CNTs	pH 6, T = 300 K	Langmuir	N/A	Electrostatic interactions	13.87	[91]

(*Continued*)

Table 7.4 Maximum adsorption capacities and interaction mechanisms of selected heavy metal ions with CNTs. (*Continued*)

				Q (mg/g)		Adsorption capacity,	
Adsorbates	**Adsorbents**	**Conditions**	**Model used**	**Experimental**	**Interaction mechanism**	**Calculate from model**	**Ref.**
	MWCNTs in the presence of humic acid (HA)	Initial concentration = 20 mg/L T = 293 K	Langmuir	N/A	Electrostatic interactions	7.776 ± 0.538	[92]
	MWCNTs/Fe_3O_4	pH 6, concentration = 30 mg L	Langmuir	19	Electrostatic interactions	38.91	[93]
	Chitosan/poly (vinyl) alcohol thin adsorptive membranes modified	pH 5.5, initial concentration = 30 mg/L, T = 20 °C	Langmuir	9.54	Ion exchange	11.1	[94]
	Amino functionalized MWCNTs As-grown CNTs	pH 5.4, Cu^{2+} ion concentration = 20 mg/L	Langmuir	N/A	Precipitation	14.4	[95]

(*Continued*)

Table 7.4 Maximum adsorption capacities and interaction mechanisms of selected heavy metal ions with CNTs. (*Continued*)

				Q (mg/g)		Adsorption capacity,	
Adsorbates	**Adsorbents**	**Conditions**	**Model used**	**Experimental**	**Interaction mechanism**	**Calculate from model**	**Ref.**
	Oxidize CNTS	pH 5.2, Cu^{2+} ion concentration = 20 mg/L	Langmuir	N/A	Electrostatic interaction	27.6	[95]
Zn(II)	Nitrogen-doped magnetic CNTs	pH 8	Langmuir	9.31 mmol/g	Chemical adsorption	N/A	[96]
	MWCNTs impregnated with di-(2-ethyl hexyl phosphoric acid) (D2EHPA) and tri-n-octyl phosphine oxide (TOPO)	pH 5, initial concentration = 500 Ig/L, adsorbent dosage = 500 mg	Experimental	4.82	Electrostatic interaction	N/A	[97]
	Oxidized CNTs sheets	pH 7, initial concentration = 1200 mg/L, T = 298 K	Langmuir	58	Chemical adsorption	74.63	[98]

(*Continued*)

Table 7.4 Maximum adsorption capacities and interaction mechanisms of selected heavy metal ions with CNTs. (*Continued*)

Adsorbates	Adsorbents	Conditions	Model used	Q (mg/g)	Interaction mechanism	Adsorption capacity,	Ref.
				Experimental		Calculate from model	
	Pristine MWCNT	pH 6.5–6.8, initial concentration = 0.08–3 mM	Langmuir	N/A	Electrostatic interaction	0.14 mmol/g	[99]
	O-MWCNT	pH 6.5–6.8, initial concentration = 0.08–3 mM	Langmuir	N/A	Electrostatic interaction	0.27 mmol/g	[99]
	SWCNTs (NaOC1)	pH 7, S/L: 0.05/100, initial concentration = 10 mg/L, T = 25°C	Langmuir	14.9	Electrostatic interaction	43.66	[100]
	MWCNTs (NaOC1)	pH 7, S/L: 0.05/100, initial concentration = 10 mg/L, T = 25°C	Langmuir	13.75	Electrostatic interaction	32.68	[100]
	SWCNTs purified with sodium hypochlorite	Initial concentration = 60 mg/L, T = 45 °C	Langmuir	16.18	Electrostatic interaction	46.94	[101]

(*Continued*)

Table 7.4 Maximum adsorption capacities and interaction mechanisms of selected heavy metal ions with CNTs. (*Continued*)

Adsorbates	Adsorbents	Conditions	Model used	Q (mg/g)	Interaction mechanism	Adsorption capacity,	Ref.
				Experimental		Calculate from model	
	MWCNTs purified with sodium hypochlorite	Initial concentration = 60 mg/L, T = 45 °C	Langmuir	15.77	Electrostatic interaction	34.36	[101]
	Multi-walled carbon nanotubes (MWCNTs) modified with chitosan	pH 7, dosage = 200 g, T = 298 K, flow rate = 5 ml/min	N/A	N/A	Electrostatic interaction	N/A	[102]
	Functionalized carbon nanotubes	pH 10, initial concentration = 1.1 mg/L, dosage = 0.09 g, agitation speed = 120 rpm, time = 120 min	Langmuir	2.42	Electrostatic interaction	1.05	[103]
Ni(II)	MWCNTs	pH 5.4, m/V = 0.8 g/L, T = 293 K	Langmuir	2.9036	Electrostatic interaction, p—p interaction	6.346 × 10—5 mol/g	[104]

(*Continued*)

Table 7.4 Maximum adsorption capacities and interaction mechanisms of selected heavy metal ions with CNTs. (*Continued*)

Adsorbates	Adsorbents	Conditions	Model used	Q (mg/g)	Interaction mechanism	Adsorption capacity,	Ref.
				Experimental		Calculate from model	
	Polyacrylic acid (PAA)-MWCNTs	pH 5.4, m/V = 0.8 g/L, T = 293 K	Langmuir	N/A	Electrostatic interaction, p—p	6.615 × 10—6 mol/g	[104]
	HNO3-treated MWCNTs	pH 6.5, initial concentration = 20 mg/L, amount of adsorbent = 0.8 g/L, T = 338 K	Langmuir	12.500	Ion exchange	17.86	[105]
	MWCNTs/iron oxide	Initial concentration = 6 mg/L, m/V = 0.75 g/L	Langmuir	N/A	Ion exchange	9.18	[106]
	MWCNTs	pH 6, m/V = 0.2 g/L	Langmuir	N/A	Electrostatic interaction	18.083	[107]
	Oxidized CNTs	pH 6, m/V = 0.2 g/L	Langmuir	N/A	Electrostatic interaction	49.261	[107]
	Nitrogen-doped magnetic CNTs	pH 8	Langmuir	8.06 mmol/g	Electrostatic interaction	N/A	[108]

(*Continued*)

Table 7.4 Maximum adsorption capacities and interaction mechanisms of selected heavy metal ions with CNTs. (*Continued*)

Adsorbates	Adsorbents	Conditions	Model used	Q (mg/g)	Interaction mechanism	Adsorption capacity,	Ref.
				Experimental		Calculate from model	
	MWCNTs impregnated with di-(2ethyl hexyl phosphoric acid) (D2EHPA) and tri-n-octyl phosphine oxide (TOPO)	pH 5, initial concentrations = 500 lg/L, adsorbent dosage = 500 mgH	Exprimental	4.78	Electrostatic interaction	N/A	[109]
	Oxidized CNTs	pH 6.55 ± 0.02, initial concentration = 20 mg/L, T = 333 K	Langmuir	8.75	Electrostatic interaction	9.80	[110]
	HNO3 oxidized CNTs	pH 8, initial concentration = 1 mg/L, room temperature	Experimental	6.89	Electrostatic interaction	N/A	[111]

(*Continued*)

Table 7.4 Maximum adsorption capacities and interaction mechanisms of selected heavy metal ions with CNTs. (*Continued*)

Adsorbates	Adsorbents	Conditions	Model used	Q (mg/g)	Interaction mechanism	Adsorption capacity,	Ref.
				Experimental		Calculate from model	
	NaClO-modified MWCNTs	pH 7.0, initial concentration = 10–80 mg/L, T = 298 K	Langmuir	N/A	Electrostatic interaction	38.46	[112, 113]
	NaClO-modified SWCNTs	pH 7.0, initial concentration = 10–80 mg/L, T = 298 K	Langmuir	N/A	Electrostatic interaction	47.86	[112, 113]
Cd(II)	Acid modified CNTs	pH 7, adsorbent dosage = 50 mg	Langmuir	2.02	Electrostatic interaction	4.35	[114]
	HNO3-modified CNTs	pH 5, initial concentration = 10 mg/L, Room Temperature	Langmuir	9.2	Ion exchange	10.86	[115]
	As-grown CNTs	pH 5.5, initial concentration = 4 mg/L	Experimental	11	N/A	N/A	[116]

(*Continued*)

Table 7.4 Maximum adsorption capacities and interaction mechanisms of selected heavy metal ions with CNTs. (*Continued*)

				Q (mg/g)		Adsorption capacity,	
Adsorbates	**Adsorbents**	**Conditions**	**Model used**	**Experimental**	**Interaction mechanism**	**Calculate from model**	**Ref.**
	H2O2 oxidized CNTs	pH 5.5, initial concentration = 4 mg/L	Experimental	2.6	N/A	N/A	[116]
	HNO3 oxidized CNTs	pH 5.5, initial concentration = 4 mg/L	Experimental	5.1	N/A	N/A	[116]
	KMnO4 oxidized CNTs	pH 5.5, initial concentration = 4 mg/L	Experimental	11	N/A	N/A	[116]
	Ethylenediamine-functionalized MWCNTs	Initial concentration = 5 mg/L, T = 45 °C	Langmuir	21.23	Physisorption chemisorption	25.70	[117]
	SWCNTs-COOH	pH 5, initial concentration = 20 mg/L, contract time, 120 min, adsorbent dosage = 50 mg/L, T = 298 ± 1 K	Langmuir	N/A	Chemisorption	55.89	[122]

(*Continued*)

Table 7.4 Maximum adsorption capacities and interaction mechanisms of selected heavy metal ions with CNTs. (*Continued*)

Adsorbates	Adsorbents	Conditions	Model used	Q (mg/g) Experimental	Interaction mechanism	Adsorption capacity, Calculate from model	Ref.
	Raw CNTs	pH 7, initial concentration = 1 mg/L CNT dosage = 50 mg	Langmuir	0.657	Electrostatic interaction	1.661	[123]
	HNO3 oxidized CNTs	pH 8, initial concentration = mg/L, Room Temperature	Experimental	7.42	Electrostatic interaction	N/A	[124]
	Pristine MWCNTs	pH 6.5–6.8, initial concentration = 0.05–1.8 mM	Langmuir	N/A	Electrostatic interaction	0.05 mmol/g	[125]
	O-MWCNTs	pH 6.5–6.8, initial concentration = 0.05–1.8 mM	Langmuir	N/A	Electrostatic interaction	0.22 mmol/g	[125]
	Alumina-decorated MWCNTs	pH 7, initial concentration = 1 mg/L; adsorbent dosage = 50 mg/L	Langmuir	0.948		27.21	[126]

7.4.9 Coagulation and Flocculation

The basic function of coagulation is to irritate colloid debris by neutralizing the forces that keep them apart. Examples of coagulants that do an excellent job of removing contaminants and metallic ions from wastewater effluent and creating the precipitate of amorphous metallic hydroxide include ferrous sulphate, aluminium, and ferric chloride [118–121].

7.5 Factors Influencing How Heavy Metal Ions Adhere to CNTs

The primary variables affecting the adsorption of heavy steel ions on CNTs floors include the initial concentration of steel ions, pH, contact time, CNT dosage, agitation speed, temperature, floor charge, ionic strength, isoelectric factor, and foreign ions. Figure 7.7 illustrates the variables influencing the adsorption of heavy metal ions on CNTs.

7.5.1 pH

For the heavy steel ions to adhere to the surface of the CNTs, the pH is crucial. At pH values greater than a factor of 0 charges (pHPZC), cation

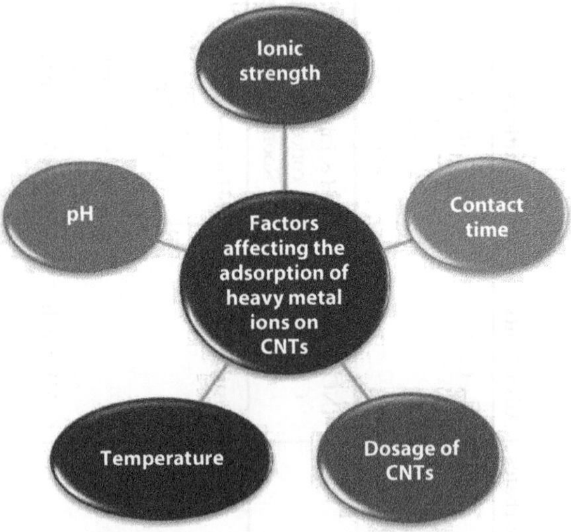

Figure 7.7 Factors influencing how heavy metal ions adhere to CNTs.

adsorption is enhanced due to the electrostatic interactions between the cation ions and the negative floor charge [127].

7.5.2 Ionic Strength

The ionic electricity of the heavy metal ions in the solution has a bearing on the adsorption of heavy metals onto the surface of CNTs. The sorption of heavy metals is typically negatively impacted by an increase in ionic electricity [128].

7.5.3 CNT Dosage

The efficiency of heavy metal absorption is also influenced by the number of CNTs used. With an increase in CNT dosage, the sorption of heavy metals can either rise or fall. Reports state that as the dose of CNTs increases, Ni2+, Cd2+, Cu2+, and Pb2+ sorption also rises [129].

7.5.4 Contact Time

The touch time also has a significant impact on the adsorption of steel ions onto CNTs. The percentage of heavy metals eliminated will typically rise as touch time increases until equilibrium is reached. Equilibrium, however, heavily depends on the steel ions' initial focus. The majority of the investigation revealed that while steel attention is minimal, equilibrium is installed rapidly [130].

7.5.5 Temperature

The impact of temperature on the sorption of heavy metals has been documented in numerous research. The rise in sorption capacity with increasing temperature is the most frequent site observation. For instance, the endothermic sorption of Pb2+ and Zn2+ would become more intense as the temperature increases [131].

7.5.6 Thermodynamic Variables

Important details concerning the sorption of heavy steel ions onto the surface of the CNTs are revealed by the thermodynamic characteristics. Measurements of entropy (DS), Gibbs power (DG), and enthalpy (DH) can be used to forecast the kind of adsorption. Based on the calculated enthalpy values, the Cd2+, Cu2+, and Pb2+ ion adsorption to SWCNTs and chemicals were stated to the bodily method [132].

7.5.7 CNT Regeneration

Regeneration of the adsorbent is not only required for recurrent use in practical packages, but it also highlights the adsorbent's affordability. CNT regeneration is significantly influenced by the pH of the regeneration fluid. Studies have shown that heavy metal ions on the surface of CNTs can be removed to regenerate them. Zn2+ and Pb2+ regeneration performs noticeably better at lower pH levels than at higher pH levels [133].

7.5.8 Isotherm Equation

The pseudo-2d order pricing equation, Langmuir, Freundlich, and Dubinin-Radushkevich (D-R) isotherm patterns are a few examples of special equations that can be used to depict the adsorption statistics. The Freundlich or Langmuir equations are typically connected with the sorption of metallic ions onto CNTs [134]. The sorption behaviour on homogeneous surfaces is relevant to the Langmuir equation.

7.5.9 Current Issues and the Need for Additional Study

The theoretical potential of CNTs as an adsorbent in water treatment has been the subject of extensive investigation in recent years. Despite their expensive costs, CNTs are anticipated to be promising adsorbents in the future due to their high adsorption compared to numerous conventional adsorbents. There are several ways discussed in the literature for the lab-scale synthesis of CNTs [135].

7.6 Conclusions

This review focused on the removal of lead, chromium, cadmium, arsenic, copper, zinc, and nickel ions from water using carbon allotropes. Numerous types of carbon allotropes, such as raw, acid-modified, and functionalized carbon allotropes, have been employed for the removal of heavy metals from water. Although raw carbon allotropes exhibited good adsorption potential for some metal ions, unmodified carbon allotropes had the greatest capacity for adsorbing the majority of heavy metals. Physical adsorption, electrostatic contact, surface complexation, and chemical interaction between the metal ions and surface functional groups are only a few of the numerous adsorption processes. The effects of process variables on the adsorption of heavy metals on the ions, such as pH, carbon allotrope dose, length, ionic strength, temperature, and surface charge Surface of carbon allotropes. The potential for CNT reuse through regeneration by metal ion desorption was also emphasized. This is a crucial consideration when deciding if carbon allotropes will eventually be used in commercial treatment facilities. However, more research is required to concentrate on the viability of producing carbon allotropes on a big scale and investigating the toxicity of carbon allotropes. Future innovation is anticipated to come from the use of carbon allotropes in industrial water treatment facilities.

Acknowledgments

The authors are grateful to HOD MATS School of Sciences, MATS University, Raipur, Chhattisgarh.

References

1. Mehndiratta, P., Jain, A., Srivastava, S., Gupta, N., Environmental pollution and nanotechnology. *Environ. Pollut.*, 2, 2, 10, 2013.
2. Nimibofa, A., Newton, E.A., Cyprain, A.Y., Donbebe, W., Fullerenes: Synthesis and applications. *J. Mater. Sci. Res.*, 7, 3, 16, 2018.
3. Mauter, M.S. and Elimelech, M., Environmental applications of carbon-based nanomaterials. *Environ. Sci. Technol.*, 42, 16, 5843–59, 2008.
4. Sardans, J., Montes, F., Peñuelas, J., Electrothermal atomic absorption spectrometry to determine As, Cd, Cr, Cu, Hg, and Pb in soils and sediments: A review and perspectives. *Soil Sediment Contam. Int. J.*, 20, 4, 447–491, 2011.
5. Binkley, J. and Simpson, J.A., Heavy metals in wastewater treatment processes, in: *Handbook of Water and Wastewater Microbiology*, pp. 597–610, 2003.
6. World Health Organization, *International Standards for Drinking Water*, WHO, Geneva, 2008.
7. Appenroth, K.-J., Definition of "heavy metals" and their role in biological systems, in: *Soil Heavy Metals*, pp. 19–29, Springer Berlin Heidelberg, Berlin, 2010.
8. Jaishankar, M., Tseten, T., Anbalagan, N., Mathew, B.B., Beeregowda, K.N., Toxicity, mechanism and health effects of some heavy metals. *Interdiscip. Toxicol.*, 7, 2, 12, 2014.

9. Sharma, P. and Dubey, R.S., Lead toxicity in plants. *Braz. J. Plant Physiol.*, 17, 18, 2005.
10. Sanders, T., Liu, Y., Buchner, V., Tchounwou, P.B., Neurotoxic effects and biomarkers of lead exposure: A review. *Rev. Environ. Health.*, 24, 1, 15–45, Jan-Mar 2009.
11. Al-Senani, G.M. and Al-Fawzan, F.F., Adsorption study of heavy metal ions from aqueous solution by nanoparticle of wild herbs. *Egypt. J. Aquat. Res.*, 44, 3, 187–194, 2018.
12. Bozbaş, S.K. and Boz, Y., Low-cost biosorbent: Anadara inaequivalvis shells for removal of Pb (II) and Cu (II) from aqueous solution. *Process. Saf. Environ. Prot.*, 103, 144–152, 2016.
13. Ihsanullah, Al Khaldi, F.A., Abusharkh, B., Khaled, M., Atieh, M.A., Nasser, M.S., Laoui, T., Agarwal, S., Tyagi, I., Gupta, V.K., Adsorptive removal of cadmium(II) ions from liquid phase using acid modified carbon-based adsorbents. *J. Mol. Liq.*, 204, 255–263, 2015.
14. Li, Q., Wu, S., Liu, G., Liao, X., Deng, X., Sun, D., Hu, Y., Huang, Y., Simultaneous biosorption of cadmium(II) and lead(II) ions by pretreated biomass of *phanerochaete chrysosporium*. *Sep. Purif. Technol.*, 34, 135–142, 2004.
15. Li, Y.H., Wang, S., Luan, Z., Ding, J., Xu, C., Wu, D., Adsorption of cadmium(II) from aqueous solution by surface oxidized carbon nanotubes. *Carbon*, 41, 1057–1062, 2003.
16. Mubarak, N.M., Ruthiraan, M., Sahu, J.N., Abdullah, E.C., Jayakumar, N.S., Sajuni, N.R., Tan, J., Adsorption and kinetic study on Sn2+ removal using modified carbon nanotube and magnetic biochar. *Int. J. Nanosci.*, 12, 6, 1350044–13500584, 2013.
17. Ihsanullah, Al Khaldi, F.A., Abusharkh, B., Khaled, M., Atieh, M.A., Nasser, M.S., Laoui, T., Agarwal, S., Tyagi, I., Gupta, V.K., Adsorptive removal of cadmium(II) ions from liquid phase using acid modified carbon-based adsorbents. *J. Mol. Liq.*, 204, 255–263, 2015.
18. Mohana, D. and Pittman, C.U., Activated carbons and low cost adsorbents for remediation of tri- and hexavalent chromium from water. *J. Hazard. Mater.*, 137, 762–811, 2006.
19. Liu, S.S., Chen, Y.Z., Zhang, L.D., Hua, G.M., Xu, W., Li, N., Zhang, Y., Enhanced removal of trace Cr (VI) ions from aqueous solution by titanium oxide–Ag composite adsorbents. *J. Hazard. Mater.*, 190, 723–728, 2011.
20. Li, Y., Jin, Z., Li, Z., A novel and simple method to synthesize SiO_2-coated Fe nanocomposites with enhanced Cr (VI) removal under various experimental conditions. *Desalination*, 288, 118–125, 2012.
21. Gupta, V.K., Agarwal, S., Saleh, T.A., Chromium removal by combining the magnetic properties of iron oxide with adsorption properties of carbon nanotubes. *Water Res.*, 45, 6, 2207–2212, 2011.
22. Barakat, M.A., New trends in removing heavy metals from industrial wastewater. *Arab. J. Chem.*, 4, 361–377, 2011.

23. Fu, F. and Wang, Q., Removal of heavy metal ions from wastewaters: A review. *J. Environ. Manage.*, 92, 407–418, 2011.
24. Marques, P.A.S.S., Rosa, M.F., Pinheiro, H.M., PH effects on the removal of Cu^{2+}, Cd^{2+} and Pb^{2+} from aqueous solution by waste brewery waste. *Bioprocess Eng.*, 23, 135–141, 2000.
25. Qu, X., Alvarez, P.J.J., Li, Q., Applications of nanotechnology in water and wastewater treatment. *Water Res.*, 47, 12, 3931–3946, 2013.
26. Chen, Q.Y., Luo, Z., Hills, C., Xue, G., Tyrer, M., Precipitation of heavy metals from wastewater using simulated flue gas: Sequent additions of fly ash, lime and carbon dioxide. *Water Res.*, 43, 2605–2614, 2009.
27. Hajiaghababaei, L., Badiei, A., Ganjali, M.R., Heydari, S., Khaniani, Y., Ziarani, G.M., Highly efficient removal and preconcentration of lead and cadmium cations from water and wastewater samples using ethylenediamine functionalized SBA-15. *Desalination*, 266, 182–187, 2011.
28. Deliyanni, E.A., Peleka, E.N., Matis, K.A., Removal of zinc ion from water by sorption onto iron-based nanoadsorbent. *J. Hazard. Mater.*, 141, 176–184, 2007.
29. Hasson, D., Sidorenko, G., Semiat, R., Low electrode area electrochemical scale removal system. *Desalination Water Treat.*, 31, 1, 35–41, July 2011.
30. Barakat, M.A., New trends in removing heavy metals from industrial wastewater. *Arab. J. Chem.*, 4, 361–377, 2011.
31. Awual, M.R., Ismael, M., Yaita, T., Safty, S.A.E., Shiwaku, H., Okamoto, Y., Suzuki, S., Trace copper (II) ions detection and removal from water using novel ligand modified composite adsorbent. *Chem. Eng. J.*, 222, 67–76, 2013.
32. Al-Rashdi, B.A.M., Johnson, D.J., Hilal, N., Removal of heavy metal ions by nanofiltration. *Desalination*, 315, 2–17, 2013.
33. Liu, C., Bai, R., Ly, Q.S., Selective removal of copper and lead ions by diethylenetriamine-functionalized adsorbent: Behaviors and mechanisms. *Water Res.*, 42, 1511–1522, 2008.
34. Peric, J., Trgo, M., Medvidovic, N.V., Removal of zinc, copper and lead by natural zeolite-a comparison of adsorption isotherms. *Water Res.*, 38, 1893–1899, 2004.
35. Bertinato, J., L'Abbe, M.R., Maintaining copper homeostasis: regulation of copper-trafficking proteins in response to copper deficiency or overload. *J. Nutr. Biochem.*, 15, 316–322, 2004.
36. Chan, Y.H., Chen, J., Liu, Q., Wark, S.E., Son, D.H., Batteas, J.D., Ultrasensitive copper (II) detection using plasmon-enhanced and photo-brightened luminescence of CdSe quantum dots. *Anal. Chem.*, 82, 3671–3678, 2010.
37. Gupta, V.K., Tyagi, I., Sadegh, H., Shahryari-Ghoshekandi, R., Makhlouf, A.S.H., Maazinejad, B., Nanoparticles as adsorbent; A positive approach for removal of noxious metal ions: A review. *Sci. Technol. Dev.*, 34, 20, 2015.
38. Mishra, A.K. and Ramaprabhu, S., Nanostructured polyaniline decorated graphene sheets for reversible CO2 capture. *J. Mater. Chem.*, 22, 9, 3708–3712, 2012.

39. Pyrzyńska, K. and Bystrzejewski, M., Comparative study of heavy metal ions sorption onto activated carbon, carbon nanotubes, and carbon-encapsulated magnetic nanoparticles. *Colloids Surf. A Physicochem. Eng. Asp.*, 362, 1, 102–109, 2010.
40. United Nations Children's Fund (UNICEF), *Progress on Household Drinking Water, Sanitation and Hygiene 2000-2017: And Hygiene I 2000-2017: Special Focus on Inequalities*, pp. 1–138, UNICEF, New York, 2019.
41. Tan, C.W., Tan, K.H., Ong, Y.T., Mohamed, A.R., Zein, S.H., Tan, S.H., Energy and environmental applications of carbon nanotubes. *Environ. Chem. Lett.*, 10, 3, 9, 2012.
42. Mauter, M.S. and Elimelech, M., Environmental applications of carbon-based nanomaterials. *Environ. Sci. Technol.*, 42, 16, 5843–5859, 2008.
43. Samonin, V.V., Nikonova, V.Y., Podvyaznikov, M.L., Carbon adsorbents on the basis of the hydrolytic lignin modified with fullerenes in producing. *Russ. J. Appl. Chem.*, 87, 2, 190–193, 2014.
44. Samonin, V.V., Nikonova, V.Y., Podvyaznikov, M.L., Sorption properties of fullerene-modified activated carbon with respect to metal ions. *Prot. Met.*, 44, 2, 190–192, 2008.
45. Tilahun, E., Sahinkaya, E., Çalli, B., A hybrid membrane gas absorption and bio- oxidation process for the removal of hydrogen sulfide from biogas. *Int. Biodeterior. Biodegrad.*, 127, 69–76, 2018.
46. Chen, X., Ok, Y.S., Mohan, D., Pittman Jr., C.U., Dou, X., The stability and removal of water-dispersed CdSe/CdS core-shell quantum dots from water. *Chemosphere*, 185, 926–933, 2017.
47. Petsriprasit, C., Namboonmee, J., Hunsom, M., Application of the electro-coagulation technique for treating heavy metals containing wastewater from the pickling process of a billet plant. *Korean J. Chem. Eng.*, 27, 3, 854–861, May 2010.
48. Choi, J. and Shin, W.S., Removal of salicylic and ibuprofen by hexadecyl-trimethylammonium-modified montmorillonite and zeolite. *Minerals*, 10, 10, 898, 2020.
49. Senthil Rathi, B. and Senthil Kumar, P., Electrodeionization theory, mechanism and environmental applications. A review, *Environ. Chem. Lett.*, 18, 1209–1227, 2020.
50. Fang, X., Xu, Z., Luo, Y., Ren, L., Hua, W., Removal of radionuclides from laundry wastewater containing organics and suspended solids using inorganic ion exchange. *Procedia Environ. Sci.*, 31, 375–381, 2016.
51. Pat-Espadas, A.M., Portales, R.L., Amabilis-Sosa, L.E., Gómez, G., Vidal, G., Review of constructed wetlands for acid mine drainage treatment. *Water*, 10, 11, 1685, 2018.
52. Teng, F.-Z., Magnesium isotope geochemistry. *Rev. Mineral. Geochem.*, 82, 1, 219–287, 2017.
53. Xuan, X. and Park, J.Y., A miniaturized and flexible cadmium and lead ion detection sensor based on micro-patterned reduced graphene oxide/carbon

nanotube/bismuth composite electrodes. *Sens. Actuators B Chem.*, 255, 2, 1220–1227, 2018.
54. Lu, C. and Liu, C., Removal of nickel (II) from aqueous solution by carbon nanotubes. *J. Chem. Technol. Biotechnol.*, 81, 12, 1932–1940, 2006.
55. Atieh, M.A., Bakather, O.Y., Tawabini, B.S., Bukhari, A.A., Khaled, M., Alharthi, M., Fettouhi, M., Abuilaiwi, F.A., Removal of chromium (III) from water by using modified and nonmodified carbon nanotubes. *J. Nanomater.*, 1–9, 2010.
56. Li, Y.-H., Wang, S., Wei, J., Zhang, X., Xu, C., Luan, Z., Wu, D., Wei, B., Lead adsorption on carbon nanotubes. *Chem. Phys. Lett.*, 357, 3–4, 263–266, 2002.
57. Dichiara, A.B., Webber, M.R., Gorman, W.R., Rogers, R.E., Removal of copper Ions from aqueous solutions via adsorption on carbon nanocomposites. *ACS Appl. Mater. Interfaces*, 7, 28, 15674–15680, 2015.
58. Huang, Y., Lee, X., Macazo, F.C., Grattieri, M., Cai, R., Mintee, S.D., Fast and efficient removal of chromium (VI) anionic species by a reusable chitosan-modified multi-walled carbon nanotube composite. *Chem. Eng. J.*, 339, 259–267, May 1, 2018.
59. Kumar, R., Khan, M.A., Haq, N., Application of carbon nanotubes in heavy metals remediation. *Environ. Sci. Technol.*, 44, 1000–1035, 2014.
60. Chang, Q. and Wang, G., Study on the macromolecular coagulant PEX which traps heavy metals. *Chem. Eng. Sci.*, 62, 4636–4643, 2007.
61. Rubio, J., Souza, M.L., Smith, R.W., Overview of flotation as a wastewater treatment technique. *Miner. Eng.*, 15, 139–155, 2002.
62. Abu-Qudais, H. and Moussa, H., Removal of heavy metals from wastewater by membrane processes: A comparative study. *Desalination*, 164, 2, 105–110, April 2004.
63. Blöcher, C., Dorda, J., Mavrov, V., Chmiel, H., Hybrid flotation-membrane filtration process for the removal of heavy metal ions from wastewater. *Water Res.*, 37, 16, 4018–26, September 2003.
64. Oikonomopoulos, I.K., Perraki, M., Tougiannidis, N., Perraki, T., Kasper, H.U., Gurk, M., Clays from Neogene Achlada lignite deposits in Florina basin (Western Macedonia, N. Greece): A prospective resource for the ceramics industry. *Appl. Clay Sci.*, 103, 1–9, January 2015.
65. Mavrov, V., Stamenov, S., Todorova, E., II, Chmiel, H., New hybrid electrocoagulation membrane process for removing selenium from industrial wastewater. *Desalination*, 201, 1-3, 290–296, November 2006.
66. Blöcher, C., Dorda, J., Mavrov, V., Chmiel, H., Hybrid flotation-membrane filtration process for the removal of heavy metal ions from wastewater. *Water Res.*, 37, 16, 4018–26, September 2003.
67. Renu, B.R., Agarwal, M., Singh, K.B., Methodologies for removal of heavy metal ions from wastewater. *Interdiscip. Environ. Rev.*, 18, 2, 124–142, October 2017.
68. Miandad, R., Barakat, M.A., Rehan, M., Plastic waste to liquid oil through catalytic pyrolysis using natural and synthetic zeolite catalysts. *Waste Manage.*, 69, 66–78, November 2017.

69. Hasson, D., Sidorenko, G., Semiat, R., Low electrode area electrochemical scale removal system. *Desalination Water Treat.*, 31, 1, 35–41, 2011.
70. Petsriprasit, C., Namboonmee, J., Hunsom, M., Application of the electrocoagulation technique for treating heavy metals containing wastewater from the pickling process of a billet plant. *Korean J. Chem. Eng.*, 27, 3, 854–861, May 2010.
71. Fu, F. and Wang, Q., Removal of heavy metal ions from wastewaters: A review. *J. Environ. Manage.*, 92, 407–418, 2011.
72. Kang, S.Y., Lee, J.U., Moon, S.H., Kim, K.W., Competitive adsorption characteristics of Co2+, Ni+2, and Cr3+ by IRN-77 cation exchange resin in synthesized wastewater. *Chemosphere*, 56, 141–147, 2004.
73. Barakat, M.A., New trends in removing heavy metals from industrial wastewater. *Arab. J. Chem.*, 4, 361–377, 2011.
74. Fu, F. and Wang, Q., Removal of heavy metal ions from wastewaters: A review. *J. Environ. Manage.*, 92, 407–418, 2011.
75. Babel, S. and Kurniawan, T.A., Low-cost adsorbents for heavy metals uptake from contaminated water: A review. *J. Hazard. Mater.*, 45, 219–243, 2003.
76. Aklil, A., Mouflihb, M., Sebti, S., Removal of heavy metal ions from water by using calcined phosphate as a new adsorbent. *J. Hazard. Mater.*, 112, 183–190, 2004.
77. Khosravi, J. and Alamdari, A., Copper removal from oil-field brine by coprecipitation. *J. Hazard. Mater.*, 166, 695–700, 2009.
78. Kurniawan, T.A., Chan, G.Y.S., Lo, W.H., Babel, S., Physicochemical treatment techniques for wastewater laden with heavy metals. *Chem. Eng. J.*, 118, 83–98, 2006.
79. Al-Rashdi, B.A.M., Johnson, D.J., Hilal, N., Removal of heavy metal ions by nanofiltration. *Desalination*, 315, 2–17, 2013.
80. Al-Rashdi, B., Somerfield, C., Hilal, N., Heavy metals removal using adsorption and nanofiltration techniques. *Sep. Purif. Rev.*, 40, 209–259, 2011.
81. Kajitvichyanukula, P., Ananpattarachaia, J., Pongpom, S., Sol–gel preparation and properties study of TiO2 thin film for photocatalytic reduction of chromium (VI) in photocatalysis process. *Sci. Technol. Adv. Mater.*, 6, 352–358, 2005.
82. Li, Y.H., Ding, J., Luan, Z., Di, Z., Zhu, Y., Xu, C., Wu, D., Wei, B., Competitive adsorption of Pb2+, Cu2+, and Cd2+ ions from aqueous solutions by multiwalled carbon nanotubes. *Carbon*, 41, 2787–2792, 2003.
83. Li, Y.H., Liu, F.Q., Xia, B., Du, Q.J., Zhang, P., Wang, D.C., Wang, Z.H., Xia, Y.Z., Removal of copper from aqueous solution by carbon nanotube/calcium alginate composites. *J. Hazard. Mater.*, 177, 876–880, 2010.
84. Ge, Y., Li, Z., Xiao, D., Xiong, P., Ye, N., Sulfonated multi-walled carbon nanotubes for the removal of copper(II) from aqueous solutions. *J. Ind. Eng. Chem.*, 20, 1765–1771, 2014.

85. Ren, X., Li, J., Tan, X., Wang, X., Comparative study of graphene oxide, activated carbon and carbon nanotubes as adsorbents for copper decontamination. *Dalton Trans.*, 42, 5266–5274, 2013.
86. Vellaichamy, S. and Palanivelu, K., Preconcentration and separation of copper, nickel and zinc in aqueous samples by flame atomic absorption spectrometry after column solid-phase extraction onto MWCNTs impregnated with D2EHPA-TOPO mixture. *J. Hazard. Mater.*, 185, 1131–1139, 2011.
87. Wang, H., Yan, N., Li, Y., Zhou, X.H., Chen, J., Yu, B.X., Gong, M., Chen, Q.W., Fe nanoparticle-functionalized multi-walled carbon nanotubes: One-pot synthesis and their applications in magnetic removal of heavy metal ions. *J. Mater. Chem.*, 22, 9230–9236, 2012.
88. Tofighy, M.A. and Mohammadi, T., Adsorption of divalent heavy metal ions from water using carbon nanotube sheets. *J. Hazard. Mater.*, 185, 140–147, 2011.
89. Moradi, O., The removal of ions by functionalized carbon nanotube: Equilibrium, isotherms and thermodynamic studies. *Chem. Biochem. Eng. Q.*, 25, 229–240, 2011.
90. Stafiej, A. and Pyrzynska, K., Adsorption of heavy metal ions with carbon nanotubes. *Sep. Purif. Technol.*, 58, 49–52, 2007.
91. Wu, C.H., Studies of the equilibrium and thermodynamics of the adsorption of Cu2+ onto as-produced and modified carbon nanotubes. *J. Colloid Interface Sci.*, 311, 338–346, 2007.
92. Sheng, G., Li, J., Shao, D., Hu, J., Chen, C., Chen, Y., Wang, X., Adsorption of copper(II) on multiwalled carbon nanotubes in the absence and presence of humic or fulvic acids. *J. Hazard. Mater.*, 178, 333–340, 2010.
93. Boro, J., Deka, D., Thakur, A.J., A review on solid oxide derived from waste shells as catalyst for biodiesel production. *Renew. Sust. Energ. Rev.*, 16, 1, 904–910, 2012.
94. Tang, W.W., Zeng, G.M., Gong, J.L., Liu, Y., Wang, X.Y., Liu, Y.Y., Liu, Z.F., Chen, L., Zhang, X.R., Characterization and mechanistic study of heavy metal adsorption by facile synthesized magnetic xanthate-modified chitosan/polyacrylic acid hydrogels. *Int. J. Environ. Res. Public Health*, 19, 17, 11123, 2022.
95. Salehi, E., Madaeni, S.S., Rajabi, L., Vatanpour, V., Derakhshan, A.A., Zinadini, S., Ghorabi, S., Monfared, H.A., Novel chitosan/poly(vinyl) alcohol thin adsorptive membranes modified with amino functionalized multiwalled carbon nanotubes for Cu(II) removal from water: Preparation, characterization, adsorption kinetics and thermodynamics. *Sep. Purif. Technol.*, 89, 309–319, 2012.
96. Li, Y.H., Luan, Z., Xiao, X., Zhou, X., Xu, C., Wu, D., Wei, B., Removal of Cu2+ ions from aqueous solutions by carbon nanotubes. *Adsorpt. Sci. Technol.*, 21, 475–485, 2003.
97. Shin, K., Hong, J., Jang, J., Heavy metal ion adsorption behavior in nitrogen-doped magnetic carbon nanoparticles: Isotherms and kinetic study. *J. Hazard. Mater.*, 190, 36–44, 2011.

98. Vellaichamy, S. and Palanivelu, K., Preconcentration and separation of copper, nickel and zinc in aqueous samples by flame atomic absorption spectrometry after column solid-phase extraction onto MWCNTs impregnated with D2EHPA-TOPO mixture. *J. Hazard. Mater.*, 185, 1131–1139, 2011.
99. Tofighy, M.A. and Mohammadi, T., Adsorption of divalent heavy metal ions from water using carbon nanotube sheets. *J. Hazard. Mater.*, 185, 140–147, 2011.
100. Cho, H.H., Wepasnick, K., Smith, B.A., Bangash, F.K., Fairbrother, D.H., Ball, W.P., Sorption of aqueous Zn[II] and Cd[II] by multiwall carbon nanotubes: The relative roles of oxygen-containing functional groups and graphenic carbon. *Langmuir*, 26, 967–981, 2010.
101. Lu, C. and Chiu, H., Adsorption of zinc(II) from water with purified carbon nanotubes. *Chem. Eng. Sci.*, 61, 1138–1145, 2006.
102. Lu, C., Chiu, H., Liu, C., Removal of zinc(II) from aqueous solution by purified carbon nanotubes: Kinetics and equilibrium studies. *Ind. Eng. Chem. Res.*, 45, 2850–2855, 2006.
103. Salam, M.A., Makki, M.S.I., Abdelaal, M.Y.A., Preparation and characterization of multi-walled carbon nanotubes/chitosan nanocomposite and its application for the removal of heavy metals from aqueous solution. *J. Alloys Compd.*, 509, 2582–2587, 2011.
104. Mubarak, N.M., Alicia, R.F., Abdullah, E.C., Sahu, J.N., Ayu Haslija, A.B., Tan, J., Statistical optimization and kinetic studies on removal of Zn2+ using functionalized carbon nanotubes and magnetic biochar. *J. Environ. Chem. Eng.*, 1, 486–495, 2013.
105. Yang, S.T., Li, J.X., Shao, D.D., Hu, J., Wang, X.K., Adsorption of Ni(II) on oxidized multi-walled carbon nanotubes: Effect of contact time, pH, foreign ions and PAA. *J. Hazard. Mater.*, 166, 109–116, 2009.
106. Mobasherpour, I., Salahi, E., Ebrahimi, M., Removal of divalent nickel cations from aqueous solution by multi-walled carbon nano tubes: Equilibrium and kinetic processes. *Res. Chem. Intermediat.*, 38, 2205–2222, 2012.
107. Chen, C.L., Hu, J., Shao, D.D., Li, J.X., Wang, X.K., Adsorption behavior of multiwall carbon nanotube/iron oxide magnetic composites for Ni(II) and Sr(II). *J. Hazard. Mater.*, 164, 923–928, 2009.
108. Kandah, M.I. and Meunier, J.L., Removal of nickel ions from water by multi-walled carbon nanotubes. *J. Hazard. Mater.*, 146, 283–288, 2007.
109. Shin, K., Hong, J., Jang, J., Heavy metal ion adsorption behavior in nitrogen- doped magnetic carbon nanoparticles: Isotherms and kinetic study. *J. Hazard. Mater.*, 190, 36–44, 2011.
110. Vellaichamy, S. and Palanivelu, K., Preconcentration and separation of copper, nickel and zinc in aqueous samples by flame atomic absorption spectrometry after column solid-phase extraction onto MWCNTs impregnated with D2EHPA-TOPO mixture. *J. Hazard. Mater.*, 185, 1131–1139, 2011.
111. Chen, C.L. and Wang, X.K., Adsorption of Ni(II) from aqueous solution using oxidized multiwall carbon nanotubes. *Ind. Eng. Chem. Res.*, 45, 9144–9149, 2006.

112. Liang, P., Liu, Y., Guo, L., Zeng, J., Lu, H., Multiwalled carbon nanotubes as solid- phase extraction adsorbent for the preconcentration of trace metal ions and their determination by inductively coupled plasma atomic emission spectrometry. *J. Anal. At. Spectrom.*, 19, 1489–1492, 2004.
113. Lu, C.Y., Liu, C.T., Su, F.S., Sorption kinetics, thermodynamics and competition of Ni2+ from aqueous solutions onto surface oxidized carbon nanotubes. *Desalination*, 249, 18–23, 2009.
114. Lu, C., Liu, C., Rao, G.P., Comparisons of sorbent cost for the removal of Ni2+ from aqueous solution by carbon nanotubes and granular activated carbon. *J. Hazard. Mater.*, 151, 239–246, 2008.
115. Ihsanullah, Khaldi, F.A.A., Abusharkh, B., Khaled, M., Atieh, M.A., Nasser, M.S., Laoui, T., Agarwal, S., Tyagi, I., Gupta, V.K., Adsorptive removal of cadmium(II) ions from liquid phase using acid modified carbon-based adsorbents. *J. Mol. Liq.*, 204, 255–263, 2015.
116. Li, Y.H., Ding, J., Luan, Z., Di, Z., Zhu, Y., Xu, C., Wu, D., Wei, B., Competitive adsorption of Pb2+, Cu2+, and Cd2+ ions from aqueous solutions by multiwalled carbon nanotubes. *Carbon*, 41, 2787–2792, 2003.
117. Li, Y.H., Wang, S., Luan, Z., Ding, J., Xu, C., Wu, D., Adsorption of cadmium(II) from aqueous solution by surface oxidized carbon nanotubes. *Carbon*, 41, 1057–1062, 2003.
118. Vukovic, G.D., Marinkovic, A.D., Colic, M., Ristic, M.D., Aleksic, R., Grujic, A.A.P., Uskokovic, P.S., Removal of cadmium from aqueous solutions by oxidized and ethylenediamine-functionalized multi-walled carbon nanotubes. *Chem. Eng. J.*, 157, 238–248, 2010.
119. Wang, H., Yan, N., Li, Y., Zhou, X.H., Chen, J., Yu, B.X., Gong, M., Chen, Q.W., Fe nanoparticle-functionalized multi-walled carbon nanotubes: One-pot synthesis and their applications in magnetic removal of heavy metal ions. *J. Mater. Chem.*, 22, 9230–9236, 2012.
120. Tofighy, M.A. and Mohammadi, T., Adsorption of divalent heavy metal ions from water using carbon nanotube sheets. *J. Hazard. Mater.*, 185, 140–147, 2011.
121. Moradi, O., The removal of ions by functionalized carbon nanotube: Equilibrium, isotherms and thermodynamic studies. *Chem. Biochem. Eng. Q.*, 25, 229–240, 2011.
122. Liang, P., Liu, Y., Guo, L., Zeng, J., Lu, H., Multiwalled carbon nanotubes as solid- phase extraction adsorbent for the preconcentration of trace metal ions and their determination by inductively coupled plasma atomic emission spectrometry. *J. Anal. At. Spectrom.*, 19, 1489–1492, 2004.
123. Cho, H.H., Wepasnick, K., Smith, B.A., Bangash, F.K., Fairbrother, D.H., Ball, W.P., Sorption of aqueous Zn[II] and Cd[II] by multiwall carbon nanotubes: The relative roles of oxygen-containing functional groups and graphenic carbon. *Langmuir*, 26, 967–981, 2010.
124. Liang, J., Liu, J., Yuan, X., Dong, H., Zeng, G., Wu, H., Wang, H., Liu, J., Hua, S., Zhang, S., Yu, Z., He, X., He, Y., Facile synthesis of alumina-decorated

multi-walled carbon nanotubes for simultaneous adsorption of cadmium ion and trichloroethylene. *Chem. Eng. J.*, 273, 101–110, 2015.

125. Cho, H.H., Wepasnick, K., Smith, B.A., Bangash, F.K., Fairbrother, D.H., Ball, W.P., Sorption of aqueous Zn[II] and Cd[II] by multiwall carbon nanotubes: The relative roles of oxygen-containing functional groups and graphenic carbon. *Langmuir*, 26, 967–981, 2010.
126. Liang, J., Liu, J., Yuan, X., Dong, H., Zeng, G., Wu, H., Wang, H., Liu, J., Hua, S., Zhang, S., Yu, Z., He, X., He, Y., Facile synthesis of alumina-decorated multi-walled carbon nanotubes for simultaneous adsorption of cadmium ion and trichloroethylene. *Chem. Eng. J.*, 273, 101–110, 2015.
127. Shadbad, M.J., Mohebbi, A., Soltani, A., Mercury(II) removal from aqueous solutions by adsorption on multi-walled carbon nanotubes. *Korean J. Chem. Eng.*, 28, 1029–1034, 2011.
128. Lu, C.Y. and Liu, C.T., Removal of nickel(II) from aqueous solution by carbon nanotubes. *J. Chem. Technol. Biotechnol.*, 81, 1932–1940, 2006.
129. Gupta, V.K., Moradi, O., Tyagi, I., Agarwal, S., Sadegh, H., Ghoshekandi, R.S., Makhlouf, A.S.H., Goodarzi, M., Garshasbi, A., Study on the removal of heavy metal ions from industry waste by carbon nanotubes: Effect of the surface modification: A review. *Crit. Rev. Environ. Sci. Technol.*, 46, 93–118, 2016.
130. Lu, C., Chiu, H., Liu, C., Removal of zinc(II) from aqueous solution by purified carbon nanotubes: Kinetics and equilibrium studies. *Ind. Eng. Chem. Res.*, 45, 2850–2855, 2006.
131. Moradi, O., The removal of ions by functionalized carbon nanotube: Equilibrium, isotherms and thermodynamic studies. *Chem. Biochem. Eng. Quart.*, 25, 229–240, 2011.
132. Wang, H.J., Zhou, A.L., Peng, F., Yu, H., Chen, L.F., Adsorption characteristic of acidified carbon nanotubes for heavy metal Pb(II) in aqueous solution. *Mater. Sci. Eng. A*, 466, 201–206, 2007.
133. Ge, Y., Li, Z., Xiao, D., Xiong, P., Ye, N., Sulfonated multi-walled carbon nanotubes for the removal of copper(II) from aqueous solutions. *J. Ind. Eng. Chem.*, 20, 1765–1771, 2014.
134. Rao, G.P., Lu, C., Su, F., Sorption of divalent metal ions from aqueous solution by carbon nanotubes: A review. *Sep. Purif. Technol.*, 58, 224–231, 2007.
135. Hu, J., Chen, C., Zhu, X., Wang, X., Removal of chromium from aqueous solution by using oxidized multiwalled carbon nanotubes. *J. Hazard. Mater.*, 162, 1542–1550, 2009.

8

Carbon Allotropes in Phenolic Compounds Removal

Manikandan Krishnamurthy and Meenakshisundaram Swaminathan

Nanomaterials Laboratory, Department of Chemistry, Kalasalingam Academy of Research and Education, Krishnankoil, India

Abstract

The phenolic compounds and their derivatives were mainly released during the agriculture farming and industrial processes including the synthesis of pharmaceuticals, paper coal conversion, metal coating, pulp, food processing, polymer and resins production, and oil refining. As a result of the above process and its wide utilization, they may exist in natural and wastewater causing a serious effect on human beings, animals and the aquatic environment. Carbon-based materials play a vital role in removing phenolic compounds through adsorption, photocatalysis and biodegradation. The present chapter summarized the use of carbon allotropes in the removal of various phenolic compounds. Among the carbon materials, activated carbon is extensively used in adsorption, photocatalysis, and microbial degradation. Relatively, the use of carbonaceous materials in dye degradation is larger than their use in phenol removal. Removal of phenolic pollutants by using cost-efficient and eco-friendly techniques is challenging at present. More research must be oriented on the utilization of these carbon materials in phenol removal as the carbon and their allotropes act as better and more efficient supports for photocatalysts to remove the organic pollutants. Since using microbes adsorbed on carbon materials to cause microbial deterioration is an easy and affordable technique, future research must be more focused on this area.

***Keywords*:** Phenols, activated carbon, graphene, graphene oxide, graphene nitride, carbon nanotubes, microbial degradation, immobilization

**Corresponding author*: m.swaminathan@klu.ac.in

Chandrabhan Verma and Chaudhery Mustansar Hussain (eds.) Carbon Allotropes and Composites: Materials for Environment Protection and Remediation, (155–172) © 2023 Scrivener Publishing LLC

8.1 Introduction

Five centuries back, the environment was not polluted so much because the natural biological processes and soil microbes were able to clean the pollutants as the concentration of pollutants was much less. After 1970, the pollution load on the earth started increasing due to the development of industrialization in various sectors all over the world, increasing population growth and multiyear droughts problems, and they are directly impacting clean water resources and their availability for household purposes. At present, environmental pollution forms the topmost priority worldwide. In 1970, the concept of green technology was proposed and implemented by all countries. The green technological processes in wastewater treatment are advanced oxidation processes and biochemical degradation. These two processes completely mineralize the pollutants in the effluents. Conventional methods like adsorption, filtration, reverse osmosis, oxidation etc. have some drawbacks or limitations in applicability, cost, and effectiveness [1]. Advanced oxidation processes (AOPs) refer to the processes in which the highly reactive hydroxyl radicals (.OH) are used for the degradation of pollutants. AOPs are: semiconductor photocatalysis (solar and UV), UV and H_2O_2 process, Fe(II/III)/H_2O_2 (photo-Fenton) process (solar and UV), ozonation process, ozone-H_2O_2 process. Among AOPs Semiconductor photocatalysis is found to be more advantageous and is widely used for effluent treatment [2, 3]. Nowadays, semiconductors are modified by different methods like doping with metals/non-metals, metal halides, anions, or coupled with other semiconductor oxides or loaded with a surface active material to enhance their efficiency and to make them solar active [4–6]. This chapter focuses on the modification of semiconductor oxides to increase their efficiency by carbonaceous materials.

About 95% of chemical compounds are made up of 'carbon compounds. The existence of an element in more than one physical state is known as allotropy and the allotropes of carbon would be classified into two types: (i) amorphous carbon allotropes and (ii) crystalline carbon allotropes. The allotropes of carbon might be either crystalline or amorphous in nature (diamond, graphite). Two well-known allotropes of carbon are Diamond and Graphite. Later eight allotropes were discovered and reported and they are: 1) diamond, 2) graphite, 3) lonsdaleite, 4) C60 (buckminsterfullerene), 5) C50, 6) C70, 7) carbon nanotubes and 8) amorphous carbon. Synthesis of these allotropes and their applications were discussed by P. S. Karthik *et al.* [7]. Some of these allotropes such as graphite, C60 (buckminsterfullerene), carbon nanotubes (CNTs) and amorphous carbon were utilized as catalysts or co-catalysts in the field of environmental remediation. All carbon-based materials (CBM) are good adsorbents. The adsorption process is highly effective among conventional technologies due to cost-effectivity and easiness. Hence, adsorption technique is extensively used to eliminate dangerous contaminants [8]. The transfer of contaminants from one phase to another, rather than their total elimination, is the adsorption process' constraint. However, the adsorption of the carbonaceous materials is also advantageous for their use in heterogeneous photocatalysis as it may enhance the photocatalytic efficiency and visible light activity.

Among the carbonaceous materials, activated carbon, CNTS, fullerenes, graphene (GR), GO, rGO and graphene nitride (g-C3N4) were studied extensively as co-catalysts for semiconductors. Figure 8.1 shows the structural models of different carbon compounds and their photocatalytic uses. As seen in the figure, these allotropes are used equally as photocatalysts for energy generation and pollutant removal.

Phenols even at low concentrations are highly toxic and carcinogenic compounds. Phenol derivatives exert negative effects on different biological processes. Phenol, if present in chlorinated water, may lead to the formation of toxic polychlorinated phenols. EPA listed the highly toxic phenolic compounds with persistence in the environment as high-priority contaminants [9–11].

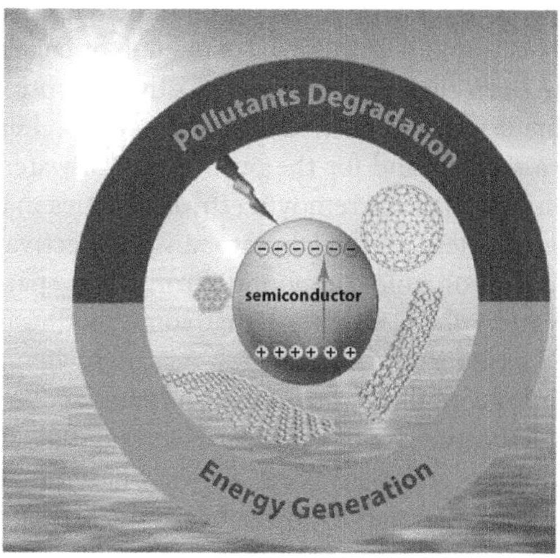

Figure 8.1 Photocatalytic applications of carbon based semiconductor composites [12].

Farming, agricultural, pharmaceutical effluents, coal conversion, paper and pulp, food processing, metal coating, polymer, and oil refining sectors are the main producers of phenolic compounds [13, 14]. Phenols are also found in natural and wastewater effluent due to their extensive utility. It's interesting to note that these compounds are formed naturally as well as by industry and human activity, which accounts for their presence in soils and sediments. Nitrophenols have been used in different sectors, which include plastics, rubber, petrochemical, paints, pesticides, pulp, and dyes production etc. Nitrophenol in water may damage kidneys, liver, skin and cause anaemia, and eye irritation [15, 16].

Adsorption and photocatalysis are the two main techniques to remove phenols from water. The molecules that have been adsorbed on the surface of the adsorbent material can be separated from the water using the adsorption method. A semiconductor creates a radical species with high reactivity during photocatalysis to destroy the impurities [17]. The highly water-soluble and stable nature of phenols makes their degradation from wastewater more challenging by conventional methods. Phenolic compounds are resistant to biodegradation [18]. This chapter discusses briefly adsorption process for phenol removal and focuses on the utility of some carbon allotropes as catalysts or co-catalysts in photocatalysis for the complete mineralization or removal of various phenolic compounds such as simple phenol, nitrophenols, chlorophenols, bisphenols and phenolic derivatives. The list of some toxic phenolic compounds is shown in Figure 8.2.

Since carbon-based materials are highly efficient adsorbents, several reviews and publications on the removal of phenols by conventional adsorption process had been reported [19–24]. Mostly adsorption technology of carbon-based materials was utilized to remove toxic chemicals or heavy metals in the wastewater and for the purification of water. Practically, we are using carbon materials for removing the impurities in the organic synthesis and water purifiers. In the laboratory, we use activated charcoal for the purification of an organic compound during its preparation to remove impurities. Traditionally, it is known that some Siddha doctors use carbon powder as a base for some medicines. As said earlier, pure adsorption technology will not completely degrade or mineralize the phenols. Hence the focus is made on the use of carbon-based materials in the combined adsorption and photocatalysis process. Carbon-based materials have also been used as support materials for microbes in the biodegradation of toxic chemicals [25].

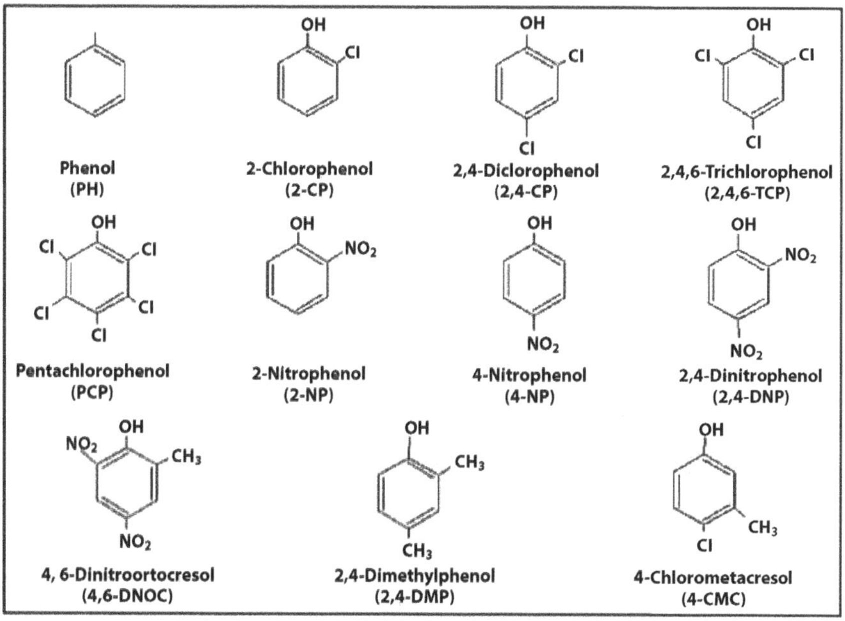

Bisphenol- A

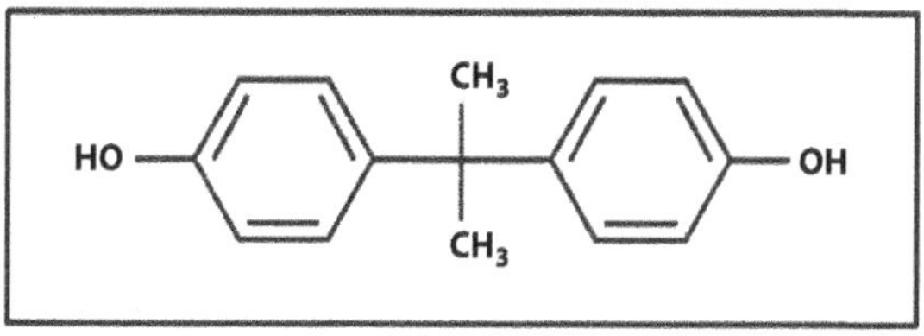

Figure 8.2 Some common toxic phenolic compounds.

8.2 Carbon Materials in Phenol Removal

8.2.1 Activated Carbon

Among the different carbonaceous materials, activated carbons (ACs) were found to successfully support and modify the photo-efficiency of semiconductor catalysts in the mineralization of organic pollutants [26]. Activated carbons are explicitly used materials because of their large surface area, effective adsorption potential and controllable pore structure. Another advantage of ACs is that they are generated easily from various sources such as coconut shells, nutshells, coal, peat, and wood. Any organic material with high carbon content can be used as a source for AC. Two variations of AC are PAC (Powdered Activated Carbon) and GAC (Granular Activated Carbon).

Activated carbon composites with metallic oxides, TiO2 and ZnO were investigated for phenol degradation [26]. AC–TiO2 has a greater photocatalytic activity relative to AC–ZnO due to the better adsorption potential of AC–TiO2 than AC–ZnO composite. Sobana *et al.* prepared activated carbon-loaded ZnO and studied the degradation of 4-acetylphenol [27]. The synergistic impact of AC was said to be the cause of the greater efficiency of this AC-ZnO composite, which was reusable. Additionally, improving the catalyst's apparent absorption is activated carbon loading. It has been reported that phenols and dyes can be removed photo-catalytically using silver ornamented on ZnO with AC (Ag/ZnO-AC) [28]. The increased efficiency of Ag/ZnO-AC was discovered through analysis of the photocatalytic activity of ZnO, Ag/ZnO-AC, and Ag/ZnO, for the elimination of nitrophenols and methyl orange using UV light. Degradation is accelerated by increased pollutant adsorption and the interaction between Ag with AC. For the removal of phenols and phenol derivatives from wastewater, this ternary catalyst is highly helpful.

Silver-doped AC-ZnO, synthesized by a calculation-electroless deposition method was reported for the removal of bisphenol-A with visible light [29]. The photocatalytic efficiency of the Ag/AC-ZnO was more than AC-ZnO and pure ZnO. In this process, superoxide radical anion is effective species for degradation. The influence of activated carbon on titania in the photodegradation of phenol and isoproturon was reported [30]. TiO_2/AC exhibits an interesting route to enhance efficiency in phenol and isoproturon degradation. This illustrates how a collaborative-synergistic effect between AC and titania can be utilized to increase efficiency. Zhou *et al.* reported the preparation of MgO/AC composites by the equal volume impregnation method for catalytic ozonation [31]. MgO/AC reveals better catalytic efficiency for phenol and COD reduction. The adsorption performance of AC with its synergistic effect showed good catalytic activity.

Activated carbon obtained from bituminous coal was reported for the preparation of titania (P25 Degussa)–carbon composite and its photocatalytic activity in phenol removal was investigated [32]. P25 Degussa with 80% anatase and 20% rutile phases of titania, was specially produced and marketed for environmental remediation. Titania-activated carbon composite increased the photodegradation of phenol, which is due to the porosity of the carbon. Bahrudin *et al.* prepared TiO_2/AC by immobilizing TiO_2 and AC using rubber-polyvinyl chloride and studied phenol removal from aqueous media. It is interesting to note that the optimized TiO_2/AC bilayer photocatalyst can remove 94% of phenol whereas TiO2 could remove only 47% in 90 min. In addition, TiO_2/AC was stable with better performance even after 10 cycles [33].

Granular activated carbon (GAC) with biochar was investigated to improve anaerobic phenol degradation. Methane generation increased during the anaerobic process through the adsorption and degradation process by biochar and GAC. The higher adsorption capacity of GAC over biochar remarkably mitigated the bio-toxicity of phenol [34].

Peroxymonosulfate (PMS) activation by AC in 2,4,6-trichlorophenol (TCP) mineralization was investigated [35]. Both UV and AC increased the degradation of TCP by activation of PMS and excellent performance of PMS/AC/UV occurred under acidic conditions. The sulfate radical was mainly responsible for TCP degradation.

8.2.2 Graphene

A review of the synthetic methods and application of semiconductor/graphene nanocomposites details phenolic compounds' degradation [36]. Large electron storage capacity and perfect electron conductivity make graphene much use as an electron acceptor and electron storage center. This reduces e−/h+ recombination pairs to enhance their photocatalytic activity. Since graphene has a lower Fermi level than the CB of many semiconductors, there is an easy electron transfer from the semiconductors to the graphene, leading to the separation of charge carriers. Singh *et al.* reported AgBr/BiOBr/graphene for phenol photo mineralization to CO_2 and H_2O within 6 hours. degradation. AgBr/BiOBr/graphene revealed excellent stability even after ten cycles [37]. Graphene functionalization to produce active sites was achieved by doping with nitrogen and the nitrogen-doped graphene was reported for the activation of the peroxydisulfate (PDS) system in the degradation of phenol [38]. Phenol mineralization was evidenced by the decrease in total organic carbon. Two intermediates hydroquinone and p-hydroxybenzoic acid were obtained during the reaction. Graphene-based TiO2-G, ZnO-G, prepared hydrothermally was reported for phenol degradation [39]. Enhancement in the photodegradation of phenol by ZnO-G and TiO2-G is caused by efficient charge separation leading to a longer lifetime of the charge carriers.

Hydrothermal synthesis of Au/BiOBr/graphene composites was reported for phenol removal [40]. Compared to BiOBr and BiOBr/graphene, Au/BiOBr/graphene had a greater photocatalytic efficiency in the breakdown of phenol. Au nanoparticles' surface plasmon resonance and smaller band gap (2.25eV) provide Au/BiOBr/graphene with better efficiency. With artificial solar light and visible light, Ag3PO4-0.02(MoS2/0.005GR) was found to be an effective catalyst for 2,4-dichlorophenol degradation [41]. MoS2 and graphene work together to boost Ag3PO4's photocatalytic activity. MoS2 and GR increased the charge separation by producing more active sites and acting as electron collectors for the interfacial electron transfer. Graphene@BiPO4 nanocomposite, developed by microwave-assisted hydrothermal method revealed higher photocatalytic efficiency than BiPO4 for phenol degradation [42]. The matched band locations of BiPO4 and graphene were the cause of the improved photocatalytic activity, which resulted in the separation of electron-hole pairs. Graphene was reported and exploited by JinwooKwon *et al.* as an adsorbent for bisphenol A. (BPA). When compared to chemically produced H-rGO, thermally produced graphene (T-rGO) has a higher adsorption capacity and selectivity for BPA. These characteristics make T-rGO an effective BPA adsorbent [43].

8.2.3 Carbon Nanotubes

Carbon nanotubes (CNTs) loaded with g-C3N4/BiVO4 were reported for phenol mineralization using solar light [44]. The stability of the g-C3N4/CNT/BiVO4 photocatalyst was excellent for several runs. A mechanism based on the Z-scheme of g-C_3N_4/CNT/$BiVO_4$ is proposed for phenol degradation. The hydrothermal method has been used to fabricate a series of reduced graphene oxide nanocomposites and ZnO nanospheres-CNT with different weight % ratios. The prepared materials were tested for photocatalytic degradation of 4-nitrophenol via the Fenton approach and using visible light [45].

Multi-walled carbon nanotubes (MWNTs) were combined with nanoscale Pd/Fe materials to create MWNT-Pd/Fe catalysts. This substance was used to remove 2,4-dichlorophenol (2,4-DCP). It was discovered that adding more chlorine atoms considerably enhanced the materials' efficiency and adsorption capacity [46]. To remove 2,4-DCP, a successful combination of physical adsorption by MWCNTs and chemical reduction by Pd/Fe nanoparticles was achieved. Significant amounts of chlorophenols were absorbed by MWCNTs.

Palladium-loaded MWCNTs were prepared by impregnation of Pd^{2+} followed by chemical reduction using NaBH4, ethanol, and H_2. The electrocatalytic performances of 4-chlorophenol (4-CP) were studied over Pd/MWCNTs [47]. Raizadaa *et al.* developed Fe doping on graphitic carbon nitride coupled with Ag_3VO_4 compound and CNT (FeCN/AV/CNT) by precipitation-deposition method. They studied the effective removal efficiency of 2,4 dimethyl phenol over the prepared catalyst under solar light illumination [48]. Sadatmansouri *et al.* fabricated [Fe (phen) 3]2[SiW12O40]·3DMF (IC–Fe) on cobalt(II) monoxide decorated

MWCNT (MWCNTs/Fe3O4/CoO) and utilized as a stable visible-light photocatalyst for 2,4,6-Trichlorophenol removal. In addition to better photocatalytic activity, the magnetic properties of the catalyst can be utilized for separation and reuse after the photoreaction [49].

8.2.4 Graphene Oxide and Reduced Graphene Oxide

Wu *et al.* synthesized Al-modified metal-organic framework MIL-68 material with rGO (MA/RG) by one-step solvothermal techniques. They have also studied their performances in p-nitro phenol (PNP) adsorption from aqueous solution. When they introduced RG into MIL-68(Al) (MA) significant morphology change of MA with increased surface area occurred. Eventually, the MA/RG-15% exhibited 64% PNP uptake, which is 123% greater than other catalysts. The major driving force for PNP removal from aqueous solution was found to be due to hydrogen bonds and $\pi - \pi$ dispersion [50].

Chengyu Zhang *et al.* have successfully made a montmorillonite-graphene oxide composite (MGC) for heavy metal ions and organic pollutants removal. The study revealed 96.82% removal efficiency obtained for PNP when MGC was used [51]. In 2020, Doan Ba Thinh *et al.* developed $MgFe_2O_4$-doped TiO_2/reduced graphene oxide (MFO-TiO_2/rGO) materials using ultrasound-hydrothermal techniques. It was directly utilized as a catalyst for p-nitrophenol mineralization. The catalytic activity of MFO-TiO_2/rGO was much higher than TiO_2/rGO, TiO_2, and MFO [52]. Nitrogen-doped reduced graphene oxide (N-RGO) with increased hydrophobicity nature along with conjugated π region materials was synthesized and the resulting material exhibited higher equilibrium adsorption for phenol and p-nitrophenol at pH of 6, and temperature of 30°C [53]. Ganguly and colleagues used graphene oxide (GO), graphene titania (GOT), and graphene iron to study the breakdown of phenol (GOI). This hybrid's photocatalytic effectiveness was successfully tested in a photoreactor under varied circumstances [54]. To effectively activate peroxymonosulfate and produce sulfate radicals for the degradation of 2,4-dichlorophenol and phenol, Hongqi Sun *et al.* produced reduced graphene oxide materials [55].

It was an effective oxidizing agent with higher oxidative potential and was found to effectively decompose the various aqueous contaminants present in the environment. It is important to note that chlorophenol and its derivatives are important organic pollutants, which may be released into the environment by human activities such as disinfection, herbicides, wood preservation and pesticides [55]. Rodríguez-López *et al.* reported the degradation of 2-Chlorophenol by using bismuth molybdate (γ-Bi_2MoO6) supported graphene oxide (GO) as a catalyst. GO reduced the bandgap of γ-Bi_2MoO_6 and thereby improving photocatalytic performances. This composite exhibited excellent photocatalytic efficiency in visible light to degrade 2-chlorophenol [56]. 2-chlorophenol removal from wastewater using Cobalt oxide-TiO_2 supported with reduced graphene oxide was reported [57]. 20 mg/L of 2-chlorophenol underwent complete mineralization in 8h under visible light with TiO_2-RGO-CoO and it maintained removal efficiency even after five runs. The catalyst, reduced graphene oxide and akageneite (Ak/rGO) was reported for the degradation of 2-chlorophenol [58]. The addition of rGO enhanced the degradation constants of 2-chlorophenol 2–5 times due to the formation of the Fe O C bond. The adsorption and synergy effect of rGO also cause an increase in the removal of 2-CP. They have found that the Ak/rGO showed enormous potential for environmental application. Gaurav Sharma *et al.* have prepared La/Co/Ni TNPs and graphene oxide (GO) supported La/Co/Ni trimetallic nanocomposite material (La/Co/Ni@GO TNC) by using microwave techniques [59]. The potential application of this material was studied in 2-chlorophenol mineralization under natural sunlight. The superior performance of La/Co/Ni@GO TNC is due to the efficient charge transfer [59].

8.2.5 Graphitic Carbon Nitride

Wang *et al.* have explored the properties of graphitic carbon nitride (g-C_3N_4) nanosheets as a visible active photocatalyst for the photodegradation of phenol in water. Synthesized graphitic carbon nitride nanosheets showed excellent efficiency of 78.9 % removal in 30 min [60].

Amanulla *et al.* designed g-CN@CuO photocatalyst for the visible light degradation of phenol. This material displayed very good photocatalytic degradation efficiency of around 87.8% within 120 min, much higher than g-CN@CuO (12%), g-CN, g-CN@CuO (4%), and g-CN@CuO (15%) [61]. Liang *et al.* have synthesized g-C_3N_4 photocatalyst with carbon vacancies by a two-step calcination process and used it for photocatalytic removal of Bisphenol (BPA). They proposed a mechanism for the above catalytic reaction and explained the contribution of carbon vacancies for an improved photocatalytic ability of catalyst in detail [62]. The 4% CuO/GCN and H2O2 integrated photo-Fenton catalyst demonstrated a good catalytic efficiency in the degradation of DMP (99%) within 120 minutes under the illumination of visible light, according to Sharma *et al.* who created the catalyst utilizing a thermal calcination method. Additionally, the comparison analysis revealed that H2O2 and Cu2+ions work together synergistically to accelerate DMP breakdown. Other phenolic compounds demonstrated notable recycle performances for the 4%CuO/GCN/H2O2 photocatalytic material. Therefore, 4%CuO/GCN/H2O2 can be used as an effective catalyst in wastewater remediation plants to remove pollutants using plenty of solar light [63].

8.2.6 Carbon Materials in the Biodegradation of Phenols

The environment-friendly techniques which can mineralize or destroy the pollutants completely are biodegradation and advanced oxidation processes. In biodegradation, microbes are capable of degrading any kind of toxic chemicals including phenols. Phenol could be used as the only carbon source by two bacterial strains, Pseudomonas cepacia and Bacillus brevis, which were found in the effluent of the phenol-formaldehyde resin production industry [64]. But the main drawback is their tolerance at high pollutant levels. Pollutants at high concentrations restrict their growth and tolerance limit. One of the strategies to overcome these drawbacks is cell immobilization. Immobilized cells have high degradation efficiency and better operational stability. We isolated soil microbes, bacillus Brevis from carbonization plant effluent and bacillus subtilis from paper mill effluent and studied the mineralisation of heterocyclics by cells immobilized on activated carbon [25].

The ability of a combined culture of Pseudomonas putida P8 and Cryptococcus elinovii H1 to degrade phenol was investigated. Compared to pure cultures, the immobilized cells can breakdown phenol up to 17 g/l more quickly. Additional storage of Pseudomonas putida P8 and Cryptococcus elinovii H1 adsorbed on activated carbon for up to 12 months was possible without a change in the effectiveness of the degradation [65]. Rhodococcus SR, a phenol-degrading microbe, was immobilized on alginate beads and granular activated carbon independently in packed bed column for phenol removal [66]. Cells grown on GAC removed more phenol (2.91 g/l/day) than cells grown on alginate (2-1 g/litre/day). The phenol-degrading strain was immobilized in the packed reactor, which had the benefits of large loading capacity and simple operation. For the treatment of phenolic wastewater, pseudomonas pictorum (NICM 2077) on activated carbon was described [67]. Excellent performance of immobilised strain in biodegradation of phenol reduces the cost and shows the simplicity of treatment of phenol. A review of the biodegradation of phenol examines numerous microbes, reactors, and different supports, including activated carbon utilized for immobilization, as well as the excellent performance of the immobilized strain in the biodegradation of phenol [68].

8.3 Conclusions

The phenolic compounds and their derivatives were mainly released during the agriculture farming and industrial processes including the synthesis of pharmaceuticals, paper coal conversion, metal coating, pulp, food processing, polymer and resins production and oil refining. As a result of the above process and its wide utilization, they may exist in natural and wastewater causing a serious effect on human beings, animals and the aquatic environment. Carbon-based materials play a vital role in removing the phenolic compounds by adsorption, photocatalysis and biodegradation.

In this book chapter, we have summarized the use of various carbon allotropes in the removal of various phenolic compounds. Among the carbon materials activated carbon is extensively used in adsorption, photocatalysis and microbial degradation. Relatively the use of carbonaceous materials in dye degradation is larger than their use in phenol removal. Removal of phenolic pollutants by using cost-efficient and eco-friendly techniques is challenging at present. More research must be oriented on the utilization of these carbon materials in phenol removal as the carbon and their allotropes act as better and more efficient supports for photocatalysts to remove the organic pollutants. Future study must concentrate more on this topic because microbial degradation by cells immobilized on carbon materials is a straightforward and economical way.

References

1. Wang, M., Cai, Y., Zhou, B., Yuan, R., Chen, Z., Chen, H., Removal of PFASs from water by carbon-based composite photocatalysis with adsorption and catalytic properties: A review. *Sci. Total Environ.*, 836, 155652, 2022.
2. Muruganandham, M., Suri, R.P.S., Sillanpää, M., Wu, J.J., Ahmmad, B., Balachandran, S., Swaminathan, M., Recent developments in heterogeneous catalyzed environmental remediation processes. *J. Nanosci. Nanotechnol.*, 14, 2, 1898–1910, 2014. https://doi.org/10.1166/jnn.2014.8718.
3. Hong, J., Cho, K.-H., Presser, V., Su, X., Recent advances in wastewater treatment using semiconductor photocatalysts. *Curr. Opin. Green Sustain. Chem.*, 36, 100644, August 2022.
4. Al Kausor, M. and Chakrabortty, D., Graphene oxide based semiconductor photocatalysts for degradation of organic dye in waste water: A review on fabrication, performance enhancement and challenges. *Inorg. Chem. Commun.*, 129, 108630, 2021.
5. Prasada, C., Liu, Q., Tang, H., Yuvaraja, G., Long, J., Rammohan, A., Zyryanov, G.V., An overview of graphene oxide supported semiconductors

based photocatalysts: Properties, synthesis and photocatalytic applications. *J. Mol. Liq.*, 297, 111826, 2020.

6. Ahmad, I., Shukrullah, S., Naz, M.Y., Bhatti, H.N., Recent progress in rare earth oxides and carbonaceous materials modified ZnO heterogeneous photocatalysts for environmental and energy applications. *J. Environ. Chem. Eng.*, 10, 3, 107762, 2022.
7. Karthik, P.S., Himaja, A.L., Singh, S.P., Carbon-allotropes: Synthesis methods, applications and future perspectives. *Carbon Lett.*, 15, 4, 219–237, 2014.
8. Sabzehmeidani, M.M., Mahnaee, S., Ghaedi, M., Heidari, H., Roy, V.A.L., Carbon based materials: A review of adsorbents for inorganic and organic compounds. *Mater. Adv.*, 2, 598, 2021.
9. Saleh, T.A., Adio, S.O., Asif, M., Dafalla, H., Statistical analysis of phenols adsorption on diethylenetriamine-modified activated carbon. *J. Clean. Prod.*, 182, 960–968, 2018.
10. Gami, A.A., Shukor, M.Y., Khalil, K.A., Dahalan, F.A., Khalid, A., Ahmad, S.A., Phenol and its toxicity. *J. Environ. Microbiol. Toxicol.*, 2, 1, 2014.
11. Anku, W.W., Mamo, M.A., Govender, P.P., Phenolic compounds in water: Sources, reactivity, toxicity and treatment methods, in: *Phenolic Compounds*, M. Soto-Hernandez, M. Palma-Tenango, M. del Rosario Garcia-Mateos (Eds.), InTech Open, United Nations SDG Publishers Compact, 2017, http://dx.doi.org/10.5772/66927.
12. Han, W., Li, Z., Li, Y., Fan, X., Zhang, F., Zhan, G., The promoting role of different carbon allotropes cocatalysts for semiconductors in photocatalytic energy generation and pollutants degradation. *Front. Chem.*, 5, 84, 1–16, 2017.
13. Bin-Dahman, O.A. and Saleh, T.A., Synthesis of carbon nanotubes grafted with PEG and its efficiency for the removal of phenol from industrial wastewater. *Environ. Nanotechnol. Monit. Manage.*, 13, 100286, 2020.
14. Anku, W.W., Mamo, M.A., Govender, P.P., *Phenolic Compounds—Natural Sources, Importance and Applications*, Intech Open, London, UK, 2017.
15. Zhang, C., Luan, J., Yu, X., Chen, W., Characterization and adsorption performance of graphene oxide–montmorillonitenanocomposite for the simultaneous removal of Pb2+ and p-nitrophenol. *J. Hazard. Mater.*, 378, 120739, 2019.
16. Khairy, M., Naguib, E.M., Mohamed, M.M., Enhancement of photocatalytic and sonophotocatalytic degradation of 4-nitrophenol by ZnO/graphene oxide and ZnO/carbon Nanotube Nanocomposites. *J. Photochem. Photobiol. A Chem.*, 396, 112507, 2020.
17. Gangulya, P., Sarkhela, R., Bhattacharya, S., Integral approach of treatment of phenolic wastewater using nano-metal coated graphene oxide in combination with advanced oxidation. *Surf. Interfaces*, 21, 100660, 2020.
18. Santana, C.M., Ferrera, Z.S., Padrón, M.E.T., Rodríguez, J.J.S., Methodologies for the extraction of phenolic compounds from environmental samples: New approaches. *Molecules*, 14, 1, 298–320, 2009.

19. Wu, Z., Yuan, X., Zhong, H., Wang, H., Zeng, G., Chen, X., Wang, H., Zhang, L., Shao, J., Enhanced adsorptive removal of p-nitrophenol from water by aluminum metal–organic framework/reduced graphene oxide composite. *Sci. Rep.*, 6, 25638, 2016.
20. Sabzehmeidani, M.M., Mahnaee, S., Ghaedi, M., Heidari, H., Roy, V.A.L., Carbon based materials: A review of adsorbents for inorganic and organic compounds. *Mater. Adv.*, 2, 598–627, 2021.
21. Zhang, J., Yu, C., Chen, Z., Luo, X., Zhao, H., Wu, F., Zeoliticimidazolate framework-8/fluorinated graphene coated SiO2 composites for pipette tip solid-phase extraction of chlorophenols in environmental and food samples. *Talanta*, 228, 122229, 2021.
22. Lee, K.M., Wong, C.P.P., Tan, T.L., Lai, C.W., Functionalized carbon nanotubes for adsorptive removal of water pollutants. *Mater. Sci. Eng. B*, 236–237, 61–69, 2018.
23. Abu-Nadaa, A. and Abdala, A., Removal of phenols and dyes from aqueous solutions using graphene and graphene composite adsorption: A review. *J. Environ. Chem. Eng.*, 9, 5, 105858, October 2021.
24. Ge, X., Wu, Z., Manzoli, M., Wu, Z., Cravotto, G., Feasibility and the mechanism of desorption of phenolic compounds from activated carbons. *Ind. Eng. Chem. Res.*, 59, 26, 12223–12231, 2020.
25. Mary, T.M.U. and Swaminathan, M., Enhanced biodegradation of thiocyanate by immobilized Bacillus brevis. *J. Indian Chem. Soc.*, 99, 100588, 2022.
26. García-Muñoz, P., Carbajo, J., Faraldos, M., Bahamonde, A., Photocatalytic degradation of phenol and isoproturon: Effect of adding an activated carbon to titaniacatalyst. *J. Photochem. Photobiol. A Chem.*, 287, 8–18, 2014.
27. Sobana, N., Muruganandam, M., Swaminathan, M., Characterization of AC–ZnO catalyst and its photocatalytic activity on 4-acetylphenol degradation. *Catal. Commun.*, 9, 2, 262–268, 2008.
28. Shintre, S.N., Wadhai, S., Thakur, P., Synthesis of Ag/ZnO-AC composite photocatalyst: Spectroscopic investigation, parameter optimization, synergistic effect and performance enhancement for cost-effective photocatalytic degradation of phenols and dyes. *Water Sci. Technol.*, 85, 9, 2663–2681, 2022.
29. Amornpitoksuk, P., Suwanboon, S., Graidist, P., Maungchanburi, S., Randorn, C., Effect of Ag loading on activated carbon doped ZnO for bisphenol A degradation under visible light. *Adv. Powder Technol.*, 29, 11, 2608–2615, 2018.
30. García-Muñoz, P., Carbajo, J., Faraldos, M., Bahamonde, A., Photocatalytic degradation of phenol and isoproturon: Effect of adding an activated carbon to titania catalyst. *J. Photochem. Photobiol. A Chem.*, 287, 8–18, 2014.
31. Zhou, L., Zhang, S., Li, Z., Liang, X., Zhang, Z., Liu, R., Yun, J., Efficient degradation of phenol in aqueous solution by catalytic ozonation over MgO/AC. *J. Water Process Eng.*, 36, 101168, 2020.
32. Velasco, L.F., Parra, J.B., Ania, C.O., Role of activated carbon features on the photocatalytic degradation of phenol. *Appl. Surf. Sci.*, 256, 17, 5254–5258, 2010.

33. Bahrudin, N.N., Evaluation of degradation kinetic and photostability of immobilized TiO_2/activated carbon bilayer photocatalyst for phenol removal. *Appl. Surface Sci. Adv.*, 7, 100208, 2022.
34. Li, Q., Gao, X., Liu, Y., Wang, G., Li, Y.Y., Sano, D., Wang, X., Chen, R., Biochar and GAC intensify anaerobic phenol degradation via distinctive adsorption and conductive properties. *J. Hazard. Mater.*, 405, 124183, 2021.
35. Ghanbari, F., Moradi, M., Gohari, F., Degradation of 2, 4, 6-trichlorophenol in aqueous solutions using peroxymonosulfate/activated carbon/UV process via sulfate and hydroxyl radicals. *J. Water Process Eng.*, 9, 22–28, 2016.
36. Kaveh, R. and Alijani, H., An overview: Recent development of semiconductor/grapheme nanocomposites for photodegradation of phenol and phenolic compounds in aqueous solution. *J. Asian Ceram. Soc.*, 9, 1, 1–23, 2021.
37. Singh, P., Sonu, Raizada, P., Sudhaik, A., Shandilya, P., Thakur, P., Agarwal, S., Gupta, V.K., Enhanced photocatalytic activity and stability of AgBr/BiOBr/grapheneheterojunction for phenol degradation under visible light. *J. Saudi Chem. Soc.*, 23, 5, 586–599, 2019.
38. Zheng, W., Xiao, X., Chen, B., A nonradical reaction-dominated phenol degradation with peroxydisulfate catalyzed by nitrogen-doped grapheme. *Sci. Total Environ.*, 667, 287–296, June 1, 2019.
39. Malekshoar, G., Pal, K., He, Q. *et al.*, Enhanced solar photocatalytic degradation of phenol with coupled graphene-based titanium dioxide and zinc oxide. *Ind. Eng. Chem. Res.*, 53, 49, 18824–18832, 2014.
40. Yu, X., Wang, L., Feng, L.-J. *et al.*, Preparation of Au/BiOBr/Graphene composite and its photocatalytic performance in phenol degradation under visible light. *J. Fuel Chem. Technol.*, 44, 937–942, 2016.
41. Peng, W.-C., Wang, X., Li, X.-Y., The synergetic effect of MoS_2 and graphene on Ag_3PO_4 for its ultra-enhanced photocatalytic activity in phenol degradation under visible light. *Nanoscale*, 6, 8311–8317, 2014.
42. Gawande, S.B., Gawande, K.B., Thakare, S.R., Manwar, N.R., Mate, V.R., Photocatalytic degradation of phenol over novel rod shaped Graphene@BiPO4 nanocomposite. *J. Phys. Chem. Solids*, 85, 132–137, 2015.
43. Kwon, J. and Lee, B., Bisphenol A adsorption using reduced graphene oxide prepared by physical and chemical reduction methods. *Chem. Eng. Res. Des.*, 104, 519–529, 2015.
44. Samsudin, M.F.R., Photocatalytic degradation of phenol wastewater over z-scheme G-C_3N_4/CNT/$BiVO_4$ heterostructure photocatalyst under solar light irradiation. *J. Mol. Liq.*, 277, 977–988, 2019.
45. Khairy, M., Naguib, E.M., Mohamed, M.M., Enhancement of photocatalytic and sonophotocatalytic degradation of 4-nitrophenol by ZnO/graphene oxide and ZnO/carbon nanotube nanocomposites. *J. Photochem. Photobiol. A Chem.*, 396, 112507, 2020.
46. Xu, J., Lv, X., Li, J., Li, Y., Shen, L., Zhou, H., Xu, X., Simultaneous adsorption and dechlorination of 2,4-dichlorophenol by Pd/Fe nanoparticles with

multi-walled carbon nanotube support. *J. Hazard. Mater.*, 225–226, 36–45, 2012.
47. Xu, J., Sheng, T., Hu, Y., Baig, S.A., Lv, X., Xu, X., Adsorption–dechlorination of 2, 4-dichlorophenol using two specified MWCNTs-stabilized Pd/Fe nanocomposites. *Chem. Eng. J.*, 219, 162–173, 2013.
48. Raizada, P., Khan, A.A.P., Singh, P., Construction of carbon nanotube mediated Fe doped graphitic carbon nitride and Ag3VO4 based Z-scheme heterojunction for H_2O_2 assisted 2,4 dimethyl phenol photodegradation. *Sep. Purif. Technol.*, 247, 116957, 2020.
49. Sadatmansouri, E., Shahrnoy, A.A., Mahjou, A.R., Synergic photocatalytic degradation of 2,4,6-trichlorophenol using magnetic multi-walled carbon nanotubes/cobalt monoxide/ionic crystal quaternary system. *J. Phys. Chem. Solids*, 151, 109897, 2021.
50. Wu, Z., Yuan, X., Zhong, H., Wang, H., Zeng, G., Chen, X., Wang, H., Zhang, L., Shao, J., Enhanced adsorptive removal of p-nitrophenol from water by aluminum metal-organic framework/reduced graphene oxide composite. *Sci. Rep.*, 6, 25638, May 16, 2016.
51. Zhang, C., Luan, J., Yu, X., Chen, W., Characterization and adsorption performance of graphene oxide–montmorillonitenanocomposite for the simultaneous removal of Pb^{2+} and p-nitrophenol. *J. Hazard. Mater.*, 378, 120739, 2019.
52. Thinh, D.B., Tien, N.T., Dat, N.M., Phong, H.H.T., Giang, N.T.H., Tai, L.T., Oanh, D.T.Y., Nam, H.M., Phong, M.T., Hieu, N.H., Improved photodegradation of p-nitrophenol from water media using ternary MgFe2O4-doped TiO2/reduced graphene oxide. *Synth. Met.*, 270, 116583, 2020,
53. Zhao, R., Li, Y., Ji, J., Wang, Q., Li, G., Wu, T., Zhang, B., Efficient removal of phenol and p-nitrophenol using nitrogen-doped reduced graphene oxide. *Colloids Surf. A Physicochem. Eng. Asp.*, 611, 125866, 2021.
54. Ganguly, P., Sarkhel, R., Bhattacharya, S., Das, P., Saha, A., Bhowal, A., Integral approach of treatment of phenolic wastewater using nano-metal coated graphene oxide in combination with advanced oxidation. *Surf. Interfaces*, 21, 10066, 2020.
55. Sun, H., Liu, S., Zhou, G., Ang, H.M., Tade, M.O., Wang, S., Reduced graphene oxide for catalytic oxidation of aqueous organic pollutants. *ACS Appl. Mater. Interfaces*, 4, 5466–5471, 2012.
56. Rodríguez-López, J., Rangel, R., Bartolo-Pérez, P., Alvarado-Gil, J.J., Cervantes-López, J.L., García, R., Ramos, A., Photocatalytic degradation of 2-chlorophenol under γ-Bi_2MoO_6/graphene oxide. *MRS Adv.*, 5, 581–589, 2020.
57. Sharma, A. and Lee, B.K., Rapid photo-degradation of 2-chlorophenol under visible light irradiation using cobalt oxide-loaded TiO_2/reduced graphene oxide nanocomposite from aqueous media. *J. Environ. Manage.*, 165, 1–10, 2016.

58. Xiao, F., Li, W., Fang, L., Wang, D., Synthesis of akageneite (beta-FeOOH)/reduced graphene oxide nanocomposites for oxidative decomposition of 2-chlorophenol by Fenton-like reaction. *J. Hazard. Mater.*, 308, 11–20, 2016.
59. Sharma, G., Kumar, A., Naushad, M., Sharma, S., Ghfar, A.A., Ahamad, T., Si, C., Stadler, F.J., Graphene oxide supported La/Co/Ni trimetallicnano-scale systems for photocatalytic remediation of 2-chlorophenol. *J. Mol. Liq.*, 294, 11160, 2019.
60. Ding, W., Liu, S., He, Z., One-step synthesis of graphitic carbon nitride nanosheets for efficient catalysis of phenol removal under visible light. *Chin. J. Catal.*, 38, 10, 1711–1718, 2017.
61. Baishnisha, A., Divakaran, K., Balakumar, V., Perumal, K.N., Meenakshi, C., Kannan, R.S., Synthesis of highly efficient g-CN@CuOnanocomposite for photocatalytic degradation of phenol under visible light. *J. Alloys Compd.*, 886, 161167, 2021.
62. Liang, X., Wang, G., Dong, X., Wang, G., Ma, H., Zhang, X., Graphitic carbon nitride with carbon vacancies for photocatalytic degradation of bisphenol A. *ACS Appl. Nano Mater.*, 2, 1, 517, 2019.
63. Sharma, K., Raizada, P., Hosseini-Bandegharaei, A., Thakur, P., Kumar, R., Thakur, V.K., Nguyen, V.-H., Singh, P., Fabrication of efficient CuO/graphitic carbon nitride based heterogeneous photo-Fenton like catalyst for degradation of 2, 4 dimethyl phenol. *Process Saf. Environ. Prot.*, 142, 63–75, 2020.
64. Arutchelvan, V., Kanakasabai, V., Nagarajan, S., Muralikrishnan, V., Isolation and identification of novel high strength phenol degrading bacterial strains from phenol-formaldehyde resin manufacturing industrial wastewater. *J. Hazard. Mater.*, 127, 238–243, 2005.
65. Morsen, A. and Rehm, H.J., Degradation of phenol by a mixed culture of pseudomonas putida and cryptococcus elinovii absorbed on activated carbon. *App. Microbiol. Biotechnol.*, 26, 283–288, 1987.
66. Pai, S.-L., Hsu, Y.-L., Chong, N.-M., Sheu, C.-S., Chen, C.-H., Continuous degradation of phenol by *rhodococcus* sp. immobilized on granular activated carbon and in calcium alginate. *Bioresour. Technol.*, 51, 37–42, 1995.
67. Annadurai, G., Balan, S.M., Murugesan, T., Design of experiments in the biodegradation of phenol using immobilized pseudomonas pictorum (NICM-2077) on activated carbon. *Bioprocess Eng.*, 22, 101–107, 2000.
68. Basha, K.M., Rajendran, A., Thangavelu, V., Recent advances in the Biodegradation of Phenol: A review. *Asian J. Exp. Biol. Sci.*, 1, 2, 219–234, 2010.

9

Carbon Allotropes in Carbon Dioxide Capturing

Elyor Berdimurodov[1*], Khasan Berdimuradov[2], Ilyos Eliboyev[1], Abduvali Kholikov[1], Khamdam Akbarov[1], Nuritdin Kattaev[1], Dakeshwar Kumar Verma[3] and Omar Dagdag[4]

[1]*Faculty of Chemistry, National University of Uzbekistan, Tashkent, Uzbekistan*
[2]*Faculty of Industrial Viticulture and Food Production Technology, Shahrisabz Branch of Tashkent Institute of Chemical Technology, Shahrisabz, Uzbekistan*
[3]*Department of Chemistry, Government Digvijay Autonomous Postgraduate College, Rajnandgaon, Chhattisgarh, India*
[4]*Institute of Nanotechnology and Water Sustainability, College of Science, Engineering and Technology, University of South Africa, Johannesburg, South Africa*

Abstract

Carbon dioxide gas is a serious problem in the modern era. Reducing its content is a crucial significant issue in the modern material and engineering sciences. Currently, carbon allotropes are recommended as excellent agents in carbon dioxide capturing. In this book chapter, the basic properties, types, synthesis methods, and mechanical, chemical and physical properties of carbon allotropes in carbon dioxide capturing were reviewed and discussed. These materials are supercapacitors for carbon dioxide. Carbon allotropes are good adsorbents for various gas: CO_2 (76.5 mg g^{-1} at 298 K), CH_4 (16.8 mg g^{-1} at 298 K) and H_2 (12.1 mg g^{-1} at 77 K). The selectivity of carbon allotropes for carbon dioxide capturing is high, confirming that this material effectively adsorbs the carbon dioxide from the gas mixtures (landfill gas, biogas and natural gas, etc.). In future research work, this chapter should be useful in the development of carbon-based materials for carbon dioxide capturing.

Keywords: Carbon allotropes, CO_2 capture, aerogels, graphene-based adsorbents, polymer-derived porous carbon, carbon-based materials, adsorption

**Corresponding author*: elyor170690@gmail.com

Chandrabhan Verma and Chaudhery Mustansar Hussain (eds.) Carbon Allotropes and Composites: Materials for Environment Protection and Remediation, (173–190) © 2023 Scrivener Publishing LLC

9.1 Introduction

9.1.1 Importance of Carbon Allotropes in Carbon Dioxide Capturing

The rise of carbon dioxide content in the atmosphere is a serious ecological and economical issue in the modern times. Scientists suggest that the removal of carbon dioxide from the atmosphere is an important aspect of material and engineering sciences. Photoelectrocatalytic, electrocatalytic and catalytic methods are also widely used in carbon dioxide capturing. In modern times, carbon allotropes are widely used in carbon dioxide capturing [1–3]. The reason for this is that the polymer-based carbon allotropes have good capacitating in carbon dioxide capturing, and good mechanical, chemical and physical properties, such as simple surface functionalization, controllable porosity, tuneable structure, end so on [4, 5]. The various types of carbon derivative materials such as mesoporous carbons, microporous carbons, polymer-derived carbons, biochar, carbon aerogels, graphene oxide and carbon nanotubes are mostly researched materials for carbon dioxide capturing. These materials functionalized with the various methods (metal oxide impregnation, hybrid/composites formation, carboxylation [-COOH], and amine modification), doping agents (S, O and N) and activating agents ($ZnCl_2$, K_2CO_3, NaOH and KOH). The various methods are applying in the syntheses of graphene based carbon allotropes, such as, *in-situ* polymerization method [30], ultrasound using [37, 38], chemical modification and the functionalized way [34, 35], low temperatures treatment method [31], sol-gel method [32], *in-situ* doping method [33], the hydrothermal method [36], simple impregnation method [39, 40], Hummer's approach, acidic treatment [32, 37]. As a result, the following type of modifications are created: polymer-GO nanocomposite, rose aerogel formation, textural features, amine functionality, metal decorated GO composite, S and N co-doped porous carbon, improved textural feature, graphite NFs activation, N-doped GO sheets, amine modified GO-aerogels, graphite oxidation, hydroxyl/carboxyl groups grafting, conversion of graphite flakes into GO, nitrogen functionalities, increased basicity, amine functionalization, incorporation of amine functionality (-NH and $-NH_2$) and amine modification. The 2D/3D assemblies containing graphene silica (template) aerogel was prepared by the one-pot facile method. The 2D graphene sheets was modified with the 3D silica (templated) aerogels to form the 2D/3D assemblies. Next, the amine was functionalized into the 2D/3D assemblies. The mesoporous cavity was rose with the carbon chain length and amine functionalization. The capacitance for carbon dioxide capture of this material was around 5 mmol/l. The heat of carbon dioxide sorption was importantly decreased to around 60 kJ/mol.

9.2 Main Part

9.2.1 Polymer-Based Carbon Allotropes in Carbon Dioxide Capturing

Various polymer-based carbon materials were suggested for carbon dioxide capture, and their various properties were investigated [6–8]. For example, the mesitylene and 1,4-benzenedimethano were modified to create the polymer-based carbon allotropes for carbon dioxide capturing by the carbonization, crafts polymerization and Friedel methods. These materials are large surface regions and good microporous polymer-based carbon nanomaterials. These properties make them become high capacitor adsorbents for carbon dioxide. They have low N-contents; this is mainly responsible for their excellent carbon dioxide adsorption [9–11].

Sang *et al.* [11] suggested carbonyl-modified hyper-cross-linked polymers based on oxygen-rich porous carbon allotropes for efficient carbon dioxide capture. Figure 9.1 shows the carbonization process of polymer-based carbon allotropes: (a) recovering characteristics; (b) selectivity for carbon dioxide and nitrogen; (c) correlation between the carbon dioxide adsorption and its flow rate; (d) values in adsorption of carbon dioxide on the carbon allotropes (e) adsorption isotherm (273 K); (f) correlation between the carbon dioxide flow and adsorption values [10]. The selectivity is also more favorable for nitrogen gas. It is also clear from the results that the 303 mg/g carbon dioxide was captured at room temperature. The oxygen content is an important factor in the high properties of these polymer-based materials for carbon dioxide capture. The values of oxygen were increased to around 30 wt% through the carbonization reaction. The obtained results confirmed that the adsorption ability surface area for carbon dioxide was 440–1769 m^2/g (Brunauer-Emmett-Teller) and microporosity was 72% to 87% for carbon dioxide capturing. The thermodynamics of carbon dioxide adsorption on the surface of these carbon allotropes show that the adsorption is exothermic processes. The rise of temperature importantly influence the adsorption capacitance of these materials because the heat of adsorption is high.

In other research work, the N-doped porous carbons (NPCs) based on triazine-based hyperpolymers were suggested as good adsorbents for carbon capture and sequestration. The obtained material is more porosity, which

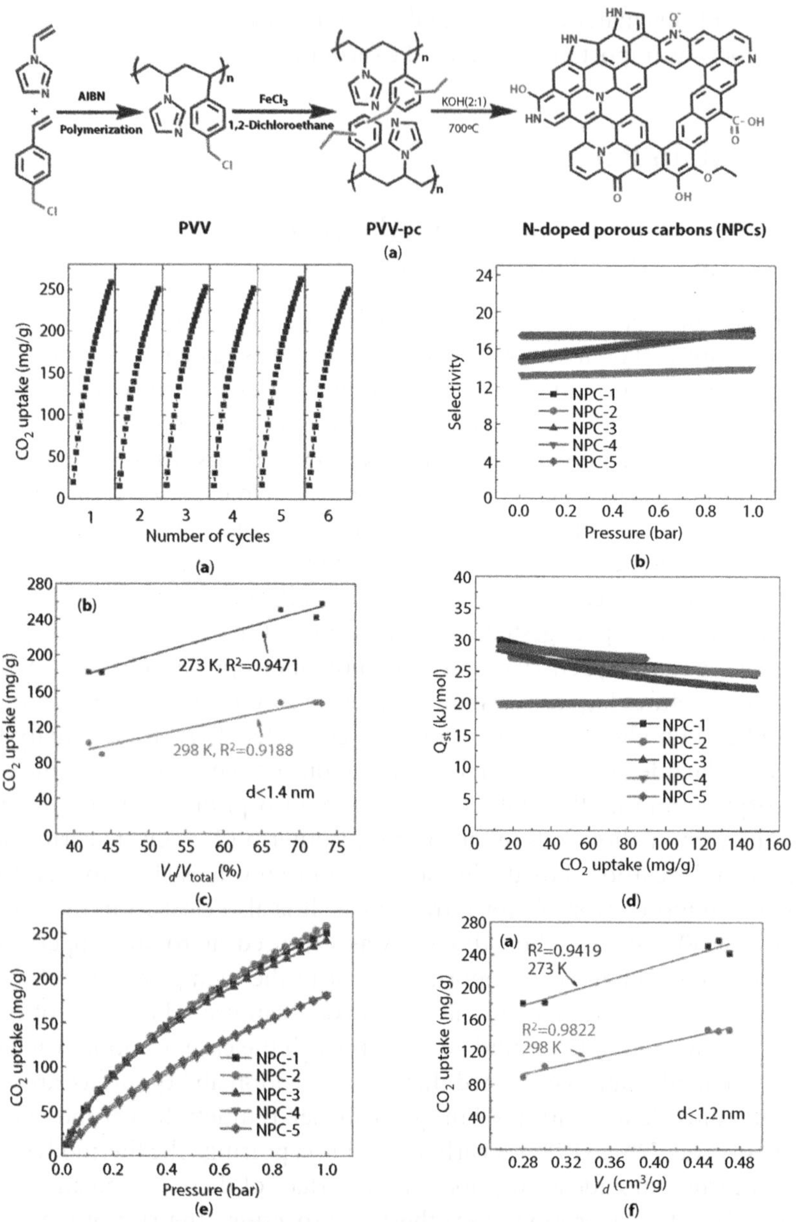

Figure 9.1 Carbonization process of polymer-based carbon allotropes: (a) recovering characteristics; (b) selectivity for carbon dioxide and nitrogen; (c) correlation between the carbon dioxide adsorption and its flow rate; (d) values in adsorption of carbon dioxide on the carbon allotropes (e) adsorption isotherm (273 K); (f) correlation between the carbon dioxide flow and adsorption values [10].

promotes high adsorption capacitance. The porosity depends on the syntheses' temperature. The NPCs were prepared at 700 °C, as a result, their porosity was extremely rose for carbon dioxide. The obtained results confirmed that the obtained NPCs adsorbed the carbon dioxide with 240 mg/g [9]. The selectivity for various gas was also checked, confirming that the carbon dioxide was most sensitively adsorbed on the surface of the investigated surface.

Jin *et al.* [12] synthesized the S, N co-doped porous carbons from polybenzoxazine for carbon dioxide capture. These materials were prepared from the formaldehyde monomers, thiourea and 4-cyanophenol by the activation, carbonization and solidification methods. Lou *et al.* [13] prepared polymer-based nitrogen-doped materials, which are more efficient for carbon dioxide capture by solvent-free methods. This material is N-doped carbons, well-developed porosities, high cross-linking degree and higher yields in the carbon dioxide capture [4]. Figure 9.2 indicates the (a) synthesis illustration of nitrogen-contained polymer-based materials, CO_2 adsorption isotherms of (b) NUT-71-F and NDPC-71-F, (c) NUT-71-S and

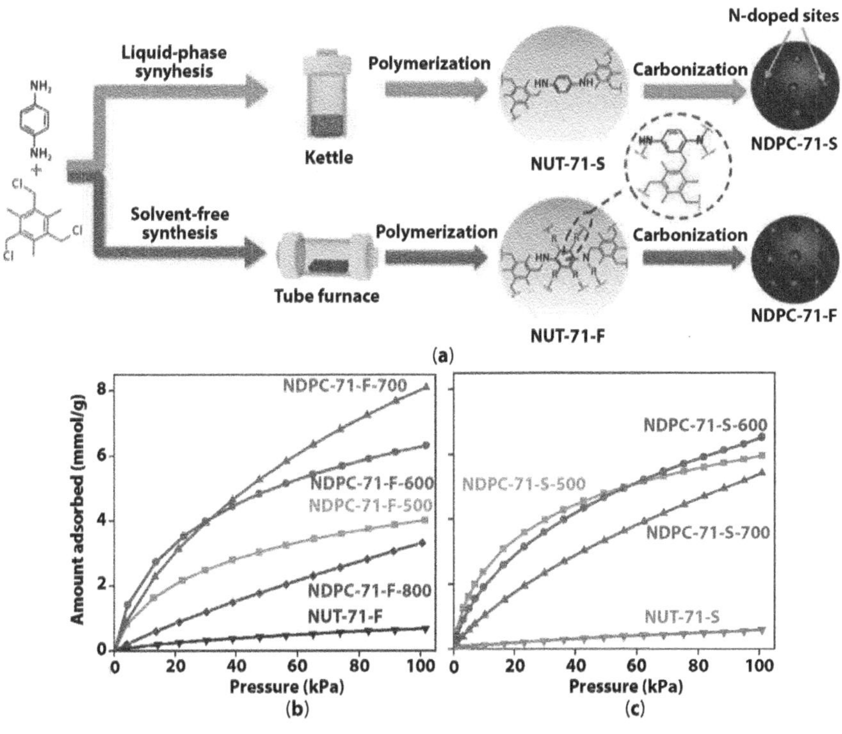

Figure 9.2 (a) Preparation of nitrogen-doped polymer-based material, CO_2 adsorption isotherms of (b) NUT-71-F and NDPC-71-F, (c) NUT-71-S and NDPC-71-S at 273 K [13].

NDPC-71-S at 273 K. It is clear from the obtained results that the nitrogen-doped polymers were prepared by the liquid-phase and solvent-free methods. In these processes, the carbonization and polymerization reactions were done. As a result, it obtained the ideal thermal stability and highly crosslinking structural carbon allotropes for carbon dioxide capturing. These nitrogen-doped materials are high microporosity. The high capacitance for carbon dioxide was 8.1 mmol/g.

Alabadi *et al.* prepared the polymer-based nitrogen-doped activated carbon (NACs) for carbon dioxide capture. Figure 9.3 indicates the (a) preparation of nitrogen-doped polymer-based material, (b) carbon dioxide adsorption on the surface of NACs, and (c) carbon dioxide adsorption at 273 K. The selectivity of these materials is also high, suggesting that they can separate carbon dioxide from nitrogen gas by around 90-95%. The volume of carbon dioxide capacitance was over 6.25 and 4.87 mmol g^{-1} at 273 K and 298 K respectively at 1.13 bar. At these high temperatures, the NAC was activated by the polyamine as the precursor. In this research work, the carbon dioxide-capturing properties of NAC-600, NAC-700 and NAC-800

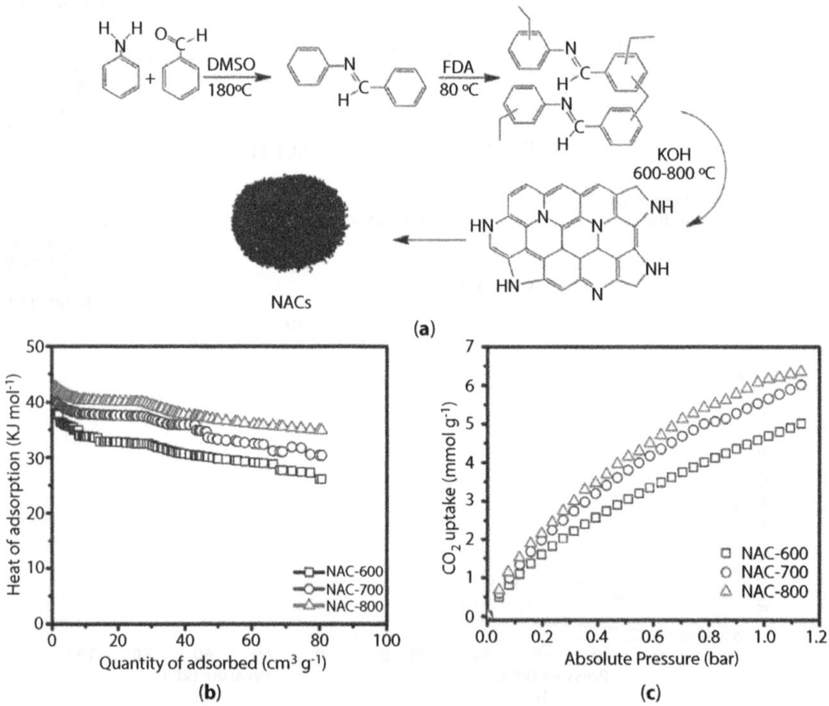

Figure 9.3 (a) Preparation of nitrogen-doped carbon dots, (b) carbon dioxide adsorption on the surface of NACs, (c) carbon dioxide adsorption at 273 K.

were investigated. Among these compounds, the NAC-800 is high carbon dioxide capacitance material with a large micropore volume (0.98 $cm^3 g^{-1}$), and good porosity behavior. In the synthesis processes, the benzaldehyde and aniline were reacted in the organic solvent to produce the imine-linked monomer. The next step is that the polyamine was synthesized from the hyper cross-linked polymerization reaction by the Friedel-Crafts methods. The reactions were carried out at 600–800 °C temperature ranges.

9.2.2 Graphene-Aerogels-Based Carbon Allotropes in Carbon Dioxide Capturing

Currently, graphene-aerogels-based carbon allotropes are performing in carbon dioxide capturing [14]. This is due to their homogenous pore dispersion and small-sized molecules [15–17]. The doping of heteroatoms, functionalization and surface modification is attributed to an increase in the adsorption capacitance of carbon dioxide capture [14, 16, 18–20]. The single-pot synthesis, hydrothermal reduction, CO_2 activation, Schiff-base reaction/pyrolysis, wet impregnation approach and sol-gel methods were used to prepare the graphene-aerogels-based carbon allotropes in carbon dioxide capturing [4]. Wang *et al.* [21] synthesized the 2D/3D modified aerogel, which is more effective for carbon dioxide capture. It was prepared to form graphene-based materials. In this work, this aerogel was prepared by the one-pot facile method. The 2D graphene sheets were modified with the 3D silica (templated) aerogels to form the 2D/3D aerogels. Next, the amine was functionalized into the 2D/3D assemblies. The mesoporous cavity rose with the carbon chain length and amine functionalization. The capacitance for carbon dioxide capture of this material was around 5 mmol/l. The heat of carbon dioxide sorption was importantly decreased to around 60 kJ/mol.

Glucose-graphene-based aerogel was prepared by CO_2 activation and hydrothermal reduction [22]. Figure 9.4 shows (a) a Pictorial representation of glucose-graphene-based aerogel (GAs) syntheses: (b) adsorption isotherm for nitrogen, (c) dependence between the adsorption and pore size, (d) methane gas adsorption character, (e) carbon dioxide adsorption character, (f) galvanostatic electrochemical analysis, (g) cyclic electrochemical analysis. It was found the following results:

(i) the glucose-graphene-based aerogel is a supercapacitor for carbon dioxide;
(ii) these carbon allotropes are good adsorbents for various gas: H_2 (12.1 mg g^{-1} at 77 K), CH_4 (16.8 mg g^{-1} at 298 K), CO_2 (76.5 mg g^{-1} at 298 K);

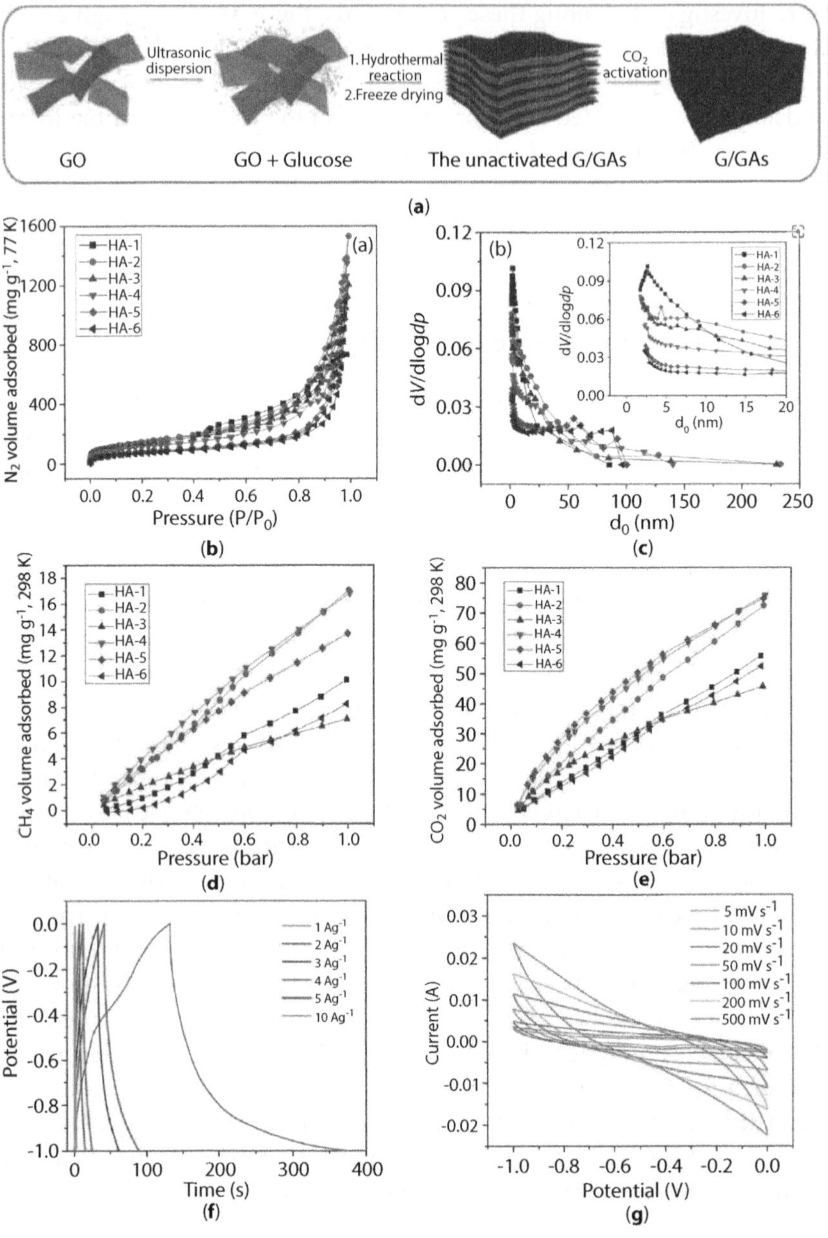

Figure 9.4 (a) Pictorial representation of glucose-graphene-based aerogel (GAs) syntheses: (b) adsorption isotherm for nitrogen, (c) dependence between the adsorption and pore size, (d) methane gas adsorption character, (e) carbon dioxide adsorption character, (f) galvanostatic electrochemical analysis, (g) cyclic electrochemical analysis [22].

(iii) the mesoporous structures, hierarchical macroporous, a high surface area (763 $m^2 g^{-1}$) and narrow mesopore size distribution are mainly responsible for its high efficiency in carbon dioxide capturing.

(iv) the high electrochemical properties and higher specific surface areas support the carbon dioxide capacitance of these materials;

(v) the selectivity of glucose-graphene-based aerogel for carbon dioxide capturing is high, confirming that this material effectively adsorbs the carbon dioxide from the gas mixtures (landfill gas, biogas and natural gas, etc.) [23].

Li *et al.* [17] synthesized the nitrogen-doped carbon aerogels (NCAs) and studied their catalyst support, energy storage, gas adsorption, heavy metal adsorption and wastewater purification. The rigidity of NCA frameworks, conductivity, N-doping, ultra-high surface area and energy storage abilities support their carbon dioxide capacity. These suggested materials were more sensitive to carbon dioxide gas. This performance depends on the change structural analysis. The CO_2 capacitance of these materials was 6.1 mmol g^{-1} at 273 K and 1 bar, 33.1 mmol g^{-1} at 323 K and 30 bar.

Jeon *et al.* [14] prepared N-doped carbon aerogels from the resorcinol/melamine by using the sol-gel method. The prepared composite is an effective adsorbent for carbon dioxide. These properties support the carbon dioxide capture ability. Figure 9.5 shows (a) the synthesis routine of N-doped carbon aerogels, (b) carbon dioxide adsorption isotherms on the surface of suggested aerogels and (c) the correlation between the pore size and adsorption values. The nitrogen-doped mainly role in carbon adsorption. The size of the pore is also the next important factor for carbon dioxide capture. It is indicated that the carbon dioxide molecules physically or chemically interacted with the nitrogen atoms in micropores. As a result, the NCAs effectively adsorbed the gas.

Chen *et al.* [16] synthesized carbon aerogels (CA) from the tetraethylenepentamine by the incipient wetness impregnation method and studied its carbon dioxide adsorption and the effect of the pore structure of supports. Figure 9.6 indicates the correlation between the carbon dioxide sorption indicating (a) temperature change and (b) component percentage [16]. It was found the following results:

(i) the adsorption kinetics of carbon dioxide on the surface of CA rose importantly with the increase in the pore size;

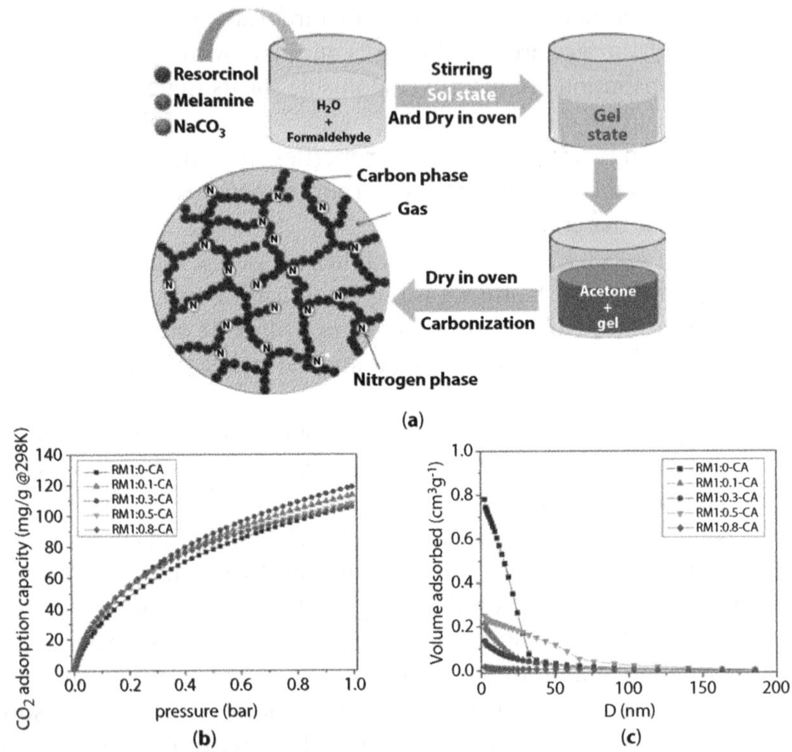

Figure 9.5 (a) Syntheses routine of N-doped carbon aerogels, (b) carbon dioxide adsorption behavior on the surface of aerogels and (c) dependence of pore size to adsorption values [14].

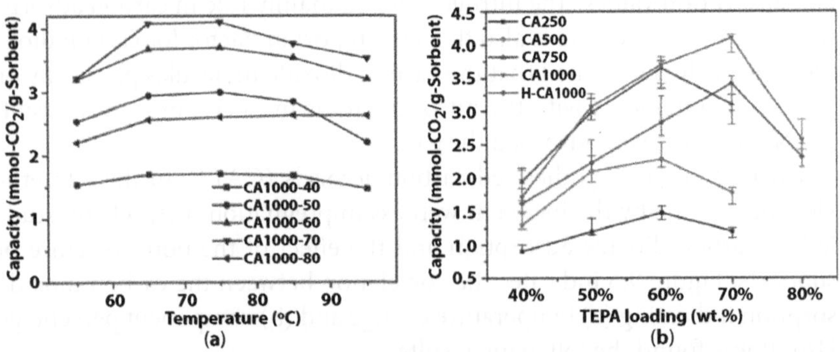

Figure 9.6 Dependence of the carbon dioxide to (a) temperature rise and (b) component percentage [16].

(ii) the values of carbon dioxide adsorption depend on the various chemical, physical and structural properties of CA;
(iii) the carbon dioxide interacted with the amine functional groups in the pore of CA;
(iv) it is determined the mesopore size of CA was 24.7 nm;
(v) the atmospheric pressure is also the next factor in carbon dioxide adsorption, and its higher values are responsible for good adsorption;
(vi) CA is regenerated material after the carbon dioxide adsorption, then it will be used again in this process;
(vii) the desorption processes increase in high temperatures [24, 25].

9.3 Functionalized Graphene-Based Carbon Allotropes in Carbon Dioxide Capturing

The graphene-based carbon allotropes were functionalized with various modifications. Their 2D and 3D materials are created, and their good mechanical features and thermal, electronic, chemical and physical properties were studied [26, 27]. Currently, the graphene-based carbon allotropes are using carbon dioxide capturing effectively, because, of their greater surface area, sheet-like morphology and high porous [4, 28]. In some research work, graphene oxide was modified to create new materials for gas storage performance [28, 29]. Various methods are applied in the syntheses of graphene-based carbon allotropes, such as the *in-situ* polymerization method [30], low-temperature treatments method [31], sol-gel method [32], *in-situ* doping method [33], modification and functionalizing [34, 35], a hydrothermal method [36], ultrasound treatment method [37, 38], simple impregnation method [39, 40], Hummer's approach, acidic treatment [32, 37]. As a result, the following type of modifications are created: polymer-GO nanocomposite, rose aerogel formation, textural features, amine functionality, metal decorated GO composite, S and N co-doped porous carbon, improved textural feature, graphite NFs activation, N-doped GO sheets, amine-modified GO-aerogels, graphite oxidation, hydroxyl/carboxyl groups grafting, conversion of graphite flakes into GO, nitrogen functionalities, increased basicity, amine functionalization, incorporation of amine functionality (-NH and $-NH_2$) and amine modification.

Liu *et al.* [38] modified the graphene oxide (GO) with tetraethylenepentamine (TEPA) to prepare the TEPA-GO by the ultrasound-mediated

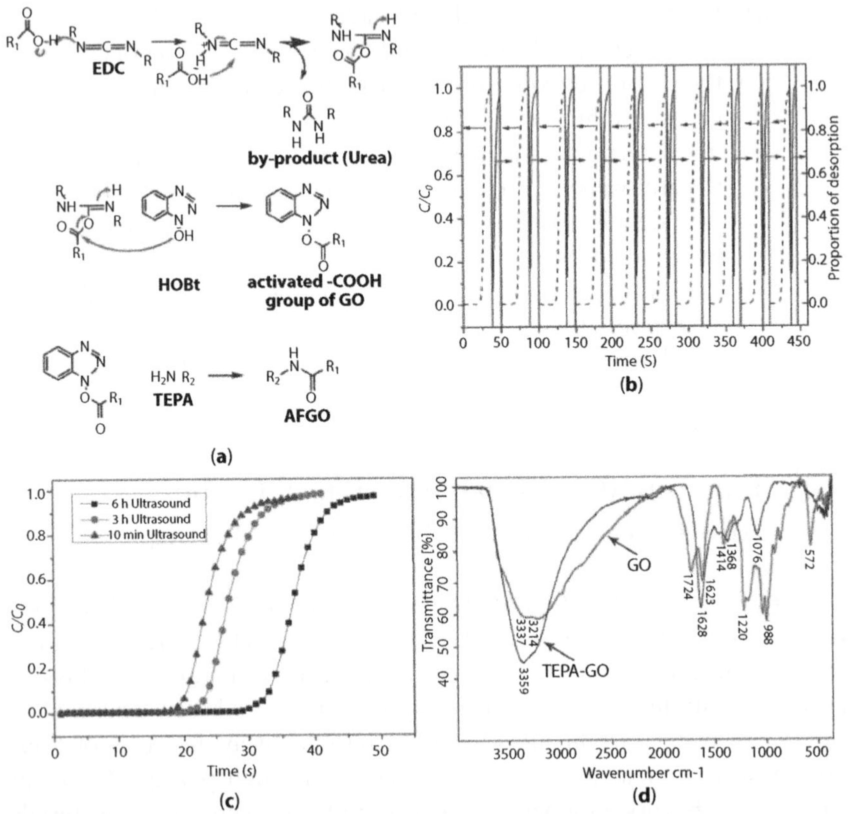

Figure 9.7 (a) Modification of graphene oxide with tetraethylenepentamine (TEPA), (b) carbon dioxide adsorption characteristics, (c) breakthrough curves of TEPA-GO, (d) infra-red spectroscopy analysis of obtained chemicals [38].

amination approach. The carbon dioxide adsorption of TEPA-GO was studied. Figure 9.7 shows (a) modification of graphene oxide with tetraethylenepentamine (TEPA), (b) carbon dioxide adsorption characteristics, (c) breakthrough curves of TEPA-GO, (d) infra-red spectroscopy analysis of obtained chemicals. The obtained results confirmed the following facts:

(i) the surface area of this material was improved by the acoustic treatment;
(ii) the elemental compositions, thermal stability, surface functionalities and textural properties of the selected adsorbents confirming the molecular structure of TEPA-GO;
(iii) the carbon dioxide capture performance of this material was high in 338 K;

(iv) the carbon dioxide was adsorbed into TEPA-GO by the chemical adsorption mechanism.
(v) the adsorption capacity and adsorption rate of graphene composites were enhanced by the 3D pore structure design;
(vi) adsorption capacity depends on the time of sonication duration, its rose with the increase of this time.

Arifin *et al.* [37] synthesized the functionalized graphene oxide (GO) with the 3-aminopropyl triethoxysilane (APTES) by the Hummer's method. Its structural and physical properties were confirmed by the surface morphology, Raman spectroscopy and Fourier transmission IR. It is indicated that the adsorption capacitance for carbon dioxide in the chemical industry.

In the following research work (Figure 9.8) [33], the gaseous mixture of CO_2, SO/SO_2 adsorption on the surface of S-doped porous carbon/rGO composites was studied. The CO_2, SO/SO_2 gas chemically electrochemically interacted with the doped S atoms. This process is electroreduction. In the pores of S-doped porous carbon/rGO composites, the reduction reaction occurred, in which the electrons are transferred through the carbon matrix. The electrons transferred from the carbon matrix to CO_2 form CO_2^{*-} as an intermediate. The reason for this is the high electron density of the π system of the matrix. As a result, the CO_2^{*-} electrostatically interacted with the positively charged carbon atoms and attached -them to S-doped atoms.

Shin *et al.* [39] enhanced the functionalized graphene-based carbon allotropes based on the nitrogen-doped agents as a good agent in carbon dioxide capturing. Figure 9.9 indicates (a) preparation of PEI-GO, (b) amount of CO_2 adsorption on GO and PEI-GO, (c) carbon dioxide

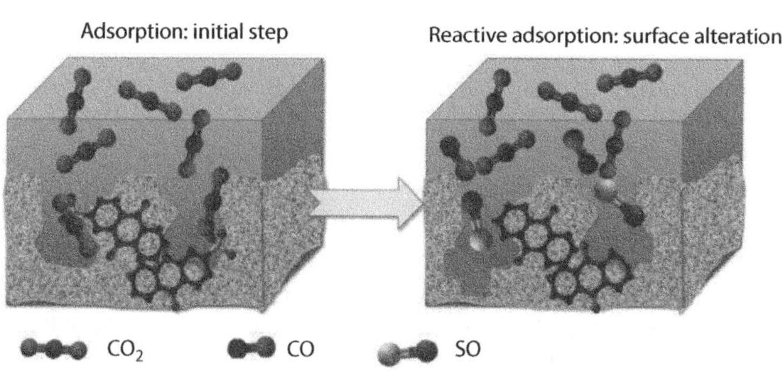

Figure 9.8 Pictorial representations of CO_2, SO/SO_2 adsorption on the surface of S-doped porous carbon/rGO composites [33].

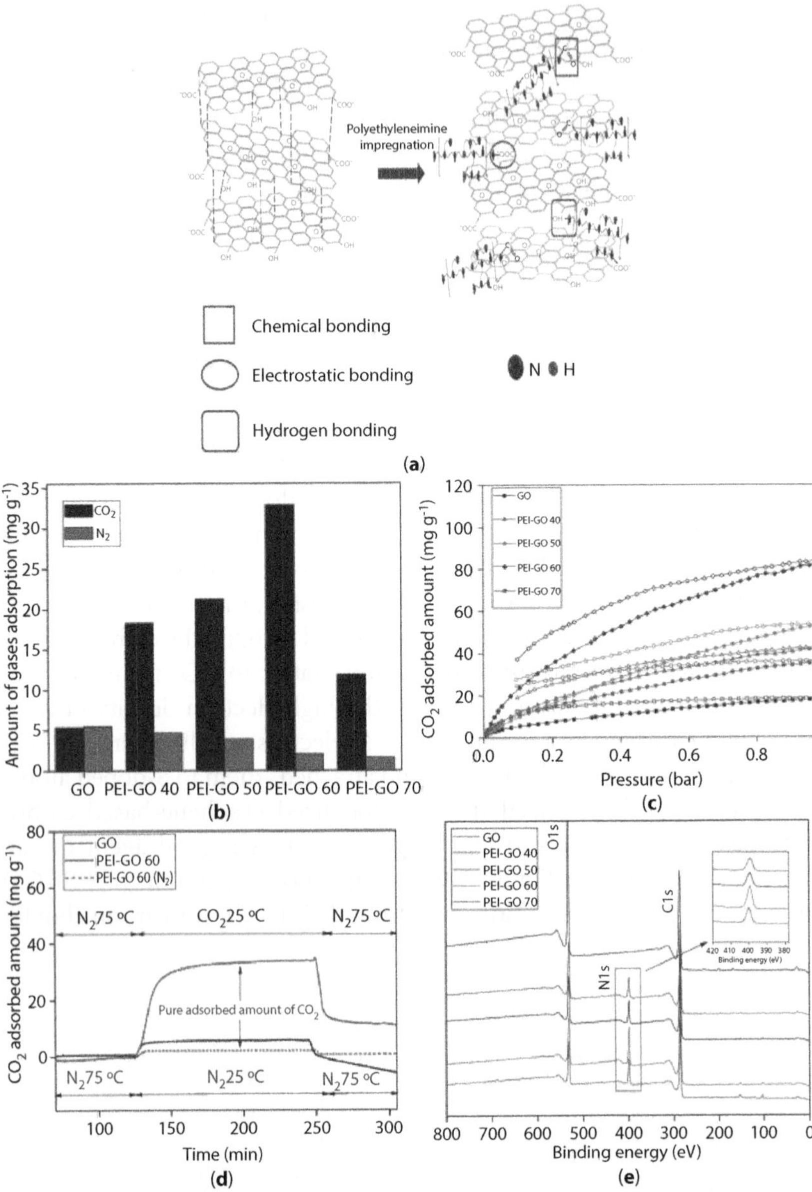

Figure 9.9 (a) Preparation of PEI-GO, (b) amount of CO_2 adsorption on GO and PEI-GO, (c) adsorption characteristics of carbon dioxide on the surface of PEI-GO, (d) correlation between the adsorption of carbon dioxide and time, (e) XPS of PEI-GO bonding the presence of C single bond N bonding at 399 eV [39].

adsorption on the PEI-GO surface, (d) correlation between the adsorption of carbon dioxide and time, (e) XPS of PEI-GO bonding the presence of C single bond N bonding at 399 eV [39]. It should be noted that the rise of polyethylenimine content is responsible for enhancing carbon dioxide-capturing capacitance. This is due to many nitrogen atoms promoting carbon dioxide adsorption. The selectivity is also high in the material for carbon dioxide. XPS results confirmed the formation of chemical bonds carbon and nitrogen atoms. The stability of this chemical bond is nearly equal to the coordination bonds. Another considerable feature is that the carbon dioxide is good donor and acceptor molecule. When the carbon dioxide interacted with the functionalized graphene, then it was activated and become good electron donor and acceptor molecule. These properties promote the adsorption of carbon dioxide on the pore of carbon functionalized graphene and also enhance the interaction tendency of carbon dioxide with the amine functional groups.

9.4 Conclusions

In this chapter, the basic properties, types, synthesis methods, mechanical, chemical and physical properties, such as the simple surface functionalization, controllable porosity, and tuneable structure of carbon allotropes in carbon dioxide capturing were reviewed and discussed. The *in-situ* polymerization method [30], ultrasound treatment method [37, 38], modification and functionalizing [34, 35], low-temperature treatments method [31], sol-gel method [32], *in situ* doping method [33], the hydrothermal method [36], simple impregnation method [39, 40], Hummer's approach, acidic treatment methods are mostly used in the syntheses of carbon allotropes in carbon dioxide capturing. These materials are supercapacitors for carbon dioxide. Carbon allotropes are good adsorbents for various gas: H_2 (12.1 mg g^{-1} at 77 K), CH_4 (16.8 mg g^{-1} at 298 K), CO_2 (76.5 mg g^{-1} at 298 K). The selectivity of carbon allotropes for carbon dioxide capturing is high, confirming that this material effectively adsorbs the carbon dioxide from the gas mixtures (landfill gas, biogas and natural gas, etc.).

References

1. Hong, L., Ju, S., Liu, X., Zhuang, Q., Zhan, G., Yu, X., Highly selective CO2 uptake in novel fishnet-like polybenzoxazine-based porous carbon. *Energ. Fuel.*, 33, 11454–11464, 2019.

2. Alabadi, A., Abbood, H.A., Li, Q., Jing, N., Tan, B., Imine-linked polymer based nitrogen-doped porous activated carbon for efficient and selective CO2 capture. *Sci. Rep.*, 6, 1–9, 2016.
3. Peng, A.-Z., Qi, S.-C., Liu, X., Xue, D.-M., Peng, S.-S., Yu, G.-X., Liu, X.-Q., Sun, L.-B., Fabrication of N-doped porous carbons for enhanced CO2 capture: Rational design of an ammoniated polymer precursor. *Chem. Eng. J.*, 369, 170–179, 2019.
4. Kamran, U. and Park, S.-J., Chemically modified carbonaceous adsorbents for enhanced CO2 capture: A review. *J. Clean. Prod.*, 290, 125776, 2021.
5. Najafabadi, A.T., Emerging applications of graphene and its derivatives in carbon capture and conversion: Current status and future prospects. *Renew. Sust. Energ. Rev.*, 41, 1515–1545, 2015.
6. Peng, A.-Z., Qi, S.-C., Liu, X., Xue, D.-M., Peng, S.-S., Yu, G.-X., Liu, X.-Q., Sun, L.-B., N-doped porous carbons derived from a polymer precursor with a record-high N content: Efficient adsorbents for CO2 capture. *Chem. Eng. J.*, 372, 656–664, 2019.
7. Geng, J., Xue, D., Liu, X., Shi, Y., Sun, L., N-doped porous carbons for CO2 capture: Rational choice of N-containing polymer with high phenyl density as precursor. *AIChE J.*, 63, 1648–1658, 2017.
8. Mahurin, S.M., Górka, J., Nelson, K.M., Mayes, R.T., Dai, S., Enhanced CO2/N2 selectivity in amidoxime-modified porous carbon. *Carbon N. Y.*, 67, 457–464, 2014.
9. Shao, L., Wang, S., Liu, M., Huang, J., Liu, Y.-N., Triazine-based hyper-cross-linked polymers derived porous carbons for CO2 capture. *Chem. Eng. J.*, 339, 509–518, 2018.
10. Shao, L., Sang, Y., Huang, J., Imidazole-based hyper-cross-linked polymers derived porous carbons for CO2 capture. *Microporous Mesoporous Mater.*, 275, 131–138, 2019.
11. Sang, Y., Chen, G., Huang, J., Oxygen-rich porous carbons from carbonyl modified hyper-cross-linked polymers for efficient CO2 capture. *J. Polym. Res.*, 27, 1–8, 2020.
12. Jin, Z., Wang, J., Zhao, R., Guan, T., Zhang, D., Li, K., Synthesis of S, N co-doped porous carbons from polybenzoxazine for CO2 capture. *New Carbon Mater.*, 33, 392–401, 2018.
13. Lou, Y.-C., Qi, S.-C., Xue, D.-M., Gu, C., Zhou, R., Liu, X.-Q., Sun, L.-B., Solvent-free synthesis of N-containing polymers with high cross-linking degree to generate N-doped porous carbons for high-efficiency CO2 capture. *Chem. Eng. J.*, 399, 125845, 2020.
14. Jeon, D.-H., Min, B.-G., Oh, J.G., Nah, C., Park, S.-J., Influence of nitrogen moieties on CO2 capture of carbon aerogel. *Carbon Lett.*, 16, 57–61, 2015.
15. Heo, Y.-J., Yeon, S.-H., Park, S.-J., Defining contribution of micropore size to hydrogen physisorption behaviors: A new approach based on DFT pore volumes. *Carbon N. Y.*, 143, 288–293, 2019.

16. Chen, L., Gong, M., Cheng, Y., Liu, Y., Yin, S., Luo, D., Effect of pore structure of supports on CO2 adsorption of tetraethylenepentamine/carbon aerogels prepared by incipient wetness impregnation method. *Pol. J. Environ. Stud.*, 28, 4127–4137, 2019.
17. Li, H., Li, J., Thomas, A., Liao, Y., Ultra-high surface area nitrogen-doped carbon aerogels derived from a schiff-base porous organic polymer aerogel for CO2 storage and supercapacitors. *Adv. Funct. Mater.*, 29, 1904785, 2019.
18. Kong, Y., Shen, X., Cui, S., Fan, M., Development of monolithic adsorbent via polymeric sol–gel process for low-concentration CO2 capture. *Appl. Energy*, 147, 308–317, 2015.
19. Alhwaige, A.A., Ishida, H., Qutubuddin, S., Carbon aerogels with excellent CO2 adsorption capacity synthesized from clay-reinforced biobased chitosan-polybenzoxazine nanocomposites. *ACS Sustain. Chem. Eng.*, 4, 1286–1295, 2016.
20. Alhwaige, A.A., Agag, T., Ishida, H., Qutubuddin, S., Biobased chitosan hybrid aerogels with superior adsorption: Role of graphene oxide in CO2 capture. *RSC Adv.*, 3, 16011–16020, 2013.
21. Wang, W., Motuzas, J., Zhao, X.S., da Costa, J.C.D., 2D/3D assemblies of amine-functionalized graphene silica (templated) aerogel for enhanced CO2 sorption. *ACS Appl. Mater. Interfaces*, 11, 30391–30400, 2019.
22. Liu, K.-K., Jin, B., Meng, L.-Y., Glucose/graphene-based aerogels for gas adsorption and electric double layer capacitors. *Polym. (Basel)*, 11, 40, 2018.
23. Oh, Y., Le, V.-D., Maiti, U.N., Hwang, J.O., Park, W.J., Lim, J., Lee, K.E., Bae, Y.-S., Kim, Y.-H., Kim, S.O., Selective and regenerative carbon dioxide capture by highly polarizing porous carbon nitride. *ACS Nano*, 9, 9148–9157, 2015. https://doi.org/10.1021/acsnano.5b03400.
24. Kong, Y., Jiang, G., Wu, Y., Cui, S., Shen, X., Amine hybrid aerogel for high-efficiency CO2 capture: Effect of amine loading and CO2 concentration. *Chem. Eng. J.*, 306, 362–368, 2016. https://doi.org/https://doi.org/10.1016/j.cej.2016.07.092.
25. Zhao, X., Hayner, C.M., Kung, M.C., Kung, H.H., Flexible holey graphene paper electrodes with enhanced rate capability for energy storage applications. *ACS Nano*, 5, 8739–8749, 2011. https://doi.org/10.1021/nn202710s.
26. Li, P. and Zeng, H.C., Hierarchical nanocomposite by the integration of reduced graphene oxide and amorphous carbon with ultrafine MgO nanocrystallites for enhanced CO2 capture. *Environ. Sci. Technol.*, 51, 12998–13007, 2017. https://doi.org/10.1021/acs.est.7b03308.
27. Low, J., Yu, J., Ho, W., Graphene-based photocatalysts for CO2 reduction to solar fuel. *J. Phys. Chem. Lett.*, 6, 4244–4251, 2015. https://doi.org/10.1021/acs.jpclett.5b01610.
28. Jiang, L. and Fan, Z., Design of advanced porous graphene materials: From graphene nanomesh to 3D architectures. *Nanoscale*, 6, 1922–1945, 2014.

29. Zhang, Y., Chi, Y., Zhao, C., Liu, Y., Zhao, Y., Jiang, L., Song, Y., CO2 adsorption behavior of graphite oxide modified with tetraethylenepentamine. *J. Chem. Eng. Data*, 63, 202–207, 2018.
30. Rodríguez-García, S., Santiago, R., López-Díaz, D., Merchán, M.D., Velázquez, M.M., Fierro, J.L.G., Palomar, J., Role of the structure of graphene oxide sheets on the CO2 adsorption properties of nanocomposites based on graphene oxide and polyaniline or Fe3O4-nanoparticles. *ACS Sustain. Chem. Eng.*, 7, 12464–12473, 2019.
31. Hsan, N., Dutta, P.K., Kumar, S., Bera, R., Das, N., Chitosan grafted graphene oxide aerogel: Synthesis, characterization and carbon dioxide capture study. *Int. J. Biol. Macromol.*, 125, 300–306, 2019.
32. Gunathilake, C.A., Ranathunge, G., Dassanayake, R.S., Illesinghe, S.D., Manchanda, A.S., Kalpage, C.S., Rajapakse, R.M.G., Karunaratne, D., Emerging investigator series: Synthesis of magnesium oxide nanoparticles fabricated on a graphene oxide nanocomposite for CO2 sequestration at elevated temperatures. *Environ. Sci. Nano*, 7, 1225–1239, 2020.
33. Seredych, M., Rodríguez-Castellón, E., Bandosz, T.J., Alterations of S-doped porous carbon-rGO composites surface features upon CO2 adsorption at ambient conditions. *Carbon N. Y.*, 107, 501–509, 2016.
34. Alghamdi, A.A., Alshahrani, A.F., Khdary, N.H., Alharthi, F.A., Alattas, H.A., Adil, S.F., Enhanced CO2 adsorption by nitrogen-doped graphene oxide sheets (N-GOs) prepared by employing polymeric precursors. *Mater. (Basel)*, 11, 578, 2018.
35. Chandra, V., Yu, S.U., Kim, S.H., Yoon, Y.S., Kim, D.Y., Kwon, A.H., Meyyappan, M., Kim, K.S., Highly selective CO2 capture on N-doped carbon produced by chemical activation of polypyrrole functionalized graphene sheets. *Chem. Commun.*, 48, 735–737, 2012.
36. Pruna, A., Cárcel, A.C., Benedito, A., Giménez, E., Effect of synthesis conditions on CO2 capture of ethylenediamine-modified graphene aerogels. *Appl. Surf. Sci.*, 487, 228–235, 2019.
37. Arifin, N.F.T., Zulkipli, N.A.N., Yusof, N., Ismail, A.F., Aziz, F., Salleh, W.N.W., Jaafar, J., Nordin, N.A.H.M., Sazali, N., Preparation and characterization of APTES-functionalized graphene oxide for CO2 adsorption. *J. Adv. Res. Fluid Mech. Therm. Sci.*, 61, 297–305, 2019.
38. Liu, Y., Sajjadi, B., Chen, W.-Y., Chatterjee, R., Ultrasound-assisted amine functionalized graphene oxide for enhanced CO2 adsorption. *Fuel*, 247, 10–18, 2019.
39. Shin, G.-J., Rhee, K., Park, S.-J., Improvement of CO2 capture by graphite oxide in presence of polyethylenimine. *Int. J. Hydrogen Energy*, 41, 14351–14359, 2016.
40. Zhao, Y., Ding, H., Zhong, Q., Preparation and characterization of aminated graphite oxide for CO2 capture. *Appl. Surf. Sci.*, 258, 4301–4307, 2012.

10

Carbon Allotropes in Air Purification

Nishtha Vaidya[1]*, Amarpreet K. Bhatia[1] and Shippi Dewangan[2]

[1]Bhilai Mahila Mahavidyalaya, Hospital Sector, Bhilai Nagar, Bhilai, Durg, Chhattisgarh, India
[2]Department of Chemistry, SW, Pukeshwar Singh Bhardiya Govt. College, Nikum, Durg, Chhattisgarh, India

Abstract
One of the most fundamental and adaptable components found in the landscape, carbon has a variable that allows it to generate different allotropes. Carbon has been employed as an energy source for the last few decades. Today, carbon plays a large and outsized role in virtually every area of applied science, including research, technology, and innovation. We should all be aware of carbon's extraordinary properties, including its many forms and allotropes, as well as how widely they are used in fields as disparate as science and environmental preservation and restoration. To cure a variety of organic and inorganic contaminants for renewable energy and environmental protection, amazingly adaptable nanomaterials, and carbon nanotubes can be used. One of the biggest environmental issues we face today is global warming, which was brought on by the improper use of fossil fuels, which resulted in the development of greenhouse gases. The majority of air pollutants, such as carbon monoxide, sulfur dioxide, nitrogen oxides, organic volatile compounds, ozone, toxic substances, and inhalable particulates (PM 2.5 and PM 10), were found as a result of incomplete combustion of the gases. These pollutants differ in their chemical makeup. A recent definition of nanoparticles of carbon. It possesses the capacity to change the world in terms of new inventions, technology, power, and accuracy as well as enable humankind will explore new depths and heights, states that one of the most significant goals of sustainable development is the protection of environmental health. The use of carbon allotropes and the elimination of environmental air contaminants and the methods used to clean them were discussed in the current book chapter. The application of carbon allotropes and their synthesis as nanoparticles, as well as the uses of

**Corresponding author*: nishthavaid@gmail.com

Chandrabhan Verma and Chaudhery Mustansar Hussain (eds.) Carbon Allotropes and Composites: Materials for Environment Protection and Remediation, (191–228) © 2023 Scrivener Publishing LLC

nanotubes for environmental air purification and their protection, are all covered in length in this chapter.

Keywords: Carbon allotropes, air purification, air pollutant, nanotubes CNTs, nanoparticles, environmental air pollutants

10.1 Introduction

Carbon is a unique and necessary element in our world. Carbon is the sixth most common element in the universe, fourth in the solar system, and 17th in the Earth's crust. Carbon's relative abundance is estimated to be between 180 and 270 parts per million [1]. All organic life in the natural world is formed of carbon and its allotropes. Every living thing on our planet, including people, animals, and plants, is made of carbon. By mass, After oxygen, carbon is the second most prevalent element in the human body. Consequently, 18% of a person's body weight is made up of carbons. Among the most notable descriptors of carbon in the fields of kinetic stability is that it has a wide variety of the broad range of metastable phases that might develop close to ambient conditions [2].

Our existence on the earth is mostly supported by carbon. The majority of elements heavier than helium were created in stars, including carbon. Numerous cycles and recycling processes have taken place on our globe, starting with clay minerals and ending with plants, animals, and air. In the form of soot and charcoal, carbon has been a common and valuable resource since early human history. It was not until the late 1800s that carbon was acknowledged as an element in nature [3]. Amorphous carbon, diamond, graphite, lonsdaleite, C60 (Buckminsterfullerene or buckyball), C540, C70, and carbon nanotubes are the eight main allotropes of carbon (CNTs; buckytube). For many years, researchers have been particularly interested in carbon and its allotropes. There have been numerous novel carbon allotropes found. Among these allotropes are carbon nanotubes (CNTs), which Iijima first discovered in 1991 [3, 5], and graphene, a one-atom-thick layer of carbon organized in a honeycomb lattice that Novoselov discovered in the year 2004 were delighted with The 2010 Nobel Prize in physics served as evidence of the value and application of carbon nanostructures in a wide range of fields [4, 6].

Environmental pollution is a widespread concern in both developing and developed nations, thus it is used not only in industries like chemistry, power, and the food industry but also in the field of environmental remediation technologies in the current situation [7]. Environmental pollution is one of the most significant types of pollution, and because it endangers both people's security and the stability of the economy, it is a major source of anxiety for everyone. Concerning this, a critical need for accessible and reliable monitoring is emerging. The use and benefits of carbon allotropes have been established, and their high potential and exceptional characteristics at the nanoscale have also been confirmed [8]. For use in the fields of pharmaceutical, medical, environmental, life sciences, and energy sciences, advanced devices that are microscopic and molecular in size are produced using nanotechnologies [9, 10]. Numerous applications of nanomaterials have been introduced in emerging nations to save the environment. Carbon allotropes such as nanosensors, nanocatalysts, nanotubes, and nanocomposites are used to improve environmental health and reduce toxins in the air, wastewater, water, soil, and food [11, 12]. Recently, the use of nanotechnology has been used to lessen air, soil, and water pollution. The most significant contaminants in the air, water, and soil are now detected by nanotechnology-based technologies, which have also led to the development of innovative waste-reducing green industrial processes [13]. Nanotechnology is crucial for reducing Iran's environmental health issues [14].

Water and air pollution are the most life-threatening environmental problems of the recent era [1, 2]. It is now essential to properly dispose of harmful waste in water and create innovative wastewater treatment methods [3]. In addition, many nations tend to find it more difficult to control outdoor air quality, and the majority of current directives only mention specific pollutants like PMs, ozone, and NO_x [4]. Pollutants in such systems have also been examined and controlled concerning indoor air quality. Pollutants and micropollutants in the environment are numerous and increasing, posing major health risks to living things [5]. Several earlier studies on renewable energy sources and solar radiation demonstrated how photocatalysts can degrade organic materials, bacteria, and viruses when UV or visible light is present [15].

Nanotechnology is used for safety applications such as gas monitoring systems, manufacturing gas leak detectors, fire, and dangerous gas detectors, ventilation control, and breath analyzers to examine how air pollution is affected by nanotechnologies. Nanotubes' capacity to absorb hazardous gases from gasoline for use in other industrial processes as well as the removal of contaminants from the environment [14].

Global warming is one of the most serious environmental issues confronting the world today. The production of greenhouse gases as a result of the incomplete combustion of fossil fuels is the primary source of air pollution today. CO, SO_2, NO_x, VOCs (volatile organic compounds), O_3, heavy metals, and respirable particulate matter are the major air pollutants (PM2.5 and PM10). CNT filters remove air pollutants from mobile sources and vehicle exhausts [16]. Many scientists have already created air and water filtration devices based on CNTs and they claim that the filters can kill the majority of bacteria in addition to blocking the smallest particles. Products containing CNTs are currently on the market and have already reached the commercial stage. Carbon nanotubes have a significant potential for effectively removing the use of mercury from the air for industrial and environmental goals [17].

10.2 Historical and Chemical Properties of Some Designated Carbon-Based Allotropes

Over the past few hundred years, carbon has been used as an energy source. Carbon has contributed significantly to nearly every field of science and technology in the contemporary era of invention. Given this element's extraordinary properties, we should all be aware of its wide range of existence and uses in the field of nanoscience.

The majority of carbon allotropes are produced via synthesis techniques. The majority of the time, the carbon allotropes are suitable for targeted drug delivery systems. Just because of the bio-friendly characteristics, The use of the allotropes in medicine delivery systems is the most promising. In addition, it offers a broad range of possible uses, such as strength applications utilizing carbon fiber and solar power as well as power and protection. Several peer-reviewed scientific studies have detailed the numerous benefits of carbon allotropes [4].

10.3 Structure and Characteristics of Carbon Allotropes

Researchers and business people are interested in the remarkable unique and adaptable features of carbon-based nanomaterials. Carbon allotropes were used in many different applications and had several amazing electrical, optoelectronic, mechanical, and chemical capabilities. These nanomaterials have mostly been studied and discussed for device applications due to their characteristics, structure, shape, bonding, charge transfer in confinement, and other factors. Additionally, graphene is referred to as the source of all carbon allotropes, making it the mother of all allotropes. Materials can be rolled, layered, and wrapped to create objects with different dimensions, such as 3-D graphite and 1-D nanotubes (Figure 10.1).

Zero-Dimensional Allotropes: Fullerene, Graphene Quantum Dots (GQDs)

One-Dimensional Allotropes: Carbon nanohorns, carbon nanofibers, carbon nanotubes (CNTs) (CNAs), Graphene Nanoribbons (GNRs)

Two Dimensional Allotropes: Graphene (GR),

Three-Dimensional Allotropes: Graphite, Diamond

There are eight different forms of carbon, including amorphous carbon, diamond, graphite, lonsdaleite, C60 (buckminsterfullerene or buckyball), C540, and carbon nanotubes (CNTs; buckytube). For many years, academics have been intensely interested in carbon and its various forms. The world of carbon and its uses was being effectively discovered through advanced research in the realm of carbon and its allotropes. The eight carbon allotropes are shown in the diagram below (Figure 10.2).

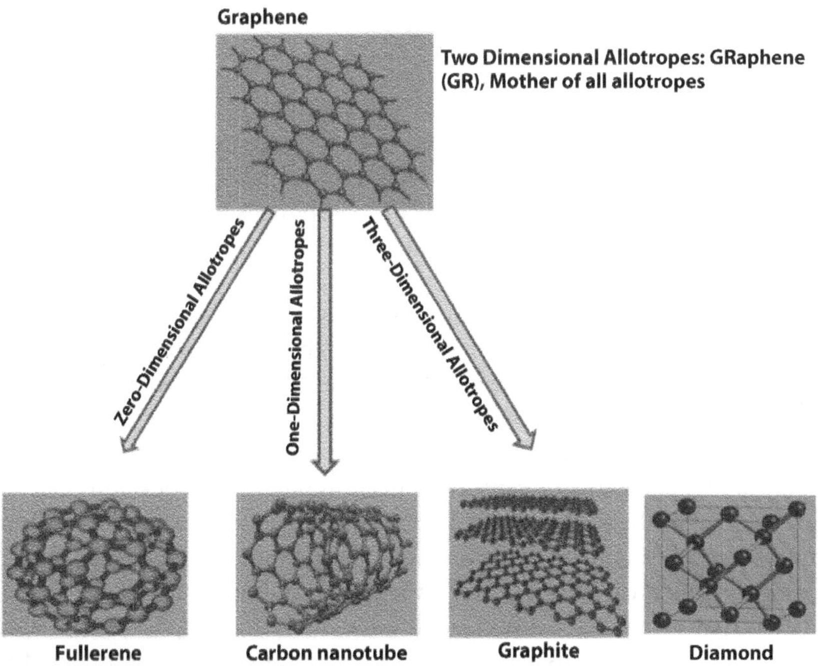

Figure 10.1 Structures of all carbon allotropes.

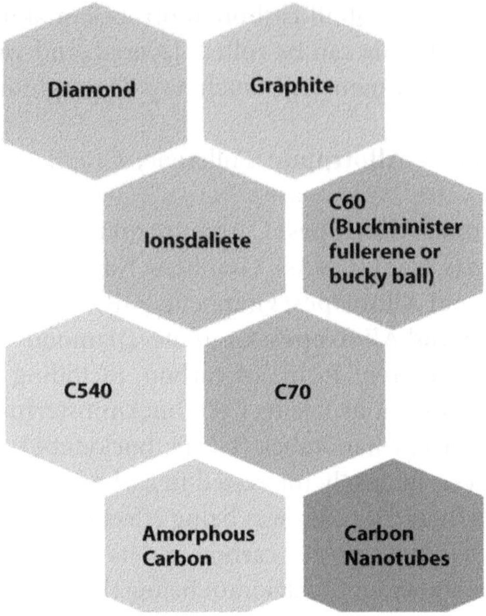

Figure 10.2 Eight allotropes of carbon.

(a) Diamond: The most prevalent 3-D allotrope of carbon is called diamond, and it has a face-centred cubic crystal structure and is meta-stable. At temperatures between 900 and 1400 degrees Fahrenheit and pressures of 5 to 6 Gpa, diamonds are molded [14]. Graphite is more stable than diamond. The inclusion of other elements, such as nitrogen and boron, affects the diamond's color [18].

The hardest naturally occurring substance is called diamond, which has individual carbon covalently linked atoms that have four neighbouring atoms and are in the sp3 hybridization state. Diamonds oxidize at higher temperatures, such as 500–700 °C in the air and 140–1700 °C in an inert environment. There is also the introduction of nanodiamonds, which are artificial arrangements of carbon atoms with characteristics and a structure similar to those of a natural diamond. Diamonds exhibit stable and complicated behavior in the presence of functional groups that contain nitrogen, which can be used for a variety of device applications. DLC is carbon that resembles diamond and contains sp2 and sp3 hybridization energy states. Diamonds have remarkable features that are connected to DLC. These properties include exceptional smoothness, high refractive index, excellent infrared transparency, great wear and corrosion resistance, strong chemical stability, and biocompatibility, as well as a low coefficient of friction [19].

(b) Graphite: Graphite was given its name in 1789 by German scientist Abraham Gottlob Werner. Each layer of atoms in carbon Occasionally called a graphene layer or sheet, it is structured in a planar honeycomb lattice with a 0.142-nm spacing. The bonds that bind each layer of graphite together are weakly unsaturated. Three more atoms are joined to the carbon atoms in graphite to create a planar array of fused hexagonal shapes. Graphite is a 3-D allotrope of sp2 hybridized carbon. It is further categorized as a collection of GR layers that are layered in the Z-direction along the XY plane [82]. The hexagonal corners of a layer of GR contain carbon atoms. Additionally, each carbon atom has three bonds visible around its borders. The van der Waals forces that exist between the layers of graphite allow them to slide past one another, allowing for the use of the material's notable thermal conductivity, electrical conductivity, and well-known lubricative feature. Low density 0.2 to 0.6 g/cm^3 graphitic foam with a thermal conductivity of around 182 W/(m • K) was reported by Klett and colleagues [83, 90]. Profitable heat sinks and heat exchangers have been made using graphitic shapes. But because covalent bonding is rigid and this is its limitation, it only has the highest heat conductivity in the Z-direction [20].

The GR layers are stacked opposite one another. However, the physical characteristics of the two types of graphite, beta (rhombohedral) and alpha (hexagonal) are similar. Graphite is a material that has a high melting point, and is resistant to high temperatures, and thermal shock, giving rise to graphite crucibles. Despite these properties, however, it is electrically desirable due to the significant electron delocalization within the carbon layers that allows valence electrons to move freely [82]. Additionally, electrodes composed of graphite powder have shown promise in a variety of experimental procedures due to their outstanding conductivity and corrosion resistance [19].

(c) Lonsdaleite: It is also known as a hexagonal diamond. Graphite-containing meteorites that have collided with the earth contain this allotrope of carbon. A meteorite was discovered in 1967, the Diablo Canyon meteorite. The interlayer links are placed in an eclipsed pattern as opposed to diamonds. On the (100) face, Lonsdaleite is found to be 58% harder than diamond. Additionally, it is far more resistant to indentation pressure than a diamond, with a resistance of about 152 Gpa as opposed to 97 Gpa [21–23].

(d) Fullerene: Buckyballs are another name for fullerenes. Fullerene, also known as Buckyball and named after the American architect Richard Buckminster Fuller, is a closed-caged structure formed by the periodic arrangement of graphene layers with the chemical formula C60. Researchers and academics from all over the world were tremendously interested in the discovery of fullerene. Before carbon soot, which was initially produced from graphite using a resistive heating technique for the first time in an inert atmosphere, fullerene was discovered [24]. Fullerene is a molecule that is entirely made of carbon atoms and has the shape of a hollow sphere, ellipsoid, or tube. Twenty hexagons and twelve pentagons make up its truncated icosahedral symmetry, which has 32 faces, 60 corners, and 90 edges. The 6:6 ring bonds between the hexagons are thought to be double bonds with shorter bond lengths, as opposed to the 6:5 bonds between the hexagon and the pentagon, which have an average bond length of 1.4. It was first discovered in 1985 at Rice University by Kroto *et al.* [25]. They received the 1996 Chemistry Nobel Prize. There are other known fullerene isomers with the carbon atoms C70, C76, C78, and C84 [9]. However, it has been discovered that C60 and C70 have different wavelengths of activity, with the former being a visible light photocatalytic material and the latter being photochemically reactive in the UV region. Open-shell fullerenes have less favorable energy characteristics. We favor structures with isolated pentagons over those with nearby, resonance-destabilizing five-membered rings. The melting temperature and bond energy per carbon atom are two additional thermodynamic parameters that are impacted by cage size [21]. The components that produce the arch across the entire surface are not distributed evenly, which increases the cage's stress. nanotubes made of carbon (Figure 10.3).

(e) Amorphous Carbon: Activated charcoal and carbon black are produced by the incomplete burning of animal and plant products. Amorphous carbon has been used in a variety of applications for the textile, plastic, and healthcare industries, as well as in the fields of gas and water filtration,

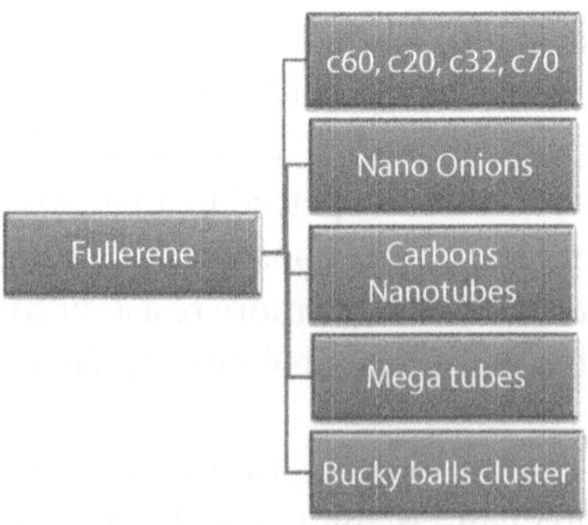

Figure 10.3 Types of fullerene.

electrical applications, and food packaging, due to the rapid advancement of a-C-reinforced nanocomposites and their superior and distinctive properties. Carbon compounds without long-range crystalline organization are referred to as amorphous carbon. Aberrations in the inter-atomic distances or inter-bonding angles of the graphite and diamond lattices do not affect the short-range crystalline order. To free the carbon cage, fullerene is thermally oxidized by increasing the temperature. Using EtOH as the carbon source, more amorphous carbon is extracted along with the residual carbon [19, 26].

(f) Carbon Nanotubes (CNT): Carbon nanotubes belong to the fullerene family and receive their name because it resembles a tube. A CNT is a cylinder-shaped, one-atom-thick portion of graphene that has been rolled up. CNTs have been developed Concerning other materials, with a length-to-diameter ratio of 132 000 000:1. CNT is divided into additional groups. A Japanese physicist named Iijima discovered carbon-based nanomaterials in 1991 [27]. The realm of carbon nanotubes with his (CNTs). The exclusive qualities of CNT have received high praise from scientists all across the world. Based on their structure and use, the CNTs are further split into two primary types, such as Multi-walled carbon nanotubes (MWCNTs) and single-walled carbon nanotubes (SWCNTs) (MWCNTs). Iijima and Ichihashi discovered SWCNTs [28], and Bethune *et al.* [29] rolled a layer of graphene into a hollow cylindrical product. can be imagined as a structure made of numerous layers of rolled graphene. MWCNTs have a structure that is comparable to multiple layers of rolled-up graphene. These long tubes include concentrically organized carbon hexagons, and the two closures are frequently covered in pentagonal structures resembling fullerenes [30–32]. On the bases of structure, CNTs can be categorized as chiral, armchair, or zig-zag [4] (Figures 10.4 and 10.5).

Due to their notable tensile strength and flexibility as well as their low coefficient of thermal expansion, which can occasionally result in ballistic transports, CNTs are relatively well-known for some distinguishing characteristics. They are renowned for their capacity for thermal insulation as well. CNTs are extremely flexible and substantially stiffer than steel. The amazing elasticity of this material has been studied using a scanning probe microscope to watch it return to its original shape after being released from tension. Because of their superior conduction properties and ability to function as metallic or semiconducting materials depending on their diameter and chirality, SWCNTs are more prevalent than MWCNTs. The valence band (VB) and conduction band (CB) meet at a degenerate point in the SWCNT's normal energy dispersion spectrum, which shows that the distance between the two bands is virtually zero and hence displays

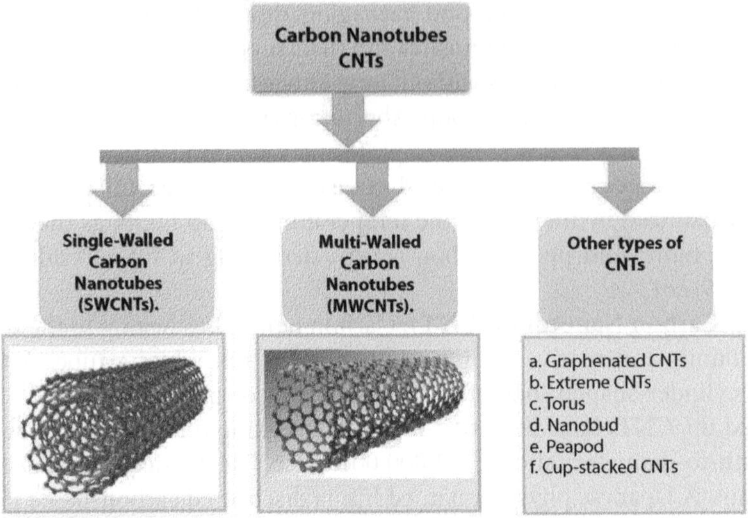

Figure 10.4 Different types of carbon nanotubes (CNT).

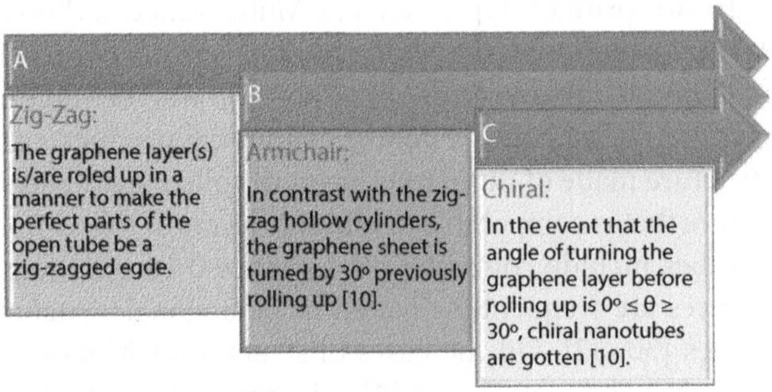

Figure 10.5 Classification of carbon nanotubes (CNT) on the bases of structure.

significant conductivity. As a result, an electron can be easily stimulated from VB to CB with very little excitation energy. Therefore, it is theoretically expected that all armchair SWCNTs will display metallic characteristics.

10.4 Uses of Carbon Nanotube Filters for Removal of Air Pollutants

The uses of nanotubes in the form of air and water pollution filters have greater attention now a day. Researchers and Scientists show a keen interest in the field of nanoscience and recently the use of allotropes of carbon made the topic of interest. The application of Nanotechnology has occurred from the conjunction of physical, chemical, and biological science and technologies. Nanoproducts such as nanofibers are also used as very thin fibers. The environmental, biological, medicinal, and energy disciplines all use nanotechnologies to create microscopic and nanoscale devices. Today, environmental protection and environmental health care practices use technology based on nanotechnologies. Other developing countries use nanoproducts for environmental health, such as nanocomposites, nanotubes, nanocatalysts, and nanosensors, to eliminate or reduce the level of soil, water, wastewater, air and food pollutants. The United States Environmental Protection Agency (US EPA) examined how nanotechnology can be used to perceive, measure, and remove pollutants from the air, water, and soil as well as determine new green industrial processes that reduce waste. The 21st century is posed with many health and environmental challenges [13, 14].

Since the atmosphere is necessary for life, its abnormalities have an impact on humans and other living things. The biggest and most accurate form of pollution in nature is air pollution [1]. One of the most significant environmental issues facing the planet today is global warming. One of the main causes of air pollution is the creation of greenhouse gases as a result of the incomplete combustion of fossil fuels [2]. The list of air contaminants is below:

Respirable particulate matter (PM 2.5 and PM 10), heavy metals, O_3, SO_2, NO_x, VOCs and CO, all have different chemical compositions, reactions, emissions, times at which they disintegrate, and capacities for long- or short-distance diffusion [3]. The incomplete burning of fossil fuels produces gaseous pollutants and alters the composition of the atmosphere in numerous ways [4]. The primary cause of air pollution in metropolitan areas is the incomplete combustion of diesel and gasoline fuels in cars, trucks, buses, and other on-road vehicles (Figure 10.6).

Noxious gases NO_x and organic volatile substances Anthropogenic sources, mobile, and stationary combustion sources are all linked to VOC pollution. The so-called volatile organic compounds are a significant class of substances that fuel combustion, incomplete combustion processes for energy production, and road transport is the main causes of emission (VOCs). This group of substances contains chemical species with an organic character, including benzene [33]. Surrounding concentrations of pollutants are associated with emissions. Several epidemiologic studies have looked into the risk of mortality connected with a rise in air pollution exposure. Air pollution affects several physiological systems (cardiovascular, respiratory, cancer, and reproductive) and affected mortality.

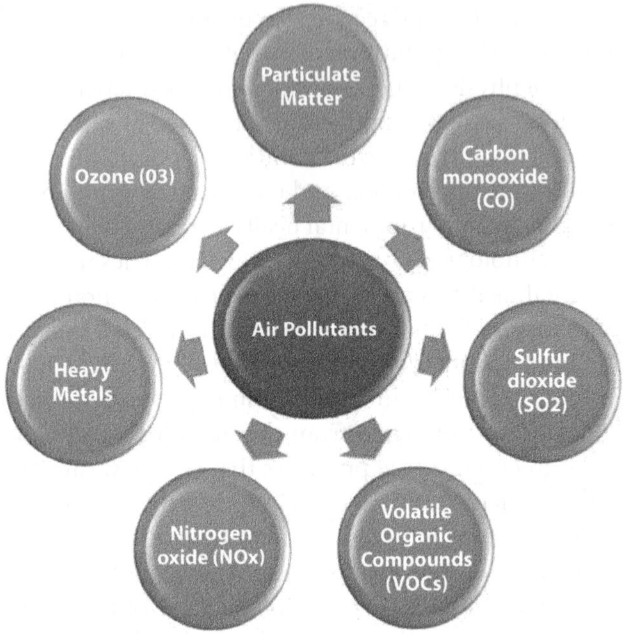

Figure 10.6 Significant causes of air pollution in cities.

A large number of deaths in urban areas was associated with the concentration of air pollutants. Most developing countries, demonstrate long-term development, yet in a few instances, there is a sign of an ecosystem that is collapsing quickly. By confirming that vehicles with high emission levels have been properly fixed, an emissions verification program has been proposed in Mexico to limit the amount of air pollution emissions from vehicles. The program's objectives are to recommend a car with a gasoline engine for less air pollution. We looked at four different types of pollution: hydrocarbons (HC), NO_x, CO, and CO_2. In the gas sorption processes, carbonaceous materials like active carbon and carbon fiber are frequently utilized [33].

CNTs have been used as environmental contaminants recently. CNTs can be employed in numerous applications thanks to their distinctive chemical and physical characteristics, such as their high BET area, such as a new kind of sorbent for environmental applications. Recently, the exploration and development of nanotube filters for purifying water and air was effectively developed and put to use to successfully filter out biological agents from water and remove bioaerosols [34, 35]. The advantages of carbon nanotube filters include the removal of air contaminants from moving sources. Use of carbon nanotubes as an air filter for membrane support of various materials with pores 10 mm in diameter. When removing air contaminants from vehicle exhaust, the prepared CNTs filters' removal efficacies were evaluated. Both research on air pollutants and measures to reduce air pollution can benefit from the deployment of CNT filters to remove air pollutants from moving sources.

10.5 Physicochemical Characterization of CNTs

Physiochemical characterization of CNTs was observed by using different techniques. CNTs characterized and examined both after and before the sorption process for air contaminants. Numerous technologies, including Brunauer, Emmett, and Teller (BET), EDS, XRD, and SEM, are utilized by most businesses to identify the precise surface area to compare physicochemical variations (Figure 10.7).

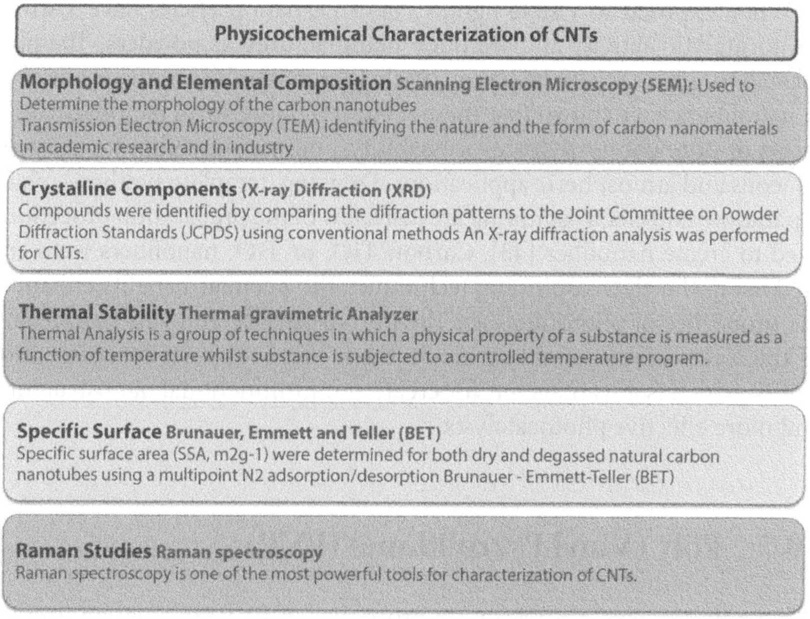

Figure 10.7 Physicochemical characterization of CNTs [33] (open access).

10.6 TiO_2 Nanofibers in a Simulated Air Purifier Under Visible Light Irradiation

The CNT/TiO_2 nanofiber may be used in air purifiers to increase the effectiveness of air treatment and air quality while using less energy [36]. In the realm of environmental science, TiO_2 is frequently utilized as a catalyst in environmental applications. The chemical features of TiO_2, including its low cost, non-toxicity, and eco-friendliness, were taken into consideration for air filtration. TiO_2, however, has limits when exposed to UV light [37]. The synthesis and modification of TiO_2 by carbon nanotubes (CNT) and their complexes, which improve the photocatalytic activity when exposed to visible light radiation, was the subject of several investigations [38, 39]. CNT/TiO_2 nanoparticles are thought to be effective photocatalysts for the oxidation of a variety of pollutants in both air and water. Because of the capacity of photocatalysts, hazardous organic compounds can be broken down into carbon dioxide and water [40, 41]. The effect of CNT on TiO_2 nanoparticles enhances photocatalytic activity. Carbon and its allotropes promote electron-hole separation, reduce charge recombination, and improve TiO_2 conductivity by acting as an electron trap [42].

When exposed to visible light, CNT/TiO_2 nanoparticles have a strong photocatalytic activity and are often manufactured as powders. The photocatalytic efficiency of the nanoparticle is high. Studies on the production of nanofibers have shown that these materials outperform nanoparticles in terms of pore connectivity, processability, and recoverability for usage in aqueous and atmospheric applications. Drawing, template synthesis, phase separation, and electrospinning are just a few of the preparation procedures used to create nanofiber [43]. Carbon/TiO_2 or TiO_2 nanofibers were created using the electrospinning technique. The peculiar benefits enhanced the properties of photocatalysts. The porous surface and vast surface area of the electrospun nanofibers made them ideal supports for adsorption ability, which is advantageous for creating multifunctional nanostructures and more effective photocatalysts.

10.7 Poly (Vinyl Pyrrolidone) (PVP)

Good adhesion and complexation properties of PVP consider an important synthetic polymer with such as Excellent physiological compatibility, biocompatibility, high hydrophilicity, low chemical toxicity, good solubility in water, and solubility in a variety of organic solvents [44]. PVP is combined with metal oxide nanoparticles, organic and inorganic chemicals, and the electrospinning process to create a variety of composite fibers [45, 46]. A straightforward method for electrospinning carbon/TiO_2 nanofibers and TiO_2 nanofibers employing poly (vinyl pyrrolidone) (PVP) as a polymeric binder was also described in previous research [43, 44, 47]. They found that carbon/TiO_2 or TiO_2 nanofiber photocatalysts were superior to carbon/TiO_2 or TiO_2 nanoparticle photocatalysts in terms of increased crystallinity, surface area, and photocatalytic activity.

10.8 VOCs

Urban and industrial settings are where anthropogenic contaminants are produced. VOCs are poisonous, and some of them are thought to be teratogenic, mutagenic, or carcinogenic. In addition, VOC emissions play a role in the greenhouse effect, stratospheric ozone depletion, and the development of smog and ozone in cities. In recent years, reducing VOCs has necessitated the air filtration system in vehicles. Typically, porous materials like charcoal and activated carbon are employed as adsorbents in automotive air purifiers that use adsorption treatment technology to filter the air. Some VOCs are captured while the adsorbent circulates the air inside an automobile. This method is straightforward and inexpensive, but its removal effectiveness is poor, necessitating repeated regeneration of the adsorbent.

Photocatalytic degradation is a new viable and advanced technology to remove indoor air level quality [48]. Using a suitable substitute photocatalyst to enhance the physical properties of TiO_2 by photocatalytic oxidation technology involves the elimination of VOC vapors under visible light irradiation [36].

Gaseous benzene serves as an important prototype for the growth of VOC photocatalytic air purification, According to studies on the visible-light-induced photocatalytic breakdown of benzene (gas phase) in a synthetic indoor air system.

10.9 Heavy Metals

The removal of mercury vapor by sorbent is a significant environmental concern due to its observable toxicity and numerous industrial applications. Environmental and industrial pollution has led to numerous health-related and other issues in nature in recent years. The primary contaminants coming from various industrial operations include heavy metals like mercury [49]. These circumstances occur in the processes of primary metal production, its production of gold and cement, the combustion of coal the refining of oil, and chemical synthesis, which may lead to significant environmental contamination. A highly hazardous heavy metal that can be created either naturally or artificially is mercury [17, 50]. Mercury pollution from human activities is the principal known source of mercury in the atmosphere. Over the past few decades, numerous legislative measures have been implemented at the national and international levels to lessen mercury poisoning in the environment. A recent global agreement to lessen mercury contamination. Mina Mata Convention calculated in 2013 that 2700–6000 metric tonnes of mercury were unintentionally emitted into the environment. The elemental form of mercury is the most significant one in terms of toxicology (Hg0). The elevated vapor pressure in the elemental mercury produced during respiration is what frequently causes intoxication through inhalation [51]. Mercury affects most of the physiological systems, particularly the [52] central nervous system [53], reproductive system [54], and respiratory system [55]. As a brand-new member and allotrope of the carbon family, carbon nanotubes (CNTs) are cylindrical macromolecules comprised of one or more layers of carbon atoms. In the realm of environmental contamination and its prevention, the electrical, thermal, and structural properties of CNTs are crucial [56]. Recent years have seen a lot of focus on the use of CNTs in adsorption-related applications to address environmental pollution issues. Some investigations looked into how well these sorbents worked with metal cations like copper and cadmium [57], lead [58], chromium [59] mercury [60], nickel [61] zinc [57] drugs [62], Endocrine-disrupting substances, organic pollutants in the aquatic environment, and benzene, toluene, and ethanol in the ambient gaseous environment. The related research discussed the process of adsorbing elemental mercury vapor by SWCNTs, and the adsorption mechanism and amount were evaluated and optimized in a gaseous atmosphere. Significant concentrations of heavy metals are present in the aqueous and gaseous streams from industrial processes. Because of the main characteristics of mercury, including its non-biodegradability and its toxicity, people may have substantial health issues. Even minor exposures to mercury can have detrimental effects on health. After exposure to various mercury compounds by inhalation, ingestion, or skin contact, neurological and behavioral problems were noted. Some of the symptoms include tremors, restlessness, memory loss, neuromuscular side effects, headaches, and cognitive and motor dysfunction. Workers may have moderate, subclinical signs of central nervous system damage after years of exposure to air containing at least 20 g/m^3 of elemental mercury. Heavy metal exposure is also impacted by the renal system. The kidney has also been impacted and reported on, with symptoms ranging from renal failure to increased protein in the urine. Adsorption, electrodialysis, ultrafiltration, and ion exchange are a few of the methods that have been developed to effectively remove heavy metal pollution [63]. The most significant and often used technology among them is adsorption; It is a quick and easy process for handling and equipment costs. Adsorption capacity, selectivity for particular chemicals, and the longevity and renewability of the adsorbents play an important part in deciding the effectiveness of removal adsorption. Due to their unique properties, heavy metal removal using nanoparticles like CNTs has received a significant deal of attention lately. Different quantities of mercury vapor can be positively adsorbed by SWCNTs. Other investigations have demonstrated that carbon nanotubes are capable of removing metals from gaseous and aqueous environments [57, 58, 61]. In the majority of investigations, the thermal desorption approach was employed more frequently than chemical methods. With the use of single-walled carbon nanotubes, heavy metals can be eliminated from the environment. Since they have sufficient absorption and desorption capabilities, single-walled carbon nanotubes would be preferable for mercury extraction and separation from the gaseous atmosphere. These sorbents were replicable and affordable in terms of "materials of the twenty-first century.

10.10 Particulate Matter (PM)

According to strong evidence, air pollution's fine particulate matter (PM) can enter the lower airways and cause negative health consequences even at concentrations below those advised by the WHO. Even though ideally, the air is seen as an essential component of human rights, air pollution has become a greater hazard to public health since the turn of the twenty-first century. Recent studies have identified air pollutants as one of the leading causes of death and disability globally is fine particulate matter (PM) with a diameter of less than 2 m (PM 2.5), for example. It was found that prolonged exposure to PM 2.5 is associated with negative health effects and natural causes of mortality. Epidemiological studies show a link between persistent exposure to high PM concentrations and the development of various respiratory disorders such as asthma, diabetic disease, and lung cancer are significantly linked to the risk of cardiovascular diseases [64]. Carbon, a component of particulate matter (PM), which is present has a dose-dependent relationship with a decline in lung capacity and function in alveolar macrophages (AM). Carbonaceous PMs are found in the lungs. The characterization was done by TEM, HRTEM, (EDX, Raman spectroscopy, and NIRFM methods. Research is a suitable way to classify PM Constituents as well. Carbon nanotubes (CNTs), an allotrope of carbon generated from both natural and man-made sources, PM levels are high both indoors and outside. CNTs are frequently exposed by people [64].

10.11 Techniques to Remove Air Pollutants and Improve Air Treatment Efficiency

Atmospheric pollution is one of the most significant environmental issues today, according to modern civilization. Humans, animals, and plant life are all harmed by air pollution elements. Forest fires, volcanic eruptions, soil degradation, and anthropogenic causes, particularly emissions brought on by human activity. Primary and secondary pollutants are two categories of air pollutants depending on where they come from. Nitrogen oxides (NOX), CO, SO_2, and particles (PM10 and PM2.5) are examples of primary pollutants. Secondary pollutants, on the other hand, are those that result from chemical reactions taking place in the atmosphere, such as tropospheric ozone (O3), HNO_3, and H_2SO_4, NO_x ($NO_x = NO+NO_2$) in the atmosphere and a rise in nitrogen oxide concentration ($NO_x= NO+NO_2$) in the atmosphere are both associated with several adverse and harmful effects. To address environmental issues including global warming, tropospheric ozone, photochemical haze, and even ozone layer depletion brought on by N2O, as well as how NOX and sulfur oxides contribute to the formation of acid rain, deNOx action has become a mandated and required activity [65–67]. Instead, when lung tissues are exposed to a high enough level of NO_x growth, it has a variety of effects on human health, including emphysema, bronchitis, etc. Another issue is how NO_x interacts and connects to other air contaminants, like volatile organic compounds (VOC). The by-products of co-creation, such as nitrous acid or peroxy acyl nitrate (PAN), maybe ten times greater harmful to people than NO_x. Ozone in the troposphere can also cause damage to the eyes and lungs [68–72]. The hourly NO_x air content is restricted to 0.1 ppm and 0.2 ppm, respectively, by several rules that have been enacted by the US Environmental Protection Agency (EPA) and the European Environmental Agency (EEA). However, the suggested upper limit of breathing NOX in urban and working environments is typically exceeded in the heart of large cities [73, 74]. Additionally, some voices call for the administrators to announce a new, 50–75 ppb short-term NO_x standard. The research community is now more interested than ever in looking into novel ways to successfully reduce NO_x pollution as a result of this worry. the primary contributor to NO_x emissions in metropolitan areas. Three distinct forms of NO_x are recognized as a result of three primary mechanisms that contribute to the production of nitrogen oxides during combustion. The manufacturing industry and the transportation industry are the sources of NO_x inlets.

1. Thermal, prompt, and fuel NO_x. Nitrogen and oxygen react at high temperatures to produce thermal NOX.
2. Prompt NOX: this takes place as a result of a quick reaction involving the radicals of nitrogen, oxygen, and hydrocarbons.
3. Fuel NOX: is produced by direct oxidation of fuel-borne N-H organo-nitrogen molecules.

The ring structures are disrupted during combustion, resulting in the formation of substances like CN, NH_2, NH_3, or HCN, which combine with O_2 to produce NO_x. There are primary and secondary techniques for reducing NOX emissions that have been developed and put into use [66, 67]. The term "primary methods" refers to techniques that remove NO_x from a combustion zone, such as a furnace, without the need for an extra reactor.

The secondary methods are used to reduce the already formed NO_x, with selective catalytic reduction (SCR) and selective non-catalytic reduction (SNCR) being the most popular ones. Other secondary methods include NOX absorption, phase separation, NO_x storage and reduction (NSR), ozone injection, and NO conversion into HNO_3 [75–77]. None of them, however, can be used to effectively reduce the NO_x that urban dwellers are exposed to. To do this, photocatalysis is a novel and effective method for the elimination of unwanted gaseous molecules. The significance and specialized application of nanomaterials in the PCO methods used to eliminate NO_x gases Through the application of photochemical oxidation processes, by using a novel and promising photocatalysis technique, the NO and NO_2 gases are converted to NO form and then expelled from the atmosphere.

Researchers are now creating novel nanomaterials to boost their photocatalytic activity to reduce NOX emissions and enhance air quality. Recent years have seen the introduction of new TiO_2-based nano compounds with the largest specific surface area (SSA), various morphologies, and chemical alterations. The smallest nanoparticles, self-ordered nano-tubular films, and nanotubes produced high photocatalytic capabilities. On the other hand, when the oxide was doped with ions, the creation of intermediate energy states between the conduction band and the valence band enhanced the photocatalytic performance of TiO_2. In contrast, visible light photocatalysts other than titanium oxide have been researched and show a good De-NOX efficiency when exposed to light with a wavelength of at least 400 nm.

10.12 Removal of NOX by Photochemical Oxidation Process

Titanium dioxide, TiO2-based photocatalysis research, and application are mostly used in the air environment. Because of its distinctive qualities, such as non-toxicity, significant oxidation power, and chemical inertness, this TiO_2 oxide is the most promising photocatalyst. Three main stable polymorphs of TiO_2 exist, with the anatase phase having the maximum photocatalytic activity [78].

The anatase phase, which is a semiconductor material, has a Due to its photocatalytic property, which is activated by UV radiation with wavelengths shorter than 387 nm, it has a broadband gap of 3.2 eV. In this region, 50 of the photon energies (1013 quanta•cm2•s, at midday) are sufficient to activate TiO_2 and produce separated electron-hole pairs, even if in the best instances UV radiation only accounts for 4-5% of the total energy of sunshine. The company Evonik produces the TiO_2-P25 product (hence referred to as P25), which is based on this titanium dioxide polymorph and is made up of 75% anatase and 25% rutile. Usually, this product is offered for sale in stores. The improved photocatalytic activity of this compound has been attributed to a synergistic interaction between the two phases. However, it has been noted that water is crucial for the production of hydroxyl radicals, or OH•, which increases the effectiveness of heterogeneous photocatalysis [79–81]. The hydroxyl groups on the photocatalyst surface are crucial since it is primarily through their interaction that NO is adsorbed [82]. For the photocatalyst deactivation, two key aspects were taken into account: water consumption on its surface (if not continually supplied) and adsorbed NO_3 species. However, cleaning the photocatalyst with water makes it simple to regenerate [83, 84]. Surface Area and Morphology: By preparing compounds with a high specific surface area, the photocatalytic activity demonstrated by titanium dioxide was easily boosted (SSA). Two basic strategies are used for this:

(i) On several surfaces with a higher specific surface area, I dispersed titanium dioxide nanoparticles.
(ii) Using several synthetic approaches to regulate the TiO_2 morphology and size.

AACA, or aerosol-assisted co-assembly of nanoparticles, appears to be a successful method for creating surfaces for larger composites. The AACA is a relatively rapid process that disperses P25 and SiO_2 nanoparticles to produce nanoporous microparticles from sprayed droplets of a colloidal solution as an aerosol (6.1 mm of particle size and 10-100 nm of pore size) (Degussa, Aerosol 200) [85].

Thermo hydrolysis, which in this case uses $TiOSO_4$ as the precursor, is promoted as an environmentally acceptable method of producing TiO_2 nanoparticles. Once attained, calcination can be used to boost the crystallinity of the nanoparticles. The pre-calcination samples performed better in terms of NOx gas removal than P25, while the latter performed better than the calcined samples. The findings reflect the elevated SSA values achieved for the pure samples (167 m^2g^1) and their decline following calcination (to 27 m^2g^1) because crystallinity increased [86, 87].

10.13 Chemically Adapted Nano-TiO_2

One of TiO_2's key shortcomings in photocatalysis applications is the requirement for UV light absorption to generate charge carriers through photogeneration. According to reports, nitrogen-doped TiO_2 exhibited photocatalytic activity when exposed to visible light. Since then, changing the band gap by doping or altering TiO_2 has received a lot of attention [85–87].

The dopant may be incorporated into the crystal lattice's interstices or employed to swap out the ions in the anionic or cationic sublattices. The introduction of a dopant distorts the crystal lattice, which results in changes to the structure of electronic bands 88, by creating exciting interstates or deforming its electronic bands. The morphology of the N-TiO_2/Fe_2O_3 samples is a more uniform, rod-like nanoparticle (10 nm in diameter; 20–60 nm in length). The Fe codoping had no noticeable influence on the morphology. The SSA for N-TiO_2/Fe_2O_3 samples (120 m2 • g1) is significantly lower than that for samples of Fe-N-TiO_2 (180 m^2g). When exposed to visible light, it is possible to see strong photocatalytic activity due to the presence of N atoms and oxygen vacancies. The presence of an iron dopant in the TiO_2 structure creates a recombination channel for the photogenerated electrons and holes, which precludes the formation of NO_2, during the photocatalytic breakdown of nitric oxide. Rh, Pt, Pd, Fe, and then Cu are the metals that are most effective at removing NOx from the atmosphere. TiO_2 that has been doped with various concentrations of Ag breaks down N_2O. The results were better than those of other investigations that had been written about in the literature.

10.14 Alternative Nanoparticulated System

TiO_2 performs superbly as a photocatalyst, without a doubt, However, its photocatalytic activity is restricted to the presence of UV light (387 nm). The photocatalytic concert of De-NOx materials using TiO_2 is dramatically reduced in certain conditions in the shadows of buildings or narrow streets where there is no direct sunlight in Central/Northern Europe or North America, where low levels of UV radiation reach the surface due to geography and gloomy weather. Additionally, DeNOx has stated that future implementation will involve a large-scale operation, thus the least expensive materials need to be taken into account. These problems have increased experts' interest in researching novel alternatives [65].

10.15 Photodegradation of NOX Evaluated for the ZnO-Based Systems

The hydrothermal method was used to create several multi-cationic layered double hydroxides (Zn/Ni/La/TiLDH). At higher temperatures, the equivalent calcined products were nickel-zinc titanium oxide phases. With increasing calcination temperature, the photocatalytic activity of NOx breakdown decreases, with activation of the De-NOX due to ZnO synthesis during calcination. It has been discovered that constructing highly oriented ZnO nanorod/nanotube arrays over a large area of N-doped titania (TiO_2xNy) nanoparticles can make zinc oxide work as a visible light photocatalyst. When compared to pure ZnO nanorods, nanotubes, P25, and TiO_2xNy, this nano-scale superstructure demonstrated high photocatalytic activity for the breakdown of NO. This increased activity was attributed to the ZnO/TiO_2xNy heterojunction structure's promotion and rise in the charge separation of the pair h+/e as well as the unique surface features of the well-aligned structure. Researchers have also looked into nanostructured photocatalysts for visible light. The first work to be published demonstrates the removal of NOx under visible light using nanoplate microspheres made of bismuth oxy bromide. The Nonaqueous Sol-gel technique was used to create the BiOBr microspheres, and the ingredients used were bismuth (III) nitrate, cetyltrimethylammonium bromide, and ethylene glycol (65).

10.16 Synthesis and Applications of Carbon Nanotubes

The first mechanism for all procedures, the growth mechanism, is where carbon nanotube production got its start. In earlier research publications, several methods for the production of nanotubes and their present state are briefly discussed. Three primary methods are often used to create carbon nanotubes [17, 88] (Figure 10.8).

Today, scientists were looking for more ways to create nanotubes for economic reasons in their research on nanomaterials. In the first arc discharge technique of this technology, an arc discharge occurs between two carbon electrodes, either with or without a catalyst, and vapor is created.

The second method for creating nanotubes is called laser ablation; in this method, a high-powered laser beam is focused on a volume of feedstock gas containing carbon (methane or carbon monoxide). Currently, arc discharge techniques often produce large numbers of impure material, whereas laser ablation produces a very small number of pure nanotubes. Chemical Vapor Deposition (CVD) is substantially more appropriate for the synthesis of Multi-walled Nanotubes (MWNTs) than low-quality Single-walled nanotubes (SWNTs). The SWNTs made via CVD have a wide range of diameters that can be difficult to manage. But this approach is relatively simple to scale up, making it perfect for commercial production [17].

(a) Arc discharge: C60 fullerenes are mostly produced using the carbon arc discharge process. Due to its widespread use and potential ease of production, This approach is the most widely used. This approach, however,

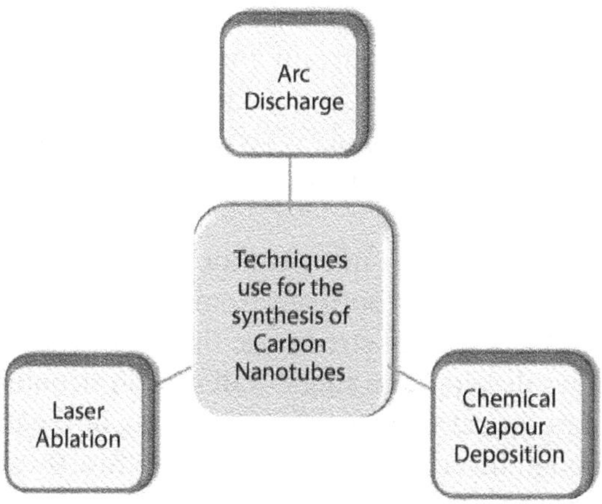

Figure 10.8 The methods used to create carbon nanotubes.

requires separating the nanotubes from the catalytic metals and soot that are present in the crude product. Two carbon rods are arranged end to end and spaced around 1mm apart in the procedure, which results in the formation of nanotubes. The container is normally filled with inert gas (helium, argon) at low pressure (between 50 and 700 bar).

According to certain recent studies in this field, nanotubes can also be made using liquid nitrogen and the arc technique. A direct current of 50 to 100 A driven by around 20 V results in a high-temperature discharge between the two electrodes. One of the carbon rods is vaporized by the discharge, and the other rod develops a tiny rod-shaped deposit. Key elements in the manufacture of high-yield nanotubes include the uniformity of the plasma arc and the temperature of the deposit form on the carbon electrode. Understanding of the development mechanism is growing, and investigations have revealed that various diameter distributions based on the helium and argon combination have been found. The thermal conductivities and diffusion coefficients of these mixes vary. These parameters affect how rapidly the carbon and catalyst molecules scatter and cool, which affects the diameter of the nanotubes formed during the arc process. This shows that temperature, carbon, and metal catalyst concentrations, and the plasma's quenching rate all have an impact on the distribution of nanotube diameters. It also shows that different-sized single-layer tubules can form and grow on metal particles [89].

(b) Laser ablation: Smalley's group 25 at Rice University first published the laser vaporization method for creating carbon nanotubes in 1995. The tool that Smalley's team used to make nanotubes using laser vaporization. At 1200 °C, a graphite target is heated. in an oven and vaporized using a continuous laser. The fundamental distinction between continuous and pulsed lasers is that the latter requires a lot more light (100 kW/cm2 as opposed to 12 kW/cm2). To maintain a pressure of 500 Torr, helium or argon gas is poured into the oven. A quickly expanding and cooling plume of extremely hot vapor is created. When the vaporized species cool to form larger clusters, such as fullerenes, small carbon molecules, and atoms quickly condense.

The catalysts also start to condense, though initially more slowly, and attach to carbon clusters to stop them from forming cage structures. 30 When catalysts adhere to cage structures, they may even open the cages. Tubular molecules develop into single-wall carbon nanotubes from these early clusters until the catalyst particles are too big or the temperature has dropped low enough that carbon can no longer diffuse through or over the catalyst particles. The nanotube stops growing when particles become so densely coated in carbon that they are unable to take in any more. In this case, the created SWNTs are grouped by van der Waals forces. There are some remarkable, although not exact, similarities between the spectrum emission of excited species during laser ablation of a composite graphite target and that of laser-irradiated C60 vapor. This implies that fullerenes are also produced by laser ablation of catalyst-filled graphite.

Laser ablation and this form of arc discharge are essentially identical since the optimal background gas and catalyst mix is the same. This could be because the conditions for the reactions must be relatively similar, and they unquestionably employ the same mechanism [91].

(c) Chemical vapor deposition: Chemical vapor deposition (CVD) synthesis can be performed by putting a carbon source in the gas phase and using an energy source, such as a plasma or a resistively heated coil, to transmit energy to a gaseous carbon molecule. The main sources of gaseous carbon were acetylene, carbon monoxide, and methane. The actual synthesis of the carbon nanotube occurs after the catalyst has first been prepared in a two-step procedure known as CVD. The catalyst is typically made by sputtering a transition metal onto a substrate and then inducing catalyst particle nucleation through thermal annealing or chemical etching. The substrate forms clusters as a result of thermal annealing, from which the nanotubes will grow. The typical temperature range for the CVD production of nanotubes is between 650 and 900 C. About 30% of yields are typical with CVD. These are the fundamental tenets guiding the CVD process. Over the past two decades, numerous CVD techniques have been created, including laser-assisted CVD, vapor phase growth, alcohol catalytic CVD, plasma-enhanced CVD, thermal chemical CVD, and CVD on aerogel support [91].

10.17 Mechanism of Technologies

(a) Adsorption techniques: Environmental remediation methods used in this method include the adsorption of gases, liquids, and organic and inorganic pollutants. It is a procedure where the adsorbate, which is present in a gaseous or liquid bulk, clings to an adsorbent's surface. Adsorbate merely adheres to the adsorbent's surface and does not diffuse inside it [92]. This method is frequently used in industrial processes and occurs naturally in the environment and ecosystem. Both CNTs as SWCNTs and MWCNTs differed based on morphologies. Compared to MWCNTs, SWCNTs have smaller inner and outer diameters [93]. The tube diameter and bundle geometry of different CNTs will have an impact on how well atoms or molecules are adsorbed [94]. CNTs with smaller sizes, greater specific surface areas, and distinctive hollow and layered structures will have higher adsorption capacities while performing the adsorption. Higher adsorption capacities will be found in CNTs with smaller diameters, larger specific surface areas, and characteristic hollow and layered architectures (Figure 10.9).

SWCNTs are less effective at adsorbing the adsorbate than MWCNTs. But since each contaminant has its adsorption capacity due to variations in the bonding connections between its molecules, different pollutants or impurities have varying adsorption capacities. The CNT adsorption mechanisms for organic compounds involve interactions between the molecules of the pollutants, including electrostatic interaction, and hydrophobic interactions with bonds, hydrogen bonds, and other molecules [96].

Additionally, the electrical and transport characteristics of nanotubes can be impacted by the numerous molecular bonding interactions [95]. Additionally, interactions between the metal ions and CNT functional groups lead to electrostatic attraction, surface precipitation, physical adsorption, and complexation-mediated heavy metal adsorption [96]. Along with that, the mechanics of carrier localization and charge distribution would be impacted by the geometric structure of CNTs. A bundle of CNTs will provide significantly stronger molecule adsorption on the surface or inside the nanotube than a single tube [95]. CNTs in bundles have two kinds of adsorption sites:

External sites (exterior and groove)
Internal sites (interior and interstitial) [94, 97].

The adoption ability for CNT in bundles is high in interstitial and groove sites, whereas the adsorption ability and charge transferred for SWCNT are weak and modest, indicating that the interior site is slightly more energetic

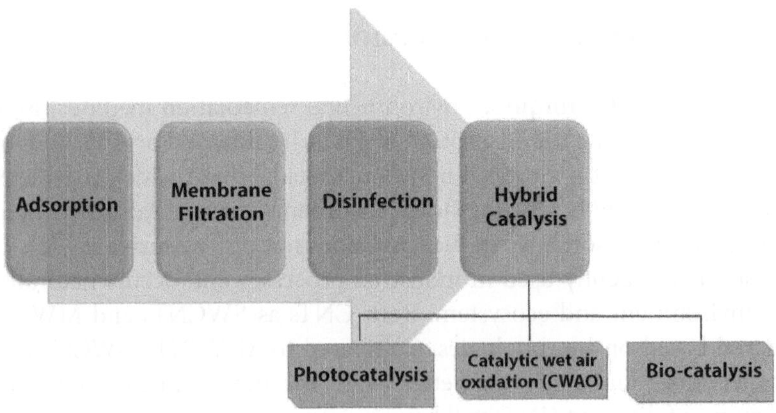

Figure 10.9 Different mechanisms involving CNT technologies.

than the outer site. Thus, it was determined that increasing the number of nanotubes employed can improve the molecule's absorption [95].

(b) Membrane Filtration: Several methods have been used in water treatment or water sanitization and purification systems, one of them is membrane filtration technologies like microfiltration, nanofiltration, forward osmosis, and reverse osmosis. This technique involves water treatment to control the quality of water to the extent that water is safe to drink. Due to the filtration mechanism used by the CNT membrane, which retains impurities from wastewater as it travels through the membrane, it differs from adsorption. Freestanding CNT membranes and composited CNT membranes are the two different forms of membrane filters. Water distillation and wastewater treatment applications use freestanding CNT membranes. also [98, 99].

Water purification, oil-water separation, and the elimination of organic and inorganic contaminants all use composited CNT membranes.

(c) Disinfection: CNT membrane has been utilized for removing byproduct fumigation in addition to purification. As previously stated, the bacterium was rendered inactive by the use of CNTs in membrane technology. The bacterium's cell wall and membrane can be damaged by CNTs, which can then invade the cell body and damage the DNA and cause protein malfunction. secondary products (reaction oxygen species, or ROS) production that causes cell body harm Their size and form have an impact on how antibacterial they are [96, 100].

Another method of the ability of CNTs to inactivate microbial cells is projected to be the generation of ROS and oxidative stress, where the detrimental ROS is formed once the CNTs reach the bacterial body.

(d) Hybrid Catalysis: The water purification process with this technology is also the most economical and environmentally beneficial. It is economical and environmentally friendly, and it delivers customized synthetic catalysts and screening. The ability to use CWAO in treatment systems has significantly expanded because of the addition of CNT as a catalyst to WAO. For various applications, the higher catalytic performance of CNTs is provided by their surface chemistry. The additional three categories for heterogeneous catalysts are non-noble metal, noble metal, and carbon CWAO [99]. Carbon nanotubes (CNTs) are attracting more attention than other catalysts due to their distinct chemical and thermal stabilities. The carbon catalyst, which produces CNT-based WAO, was first introduced in recent years. Oxygen, nitrogen, and sulfur heteroatoms are highly reactive with the surface of carbon [101] (Figure 10.10).

(e) Bio-catalysis: By using enzymes as the catalyst, bio-catalysis, a branch of applied biology and biotechnology, transforms organic substances into

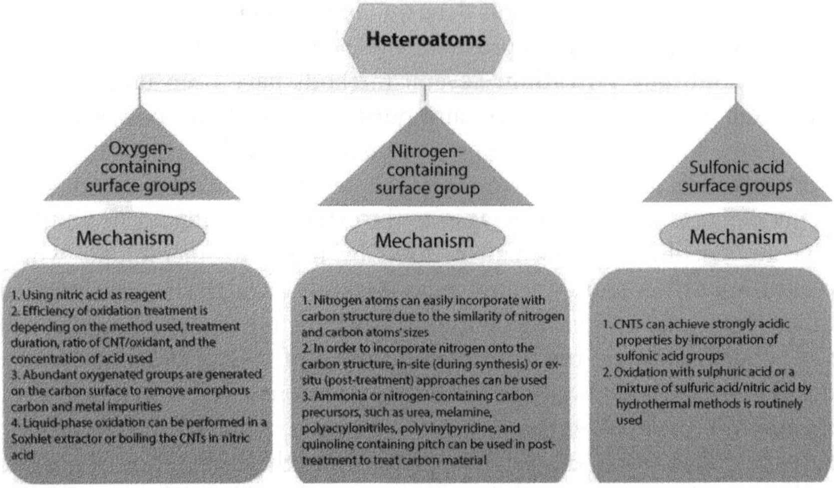

Figure 10.10 Common heteroatoms and their mechanisms in tuning CNTs properties.

other products with the least amount of waste. The enzymes are delicate; if pH or temperature conditions were unfavorable for the enzyme, they became denatured and were no longer viable for survival. Enzymes are immobilized using CNTs in biocatalysis to increase their catalytic stability, selectivity, and reusability. Adsorption to create ester in a non-aqueous media can be used to immobilize lipase on CNT. However, there are still challenges with the enzyme immobilization on CNT, including enzyme structural changes (covalent linkage), leaching (physical adsorption), and low-cost efficiencies, which limit its potential for widespread use in the treatment of sewage and water [102, 103]. CNT was regarded as a superior technology that can aid in environmental remediation. However, the graphic outlines the benefits and drawbacks of CNTs.

(f) Wet air oxidation (WAO): Wet air oxidation is one of the greatest methods for breaking down dangerous, toxic, and highly concentrated organic contaminants in aqueous solutions (WAO). It is possible for organic molecules to finally oxidize in WAO to CO2, H2O, and a few other innocuous byproducts without the emission of HCl, NOx, SOx, dioxins, or fly ash. The functionalized carbon materials MWCNTs, CFs, and graphite work well in the CWAO of phenol. These materials' successful catalytic performance depends on the carboxylic acid groups on their surfaces, which also aid in the production of active radical species by the CWAO of phenol. In the CWAO over MWCNTs, phenol first undergoes oxidation into hydroquinone and then into p-benzoquinone. Consequently, when the ring is broken, carboxylic acids are produced. The majority of carboxylic acids can instantly undergo oxidation, resulting in the production of carbon dioxide and water [104]. Advantages and disadvantages of CNTs are described in Figure 10.11.

Environmental endocrine disruptors (EDCs) are increasingly recognized as a significant global problem. We are exposed to this organic pollution at work, at home, on the farm, in the food we eat, and in the water, we drink. Due to the widespread use of EDC analogues in industry, there are emissions of EDCs into the environment. Unfortunately, even in low quantities, EDCs can accumulate in the environment and result in a variety of health issues for the general population. successfully creating a carbon nanotube with a biological polymer modification. EDCs contain both organic and synthetic mechanisms. Bisphenol A (BPA), the most prominent synthetic component of EDCs, is a polar molecule with two non-saturated phenolic and acetone connections. The solubility and stability of the BPA components are both improved by this complex structure [8]. BPA is employed in a variety of sectors, including packaging, the manufacture of water pipes, toys, medical equipment, and plastics. However, there are two basic methods for people and other species to be exposed to BPA: direct exposure and indirect exposure [9–11]. BPA can cause a variety of health

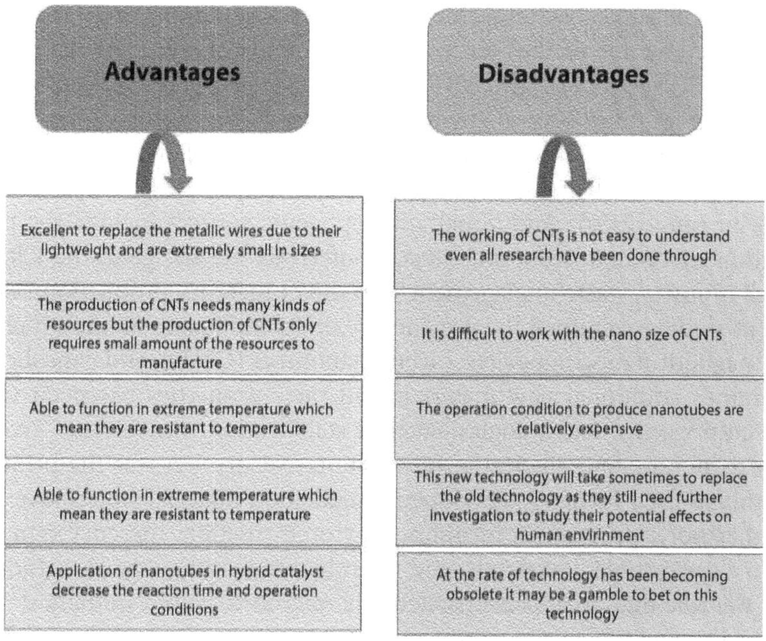

Figure 10.11 Advantages and disadvantages of CNTs.

issues, including endocrine system disruption, testicular cancer, brain cancer, neurobehavioral issues, kidney and liver damage, the development of anomalies, and even the encouragement of cancer cells in people are all examples of cancers that can occur. Therefore, in both industrialized and developing nations, the use of this component is restricted or prohibited. The adsorption procedure is one of the most widely used techniques for removing BPA from water and wastewater effluent. Low amounts of organic contaminants are reduced via adsorption. Recently, BPA was reduced using nanotubes as a selective adsorbent, but these complexes did not distribute properly in the solution. The other significant issue with nanotubes was the absence of surface functional groups.

Recently, nanotubes and biopolymers were mixed to increase efficiency. Chitosan is a well-known biopolymer that is safe for the environment and has been utilized to increase the adsorption capacity and effectiveness of other adsorbents. Further studies led to the decision to synthesize magnetic multi-wall carbon nanotubes and biopolymer-modified synthetic biopolymer nanotubes [105].

One of the major global problems is mercury poisoning because of its dangerous effects on human health. In 2015, the world's anthropogenic sources released about 2220 tonnes of mercury into the atmosphere. Mercury toxicity in soil, water, and the air is connected to potential toxicity to people and ecosystems. Mercury (Hg), a hazardous heavy metal, has been listed as one of the "ten major compounds of concern". According to a UNEP report (UNEP, 2018). Higher mercury (Hg) concentrations have been found in several environmental media as a result of the industrial usage of coal. 24% of the estimated emissions from anthropogenic sources come from stationary combustion of fossil fuels, principally from coal burning (21%). Additional anthropogenic sources include the production of cement, iron, and steel, nonferrous metals, gold, the Chlor-alkali industry, waste management, and direct mercury production. More recent studies examined the use and use of several nanomaterials, including magnetic nanocomposites, carbon nanotubes (CNTs), and nanosheets. Examines additional new materials and their properties based on various cutting-edge technologies, such as stabilization/containment, oxidation/reduction, and adsorption/desorption. Adsorption was the primary technique used by newly developed remediation materials to remove Hg. Furthermore, procedures including phytoremediation, mercury removal from algae, artificial wetlands, and microbial reduction were also studied, and the role that organisms, particularly microbes, played in these strategies is shown [106].

10.18 Conclusion

Nanotoxicology is the study of how biocompatible nanoparticles are; not all nanomaterials are biocompatible. The impacts of nanoparticles on biological bodies are being studied in cutting-edge science. A hard and interesting way to learn more about different materials and their hazardous effects is through the study of nanotoxicology. Because NOx gases present in the atmosphere have a variety of negative and hazardous impacts, reducing NOX concentration is now a requirement for environmental safety. A novel and advantageous method to reduce NOX emissions is photocatalysis. To directly eliminate the NOX that urban center dwellers breathe in, some techniques have been devised and put into practice. Our cities are implementing new anti-pollution building materials that can create the photocatalytic oxidation (PCO) of NOX pollutants. NO, and NO2 gases are transformed into NO3 via the PCO process, which removes them from the atmosphere. New materials are created from a nanoscience perspective to improve their photocatalytic activity for De-NOX procedures.

The most promising photocatalyst is an oxide called TiO_2, which is non-toxic, has a high oxidation power, and is chemically inert. With a greater specific surface area (SSA), a distinct shape, and chemical alterations, new advanced TiO_2 nanoparticles are created. Different substrates, including the TiO_2 nanoparticles, were distributed using organic and carbon fibers, mesoporous materials, clay composites, and nanoporous microparticles to increase the SSA and, consequently, the photocatalytic activity toward De-NOX processes.

The use of activated carbon (AC)/TiO_2 composites seems to be the most promising strategy since AC not only enhances the photocatalytic process but also makes NO molecules easier to adsorb. Better performance against NOX removal is generally attributed to the recombination reaction's slower rate. Alternatives to TiO_2 have been reported, including photocatalysts for visible light. Bismuth oxyhalides and ZnO/TiO_2xNy composites are among the best photocatalysts that operate under visible light irradiation at wavelengths greater than 400 nm.

A new sophisticated nano photocatalyst must be produced using the new additives in large quantities at the lowest cost before they can be used in the built environment. There are a lot of fascinating nanomaterials out there that could be used to make commercial DeNOX products. The new trials must be used to deepen our comprehension of the PCO mechanism, enhance visible-light photochemical activity, develop novel cement-based materials that combine these state-of-the-art photocatalytic additives, and show that scaling up their production is feasible. CNT had a balancing effect on the enhancement of photocatalytic activity for CNT/TiO2 nanofibers by increasing specific surface area, lowering band gap energy, and minimizing electron recombination. These findings imply that innovative and promising CNT/TiO2 nanofibers are material for use in air purification systems.

References

1. Nasir, S., Hussein, M.Z., Zainal, Z., Yusof, N.A., Carbon-based nanomaterials/allotropes: A glimpse of their synthesis, properties and some applications. *Materials*, 11, 295, 2018.
2. Sederberg, D., *Allotropes of Carbon: It's All in the Way You're Put Together*, Content Area: Science, Chemistry, Grade Level: 8-12, Physical Science Across Scale, Purdue University, ICP Draft Date: July 2009.
3. Greenwood, N. and Earnshaw, A., *Chemistry of the Elements*, Indiana Department of Education (2009). Indiana Standards and Resources, Pergamon Press, New York, 1984, Retrieved May 1, 2008, from http://www.indianastandards.org/index.asp.
4. Karthik, P.S., Himaja, A.L., Singh, S.P., Carbon-allotropes: Synthesis methods, applications, and future perspectives. *Carbon Lett.*, 15, 4, 219–237, 2014.
5. Iijima, S., Helical microtubules of graphitic carbon. *Nature*, 354, 56, 1991. http://dx.doi.org/10.1038/354056a0.
6. Novoselov, K.S., Geim, A.K., Morozov, S.V., Jiang, D., Zhang, Y., Dubonos, S.V., Grigorieva, I.V., Firsov, A.A., Electric field effect in atomically thin carbon films. *Science*, 306, 666, 2004. http://dx.doi. org/10.1126/science.1102896.
7. Chung, J.H., Hasyimah, N., Hussein, N., Application of carbon nanotubes (CNTs) for remediation of emerging pollutants-a review. *Trop. Aquat. Soil Pollution*, 2, 1, 13–26, 2022.
8. Villarreal, C.C., Pham, T., Ramnani, P., Mulchandani, A., Carbon allotropes as sensors for environmental monitoring. *Curr. Opin. Electrochem.*, 9, 11, July 21, 2017.
9. Ray, P.C., Yu, H., Fu, P.P., Toxicity and environmental risks of nanomaterials: Challenges and fkuture needs. *J. Environ. Sci. Health C, Environ. Carcinog. Ecotoxicol. Rev.*, 27, 1, 1–35, 2009. https://doi.org/10.1080/10590500802708267.
10. Zhang, W., Zhang, D., Liang, Y., Nanotechnology in remediation of water contaminated by poly- and perfluoroalkyl substances: A review. *Environ. Pollut.*, 247, 266–276, 2019. https://doi.org/10.1016/j.envpol.2019.01.045.
11. Seferos, D.S., Giljohann, D., Hill, H.D., Prigodich, A.E., Mirkin, C.A., Nanoflares: Probes for transfection and mRNA detection in living cells. *J. Am. Chem. Soc.*, 129, 50, 15477–15479, 2007. https://doi.org/10.1 021/ja0776529.
12. Wu, Y., Pang, H., Liu, Y., Wang, X., Yu, S., Fu, D., Chen, J., Wang, X., Environmental remediation of heavy metal ions by novel-nanomaterials: A review. *Environ. Pollut.*, 246, 608–620, 2019. https://doi.org/10.1016/j.envpol.2018.12.076.

13. Kumar, A. and Jakhmola, A., RNA-mediated fluorescent Q-PbS nanoparticles. *Langmuir*, 23, 6, 2915–2918, 2007. https://doi. org/10.1021/la0628975.
14. Aghalari, Z., Dahms, H.-U., Sillanpää, M., Aghalari, Z. *et al.*, Investigating the effectiveness of nanotechnologies in environmental health with an emphasis on environmental health journals. *Life Sci. Soc. Policy*, 17, 8, 2021.
15. Muscetta, M. and Russo, D., Photocatalytic applications in wastewater and air treatment: A patent review (2010–2020). *Catalysts*, 11, 7, 834. https://doi. org/10.3390/catal11070834
16. Romero-Guzmán, L., Reyes-Gutiérrez, L.R., Romero-Guzmán, E.T., Savedra-Labastida, E., Carbon nanotube filters for removal of air pollutants from mobile sources. *JMMCE*, 6, 105–118, 2018.
17. Golbabaei, F., Ebrahimi, A., Shirkhanloo, H., Koohpaei, A., Faghihi-Zarandi, A., Single-walled carbon nanotubes (SWCNTs), as a novel sorbent for determination of mercury in air. *Glob. J. Health Sci.*, 8, 7, 273–280, 2016.
18. Kaiser, W. and Bond, W.L., Nitrogen, a major impurity in common type I diamond. *Phys. Rev.*, 115, 857, 1959. http://dx.doi.org/10.1103/ PhysRev.115.857.
19. Selvam, A., Sharma, R., Sutradhar, S., Chakrabarti, S., *Synthesis of carbon allotropes in nanoscale regime*, Springer Nature, Singapore Pte Ltd, 2021.
20. Chung, D.D.L., Review graphite. *J. Mater.*, 37, 8, 1475–1489, 2002.
21. Krueger, A., *Carbon Materials and Nanotechnology*, pp. 33–122, Wiley-VCH, Weinheim, 2010.
22. Yulong, Y. and Xinsheng, P., Recent advances in carbon-based dots for electroanalysis. *Analyst*, 141, 2619–2628, 2016.
23. Namdari, P., Negahdari, B., Eatemadi, A., Synthesis properties and biomedical applications of carbon-based quantum dots: An updated review. *Biomed. Pharmacother.*, 87, 209–222, 2017.
24. Sun, H., Wu, L., Wei, W., Qu, X., Recent advances in graphene quantum dots for sensing. *Mater. Today*, 16, 433–442, 2013.
25. Kroto, H., Heath, J., O'Brien, S. *et al.*, C60: Buckminsterfullerene. *Nature*, 318, 162–163, 1985.
26. Zhang, F., Hou, P.X., Liu, C., Cheng, H.M., Epitaxial growth of single-wall carbon nanotubes. *Carbon*, 102, 181–197, 2016.
27. Iijima, S., Helical microtubules of graphitic carbon. *Nature*, 354, 56–58, 1991.
28. Ijima, S. and Ichihashi, T., Single-shell carbon nanotubes of 1-nm diameter. *Nature*, 363, 6430, 603–605, 1993.
29. Bethune, D.S., Kiang, C.H., De Vries, M.S., Gorman, G., Savoy, R., Vazquez, J., Beyers, R., Cobalt-catalyzed growth of carbon nanotubes with single-atomic-layer walls. *Nature*, 363, 6430, 605–607, 1993.
30. Tian, P., Tang, L., Teng, K.S., Lau, S.P., Graphene quantum dots from chemistry to applications. *Mater. Today Chem.*, 10, 221–258, 2018.
31. Ouyang, M., Huang, J.L., Lieber, C.M., Fundamental electronic properties and applications of single-walled carbon nanotubes. *Acc. Chem. Res.*, 35, 1018–1025, 2002.

32. Ouyang, M., Huang, J.L., Lieber, C.M., Scanning tunneling microscopy studies of the onedimensional electronic properties of single-walled carbon nanotubes. *Annu. Rev. Phys. Chem.*, 53, 201–220, 2002.
33. Romero-Guzmán, L., Reyes-Gutiérrez, L.R., Romero-Guzmán, E.T., Savedra-Labastida, E., Carbon nanotube filters for removal of air pollutants from mobile sources. *JMMCE*, 6, 105–118, 2018.
34. Brady-Estévez, A.S., Kang, S., Elimelech, M., A single-walled-carbon-nanotube filter for removal of viral and bacterial pathogens. *Small*, 4, 481–484, 2008. https://doi.org/10.1002/smll.200700863.
35. Guan, T. and Yao, M., Use of carbon nanotube filter in removing bioaerosols. *J. Aerosol Sci.*, 4, 611–620, 2010. https://doi.org/10.1016/j.jaerosci.2010.03.002.
36. Wongaree, M., Chiarakorn, S., Chuangchote, S., Sagawa, T., Photocatalytic performance of electrospun CNT/TiO2 nanofibers in a simulated air purifier under visible light irradiation. *Environ. Sci. Pollut. Res.*, 23, 21395–21406, 2016.
37. Tongon, W., Chawengkijwanich, C., Chiarakorn, S., Multifunctional Ag/TiO2/MCM-41 nanocomposite film applied for indoor air treatment. *Build. Environ.*, 82, 481–489, 2014.
38. Kim, C.H., Kim, B.-H., Yang, K.S., TiO2 nanoparticles loaded on graphene/carbon composite nanofibers by electrospinning for increased photocatalysis. *Carbon*, 50, 2472–2481, 2012.
39. Dai, K., Zhang, X., Fan, K., Zeng, P., Peng, T., Multiwalled carbon nanotube-TiO2 nanocomposite for visible-light-induced photocatalytic hydrogen evolution. *J. Nanomater.*, 2014, 1–8.
40. Park, H., Park, Y., Kim, W., Choi, W., Surface modification of TiO2 photocatalyst for environmental applications. *J. Photochem. Photobiol. C Photochem. Rev.*, 2012, 1–20, 2012.
41. Takenaka, S., Arike, T., Matsune, H., Kishida, M., Coverage of carbon nanotubes with titania nanoparticles for the preparation of active titania-based photocatalysts. *Appl. Catal. B*, 125, 358–366, 2012.
42. Ma, L., Chen, A., Zhang, Z., Lu, J., He, H., Li, C., A new fabrication method of uniformly distributed TiO2/CNTs composite film by insitu chemical vapor deposition. *Mater. Lett.*, 96, 203–205, 2013.
43. Chuangchote, S., Sagawa, T., Yoshikawa, S., Electrospinning of poly(-vinyl pyrrolidone): Effects of solvents on electrospinnability for the fabrication of poly(p-phenylene vinylene) and TiO2 nanofibers. *J. Appl. Polym. Sci.*, 114, 2777–2791, 2009.
44. Chuangchote, S., Sagawa, T., Yoshikawa, S., Electrospun TiO2 nanowires for hybrid photovoltaic cells. *J. Mater. Res.*, 26, 17, 2316–2458, 2011.
45. Kadirova, Z.C., Hojamberdiev, M., Katsumata, K.-I., Isobe, T., Matsushita, N., Nakajima, A., Okada, K., Photodegradation of gaseous acetaldehyde and methylene blue in aqueous solution with titanium dioxide-loaded activated carbon fiber polymer materials and aquatic plant ecotoxicity tests. *Environ. Sci. Pollut. Res.*, 21, 4309–4319, 2014.

46. Atchison, J.S., Zeiger, M., Tolosa, A., Funke, L.M., Jackela, N., Presser, V., Electrospinning of ultrafine metal oxide/carbon and metal carbide/carbon nanocomposite fibers. *RSC Adv.*, 5, 35683–35692, 2015.
47. Song, C.G., Koppala, S.K., Yoon, J.W., Characterization of electrospun TiO2 nanofibers and its enhanced photocatalytic property under solar light irradiation. *J. Ceram. Process. Res.*, 14, 6, 653–657, 2013.
48. An, T., Chen, J., Nie, X., Li, G., Zhang, H., Liu, X., Zhao, H., Synthesis of carbon nanotube-anatase TiO2 sub-micrometer-sized sphere composite photocatalyst for synergistic degradation of gaseous styrene. *ACS Appl. Mater. Interfaces*, 4, 5988–5996, 2012.
49. Zamir, S.M., Halladj, R., Nasernejad, B., Removal of toluene vapors using a fungal biofilter under intermittent loading. *Process Saf. Environ. Prot.*, 89, 1, 8–14, 2011. http://dx.doi.org/10.1016/j.psep.2010.10.001.
50. Tchounwou, P.B., Environmental exposure to mercury and its toxicopathologicimplications for public health. *Environ. Toxicol.*, 18, 149–175, 2003. http://dx.doi.org/10.1002/tox.10116.
51. Selin, N.E., Science and strategies to reduce mercury risks: A critical review. *J. Environ. Monit.*, 13, 2389–2399, 2011. http://dx.doi.org/10.1039/c1em10448a.
52. Yard, E.E., Horton, J., Schier, J.G., Caldwell, K., Sanchez, C., Lewis, L., Gastañaga, C., Mercury exposure among artisanal gold miners in Madre de Dios, Peru: A cross-sectional study. *J. Med. Toxicol.*, 8, 4, 441–448, 2012. http://dx.doi.org/10.1007/s13181-012-0252-0.
53. Carocci, A., Rovito, N., Sinicropi, M.S., Genchi, G., Mercury toxicity and neurodegenerative effects. *Rev. Environ. Contam. Toxicol.*, 229, 1–18, 2014. http://dx.doi.org/10.1007/978-3-319-03777-6_1.
54. Martinez, C.S., Escobar, A.G., Torres, J.G., Brum, D.S., Santos, F.W., Alonso, M.J., Wiggers, G.A., Chronic exposure to low doses of mercury impairs sperm quality and induces oxidative stress in rats. *J. Toxicol. Environ. Health A*, 77, 1, 143–154, 2014. http://dx.doi.org/10.1080/15287394.2014.867202.
55. Luo, G.H., Yao, H., Xu, M., Cui, X.W., Chen, W.X., Gupta, R., Xu, Z., Carbon nanotube-silver composite for mercury capture and analysis. *Energ. Fuel.*, 24, 1, 419–26, 2009. http://dx.doi.org/10.1021/ef900777v.
56. Hirlekar, R., Carbon nanotubes and applications: A review. *Asian J. Pharm. Clin. Res.*, 2, 17–27, 2009.
57. Vellaichamy, S. and Palanivelu, K., Preconcentration and separation of copper, nickel and zinc in aqueous samples by flame atomic absorption spectrometry after column solid-phase extraction onto MWCNTs impregnated with D2EHPA-TOPO mixture. *J. Hazard. Mater.*, 185, 2, 1131–9, 2011.
58. Moradi, O., Zare, K., Yari, M., Interaction of some heavy metal ions with single walled carbon nanotube. *Int. J. Nano Dimens.*, 1, 3, 203–20, 2011.
59. Jung, C., Heo, J., Han, J., Her, N., Lee, S.J., Ohd, J., Hexavalent chromium removal by various adsorbents: Powdered activated carbon, chitosan, and single/multi-walled carbon nanotubes. *Sep. Purif. Technol.*, 106, 63–71, 2013.

60. Pitoniak, E., Nanostructured silica-gel doped with TiO2 for mercury vapor control. *J. Nanopart. Res.*, 5, 281–292, 2013.
61. Adolph, M.A., Xavier, Y.M., Kriveshini, P., Rui, K., Phosphine functionalized multi-walled carbon nanotubes: A new adsorbent for the removal of nickel from aqueous solution. *J. Environ. Sci.*, 24, 6, 1133–1141, 2012.
62. Fazelirad, H., Ranjbar, M., Taher, M.A., Sargazi, G., Preparation of magnetic multi-walled carbon nanotubes for an efficient adsorption and spectrophotometric determination of amoxicillin. *J. Ind. Eng. Chem.*, 21, 889–892, 2015.
63. Pérez-López, B. and Merkoçi, A., Carbon nanotubes and graphene in analytical sciences. *Microchim. Acta*, 179, 1-2, 1–16, 2012. http://dx.doi.org/10.1007/s-012-0871-9.
64. Kolosnjaj-Tabi, J., Just, J., Hartman, K.B., Laoudi, Y., Boudjemaa, S., Alloyeau, D., Szwarc, H., Wilson, L.J., Moussa, F., Anthropogenic carbon nanotubes found in the airways of parisian children. *EBioMedicine*, 9, 2, 11, 1697–1704, 2015.
65. Balbuena, J., Cruz-Yusta, M., Sánchez, L., Nanomaterials to combat NOX pollution. *J. Nanosci. Nanotechnol.*, 15, 9, 6373–6385, 2015.
66. Chen, M. and Chu, J., NOx photocatalytic degradation on active concrete road surface—From experiment to real-scale application. *Clean. Prod.*, 19, 1266, 2011.
67. Lasek, J., Yu, Y., Wu, J.C.S., Removal of NOx by photocatalytic processes. *J. Photoch. Photobio. C*, 14, 29, 2013.
68. Newman, M.C. and Clement, W.H., *Ecotoxicology: A Comprehensive Treatment*, CRC Press, Boca Raton-Florida, 2008.
69. Chen, B., Hong, C., Kan, H., Exposures and health outcomes from outdoor air pollutants in China. *Toxicology*, 198, 291, 2004.
70. Morrow, P.E., Toxicological data on NOx: An overview. *J. Toxicol. Environ. Health*, 13, 205, 1984.
71. Last, J.A., Sun, W.M., Witschi, H., Ozone, NO, and NO2: Oxidant air pollutants and more. *Environ. Health Perspect.*, 102, 179, 1994.
72. Cormier, S.A., Lomnicki, S., Backes, W., Dellinger, B., Origin and health impacts of emissions of toxic by-products and fine particles from combustion and thermal treatment of hazardous wastes and materials. *Environ. Health Perspect.*, 114, 810, 2006.
73. Frampton, M.W. and Greaves, I.A., NOx - NOx: Who's there? *Am. J. Respir. Crit. Care Med.*, 179, 1077, 2009.
74. Williams, M.L. and Carslaw, D.C., New directions: Science and policy – out of step on NOX and NO2? *Atmos. Res.*, 45, 3911, 2011.
75. Normann, F., Andersson, K., Leckner, B., Johnsson, F., Emission control of nitrogen oxides in the oxy-fuel process. *Prog. Energy Combust. Sci.*, 35, 385, 2009.
76. Roy, S., Hegde, M.S., Madras, G., Catalysis for NOx abatement. *Appl. Energy*, 86, 2283, 2009.
77. Pârvulescu, V.I., Grange, P., Delmon, B., Catalytic removal of NO. *Catal. Today*, 46, 233, 1998.

78. Augustynski, J., The role of the surface intermediates in the photoelectrochemical behaviour of anatase and rutile TiO_2. *Electrochim. Acta*, 38, 43, 1993.
79. Dalton, J.S., Janes, P.A., Jones, N.G., Nicholson, J.A., Hallam, K.R., Allen, G.C., Photocatalytic oxidation of NOx gases using TiO2: A surface spectroscopic approach. *Environ. Pollut.*, 120, 415, 2002.
80. Yu, Q.L. and Brouwers, H.J.H., Indoor air purification using heterogeneous photocatalytic oxidation. Part I: Experimental study. *Appl. Catal. B Environ.*, 92, 3–4, 454–461, 2009.
81. Rosseler, O., Sleiman, M., Montesinos, V.N., Shavorskiy, A., Keller, V., Keller, N., Litter, M.I., Bluhm, H., Salmeron, M., Destaillats, H., Chemistry of NOx on TiO2 surfaces studied by ambient pressure XPS: Products, effect of UV irradiation, water, and coadsorbed K+. *J. Phys. Chem. Lett.*, 4, 536, 2013.
82. Li, S.C., Jacobson, P., Zhao, S.L., Gong, X.Q., Diebold, U., Trapping nitric oxide by surface hydroxyls on rutile TiO2(110). *J. Phys. Chem. C*, 116, 1887, 2012.
83. Ai, Z., Ho, W., Lee, S., Zhang, L., Efficient photocatalytic removal of no in indoor air with hierarchical bismuth oxybromide nanoplate microspheres under visible light. *Environ. Sci. Technol.*, 43, 4143, 2009.
84. Ibusuki, T., *Photocatalysis, Science and Technology*, M. Kaneko and I. Okura (Eds.), pp. 143–155, Springer, New York, 2002.
85. Kima, S.K., Chang, H., Cho, K., Kil, D.S., Cho, S.W., Jang, H.D., Choi, J.W., Choi, J., Enhanced photocatalytic property of nanoporous TiO2/SiO2 micro-particles prepared by aerosol assisted co-assembly of nanoparticles. *Mater. Lett.*, 65, 21–22, 3330–3332, 2011.
86. Di Paola, A., Bellardita, M., Palmisano, L., Amadelli, R., Samiolo, L., Preparation and photoactivity of nanocrystalline TiO2 powders obtained by thermohydrolysis of TiOSO4. *Catal. Lett.*, 143, 844, 2013.
87. Ao, C.H. and Lee, S.C., Enhancement effect of TiO2 immobilized on activated carbon filter for the photodegradation of pollutants at typical indoor air level. *Appl. Catal. B Environ.*, 44, 191, 2003.
88. Pérez-López, B. and Merkoçi, A., Carbon nanotubes and graphene in analytical sciences. *Microchim. Acta*, 179, 1–2, 1–16, 2012. http://dx.doi.org/10.1007/s-012-0871-9.
89. Ando, Y. and Zhao, X., Synthesis of carbon nanotubes by arc-discharge method. *New Diam. Front. Carbon Technol.*, NDFCT 511, 16, 123–137, 2006.
90. Pérez-López, B. and Merkoçi, A., Carbon nanotubes and graphene in analytical sciences. *Microchim. Acta*, 179, 1–16, 2012.
91. Karna, S.K., *Properties and Applications of Carbon Nanotubes Project Work FY 3114.* https://dvikan.no/ntnu-studentserver/reports/Report%20Carbon%20nanotubes.pdf
92. Artioli, Y., Adsorption, in: *Encyclopedia of Ecology*, pp. 60–65, 2008, https://dx.doi.org/10.1016/B978-008045405-4.00252-4.
93. Carrales-Alvarado, D.H., Leyva-Ramos, R., Rodríguez-Ramos, I., Mendoza-Mendoza, E., Moral-Rodríguez, A.E., Adsorption capacity of different types

of carbon nanotubes towards metronidazole and dimetridazole antibiotics from aqueous solutions: Effect of morphology and surface chemistry. *Environ. Sci. Pollut. Res.*, 27, 17123–17137, 2020.
94. Ren, X., Chen, C., Nagatsu, M., Wang, X., Carbon nanotubes as adsorbents in environmental pollution management: A review. *Chem. Eng. J.*, 170, 395–410, 2011. https://doi.org/10.1016/j.cej.2010.08.045.
95. Zhao, J., Buldum, A., Han, J., Lu, J.P., Gas molecule adsorption in carbon nanotubes and nanotube bundles. *Nanotechnology*, 13, 195–200, 2002.
96. Song, B., Xu, P., Zeng, G., Gong, J., Zhang, P., Feng, H., Liu, Y., Ren, X., Carbon nanotube-based environmental technologies: The adopted properties, primary mechanisms, and challenges. *Rev. Environ. Sci. Biotechnol.*, 17, 571–590, 2018.
97. Arora, B. and Attri, P., Carbon nanotubes (CNTs): A potential nanomaterial for water purification. *J. Compos. Sci.*, 4, 3, 135, 2020.
98. Liu, X., Zhang, S., Pan, B., Potential of carbon nanotubes in water treatment: A review, in: *Recent Progress in Carbon Nanotube Research Book 2*, Journal of Environmental Sciences, InTech, 2012, http://dx.doi.org/10.5772/51332.
99. Das, R., Abd Hamid, S.B., Ali, M.E., Ismail, A.F., Annuar, M.S.M., Ramakrishna, S., Multifunctional carbon nanotubes in water treatment: The present, past and future. *Desalination*, 354, 160–179, 2014.
100. Liu, D., Mao, Y., Ding, L., Carbon nanotubes as antimicrobial agents for water disinfection and pathogen control. *J. Water Health*, 16, 171–180, 2018.
101. Rocha, R.P., Soares, O.S.P.G., Figueiredo, J.L., Pereira, M.F.R., Tuning CNT properties for metal-free environmental catalytic applications. *C-Journal Carbon Res.*, 2, 2–18, 2016. https://doi.org/10.3390/c2030017.
102. Pavlidis, I.V., Tsoufis, T., Enotiadis, A., Gournis, D., Stamatis, H., Functionalized multi-wall carbon nanotubes for lipase immobilization. *Adv. Eng. Mater.*, 2, B179–B183, 2010. https://doi.org/10.1002/adem.200980021.
103. Neupane, S., Patnode, K., Li, H., Baryeh, K., Liu, G., Hu, J., Chen, B., Pan, Y., Yang, Z., Enhancing enzyme immobilization on carbon nanotubes via metal-organic frameworks for large-substrate biocatalysis. *ACS Appl. Mater. Interfaces*, 11, 12133–12141, 2019. https://doi.org/10.1021/acsami.9b01077.
104. Wang, J., Fu, W., He, X., Yang, S., Zhu, W., Catalytic wet air oxidation of phenol with functionalized carbon materials as catalysts: Reaction mechanism and pathway. *J. Environ. Sci.*, 26, 8, 1741–1749, 2014.
105. Mohammadi, A.A., Dehghani, M.H., Mesdaghinia, A., Yaghmaian, K., Es'haghi, Z., Adsorptive removal of endocrine disrupting compounds from aqueous solutions using magnetic multi-wall carbon nanotubes modified with chitosan biopolymer based on response surface methodology: Functionalization, kinetics, and isotherms studies. *Int. J. Biol. Macromol.*, 15, 155, 1019–1029, 2020.
106. Wang, L., Hou, D., Cao, Y., Ok, Y.S., Tack, F.M.G., Rinklebe, J., O'Connor, D., Remediation of mercury contaminated soil, water, and air: A review of emerging materials and innovative technologies. *Environ. Int.*, 134, 105281, 2020.

11

Carbon Allotropes in Waste Decomposition and Management

Swati Sahu*, Gajendra Singh Rathore and Sanjay Tiwari

Photonics Research Laboratory, School of Studies in Electronics & Photonics, Institute of Renewable Energy Technology & Management, Pt. Ravishankar Shukla University, Raipur, India

Abstract

Graphene, activated carbon, and carbon nanotubes (CNTs) are three examples of carbon nanomaterials (CNMs) that hold considerable potential for use in environmental and energy applications. Although there are several methods for producing CNMs, green techniques that utilize mainly energy-saving technology or low-cost basic materials have lately gained appeal. According to recent research, transforming low-valued waste products into high-valued CNMs helps to develop materials processing and recycling technology. With a focus on graphene, activated carbon, and CNTs, this chapter explores the biological synthesis potential for producing carbon allotropes from waste. This chapter makes an effort to give a thorough understanding of every facet of pyrolysis, making sure that all cutting-edge methods are approached from the requirements management elements that affect the final result. One of the most significant applications of pyrolysis is waste treatment. The target products vary depending on the phase and quantity of the chemical precursor, thermal temperature, catalyst activity, and the presence of post-pyrolysis phases during thermal treatment, even though the essential principles of the process are the identical in all cases. Particular significance is given to the carbons produced during breakdown processes in their varied forms and characteristics. Applications for carbon allotropes are also being studied.

***Keywords*:** Carbon nanomaterials, carbon-allotropes, waste materials, pyrolysis, decomposition, recycling, synthesis, graphene, activated carbon, carbon nanotubes (CNTs)

**Corresponding author*: swati.luck05@gmail.com

Chandrabhan Verma and Chaudhery Mustansar Hussain (eds.) Carbon Allotropes and Composites: Materials for Environment Protection and Remediation, (229–256) © 2023 Scrivener Publishing LLC

11.1 Introduction

All significant industrial sectors, including drug distribution, agriculture, and the food sector, have undergone radical change as a result of nanomaterials [1]. Chemical synthesis processes used to make nanomaterials include evaporation, dual emulsion, salting off, emulsifying diffusion, as well as solution displacement/precipitation. However, the manufacture of nanomaterials on an industrial scale has introduced a brand-new type of contamination into the environment. It is vitally desirable to limit the amounts of chemical waste discharged into the environment by developing innovative, practical techniques to overcome the shortcomings of chemical treatments while boosting production and decreasing cost [2].

CNTs, fullerenes, carbon nanofibers, as well as graphene, are some of the most crucial uses for carbon nanostructures [3]. Due to their remarkable uses in aircraft, transportation, and energy, multiwall carbon nanotubes (MWCNTs) and carbon nanofibers (CNFs) have gained prominence [4–6]. Graphite-based CNTs have lengths of several micrometers and nanometer-scale diameters [7]. Because there are many different ways to stack graphene sheets, including platelet, ribbon, and herringbone arrangements, CNFS displays cylindrical morphologies [8]. The two main distinctions between CNF and CNT are the growth structure and hollowness of the materials. Graphene sheets in CNTs must be rolled into tubes that are perpendicular to the growth axis, unlike graphite sheets in CNF, which can be parallel to the development axis. The outstanding mechanical strength, thermal conductivity, and electrical conductivity of these nanostructured materials are just a few of their exceptional qualities [9, 10]. Some industries that could profit from these qualities include construction, detectors, field emission displays (FEDs) like a traditional cathode ray tube (CRT), radiation source materials, conductive and high-strength polymers, battery storage and conversion devices, solar cells, monitoring devices, and direct methanol fuel cells (DMFCs) [11–14]. Using carbon nanostructures, the automobile sector has also employed these materials to produce very powerful engines [15, 16].

11.2 Management Methods for Waste

Plastics come in a variety of varieties, with some of them being more easily recycled than others. However, contaminants, such as grit, papers, paints, additives, metals, foil, insecticides, foreign objects, and mostly oxidized polymers that are easily visible in HDPE and PET bottles recovered from pavements have a major detrimental effect on recycling quality. In particular, plastic containers, infant bottles, and especially materials for medicinal purpose account for more than 40% of all plastic generated being used as single-use packaging, according to a survey. As a result of the excessive usage of plastic items, plastic particles are easily detected in the wind, rainfall, dust, frost, and tap as well as bottled water, salt, coffee, tea, alcohol, and foodstuff. The increased usage of single-use bags, containers, personal hygiene items, and personal protective equipment as a result of the transmission concern and international administrative rules has increased worldwide plastic trash production. Waste plastic management is crucial to dispose of plastics properly or to create new products. Table 11.1 provides some of the main methods for managing waste plastics [17].

Table 11.1 Different solid waste management techniques: benefits and drawbacks.

Waste management technique	Benefits	Drawbacks
Pyrolysis	I. Suitable for thermoplastic waste management II. Lowering greenhouse gas emissions III. Procession of a wide range of feedstock IV. High-energy recuperation	I. The evolution of technology II. Relatively low-cost III. When compared to landfilling, it is somewhat costly IV. Thermoplastic effectively treated
Landfilling	I. Low-cost II. Manage a big amount of garbage III. Shortened transportation distance IV. Nonrecyclable waste	I. Toxic chemical production II. Occupy large volume III. Heavy metal leaching into the environment, soil toxicity IV. Deforestation
Incineration	I. Waste volume minimization II. Combustible toxin destruction III. Energy recovery IV. Smallest proportion of land occupied	I. Costly II. High energy input III. Pollute the atmosphere IV. Not all materials are applicable

11.2.1 Landfilling

Municipal waste management still primarily involves landfilling. Landfills are one of the most popular methods of rubbish disposal despite having a very significant ability to harm the ecosystem. Frequent landfill monitoring is required to detect and describe environmental dangers associated with landfills. The most conventional technique of trash disposal in most countries is landfilling. But by greatly slowing down the microbial degradation of plastic over a protracted time, the presence of additives like stabilizers and antioxidants encourages the disposal of plastic polymers in landfills. Landfilling does not appear to be an acceptable option for handling waste plastic because of the long-term hardness of plastics, their sluggish kinetics for biodegradation, the finite number of landfill sites that are accessible, the total employment of beginning materials, and health hazards resulting from groundwater pollution by leeching and inclusion of dangerous compounds as well as their propensity to seep from landfills. Therefore, its use needs to be limited [18–20].

11.2.2 Incineration

Waste disposal at a high enough temperature is the procedure of incinerating (rapid oxidation). Calcination or controlled-flame combustion are other names for the process, which removes organic components from recyclable materials. New methods for producing energy are being investigated for this process of combustion. Additionally, incineration is regarded as a volume-reducer to lower the expense of landfilling. Industrial furnaces and boilers are two often used incineration techniques. With the aid of thermal energy, the industrial furnace is one type of technology used to treat hazardous waste. Boilers are included, along with concrete, limestone, aggregate, phosphate, coke burners, blast furnaces, and metallurgical furnaces. These organic compounds exhibit resistance to environmental degradation via ecological, chemical, and photochemical mechanisms [21].

11.2.3 Mechanical Recycling

Recycling old polymers is a more efficient waste management option than landfilling or incineration. Mechanical recycling is required for the circular economy to function. Metals, lithium-ion batteries, and polymers are just a few of the materials that can be employed with them. Materials that have outlived their lifespan can be processed by collecting, sorting, shredding, melting, and transforming them into recycled products for a new process. The product, which was created from recycled resources, enters its new usage phase before being disposed of. Upcycling and downcycling plastic wastes are the two types of mechanical recycling [22–25].

11.2.3.1 Downcycling Method

Downcycling is used for less usable polymers and is of poorer quality than the original material. Plastics lose some fundamental attributes, such as optical clarity, mechanical integrity, and other traits that make them suitable for their original applications in most cases, therefore unlike some of the other substances such as metal and glassware, biodegradable polymers cannot continue to remain in place after recycling. Grouping, isolating, cleaning, washing, freeze drying, chipping/sizing, filling, coloring, accumulating, and pelletization are some of the physical procedures employed in this process [26, 27].

11.2.3.2 Upcycling Method

Upcycling, often referred to as innovative reuse, is the process of using waste plastic to create new materials or higher-quality products for aesthetic or environmental purposes. The approach involves the chemical recycling of both single and mixed waste plastics. The key advantage of this technology is its simplicity and low cost. The fact that many solid waste plastics cannot be recycled because of impurities is a drawback of this technique [28, 29].

11.3 Process of Pyrolysis: Waste Management to the Synthesis of Carbon Allotropes

Pyrolysis is the heat breakdown of organic waste in the absence of oxygen to produce a complex variety of molecular structures that yield useful end-products including condensable oily fluids, explosive gases, and carbon-containing char remnants. Due to the endothermic nature of the process, an external heat source is required. This may frequently be produced since the lightweight gaseous proportion generally has a relatively high-calorie content to give the heat essential for the pyrolysis technique and, if necessary, the waste being pre-dried. Processing conditions can include self-generated inert gas, vapors, or vacuum, with temperatures generally ranging from 400°C to 900°C. The deteriorating process has the possibility of being advantageous in all aspects, and the processing parameters may be altered to increase the output of a specific percentage. The liquid portion may be processed to produce a range of compounds for use whether as refined products or as a chemical feedstock, while the char can be burnt as a solid fuel, converted into carbon black, or converted to activated carbon [30].

Only pyrolysis, out of all the existing waste treatment methods, offers a practical way to recover usable products from the entire waste stream without first separating the organic components. All organic waste, comprising household waste, garden garbage, paper, and polymers, may be processed in tandem. It should be noted that the oiled component is abundant in compounds that include oxygen and that the liquid output of biomass pyrolysis involves significant amounts of water, both of which may lower the product's value as a biofuel. Additionally, the greasy result typically has a high acidity level and is unstable chemically when stored [31]. As a result, to maximize the output of usable compounds, extensive further processing is needed.

Carbon is produced mostly by synthetic methods, unlike other raw elements for manufacturing like metals. In reality, some carbon allotropes, such as Gas chromatographs, are completely artificial. Pyrolysis, which includes burning a different organic predecessor to extreme temps in an inert atmosphere, sometimes with the aid of a catalyst, is the foundation of the bulk of carbon generation methods. This causes the precursor to thermally decompose, releasing non-carbon atoms in a variety of shapes. Due to carbon's high thermal stability, some solid carbon is always created as a remnant or in the form of smoke (in the absence of oxidants). This reactive solid carbon may adopt a range of microstructural topologies depending on the type of precursor, how it decomposes, the amount of pressure used, the pathways by which the larger carbon compounds form, and the high thermal stability of the reaction products [32, 33]. For many years, both crystalline and disordered carbons have been produced using this method [34, 35]. The pyrolysis procedure and precursors have been improved with the development of nanotechnology to produce one- and two-dimensional carbon structures on the nanoscale. Pyrolysis is the basis for the modern chemical vapor deposition (CVD) process used to create carbon nanomaterials [36, 37]. It's fascinating that the major attraction of carbon in nanotechnology is how easily different carbon nanoforms may be applied to a range of substrates [38]. Pyrolysis process optimization must be prioritized by everyone involved in the development of carbon materials and associated technologies. Waste tyres may be pyrolysis and

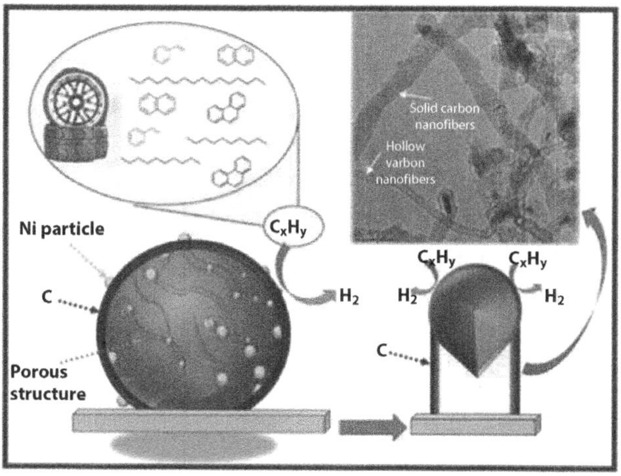

Figure 11.1 Schematic presentation of tyre pyrolysis oil design compound structure for the development of carbon nanoparticles [39] (reproduced with permission).

catalyzed to create valuable carbon nanomaterials, according to Zhang *et al.* (Figure 11.1) [39].

11.4 Synthesis Methods to Produce Carbon-Based Materials From Waste Materials

CNTs, graphite, graphene, fullerene, and AC are among the carbon allotropes in this category (AC). CNTs are of great interest to academics due to their unique properties and potential for a variety of applications. The tensile strength of CNTs, a kind of nanomaterial, is more than 100 times stronger than that of hardened steel, and they also exhibit exceptional electrical properties. They are materials that are both physically and chemically stable. CNTs are commonly employed in composite materials as conductive polymers or to boost durability. Additionally, they are employed in medical instruments, conductive coatings paint semiconductors and microelectronic transistors, energy storage, and nanosensors for ecological usage [40–43].

11.4.1 Catalytic Pyrolysis

There are three primary processes used to make CNTs: arc discharge, pulsed laser deposition, and CVD (Table 11.2). CNTs are nevertheless produced

Table 11.2 Synthesis methods for carbon-based materials comparison.

Synthetic method	Advantages	Disadvantages
Pyrolysis catalyzed	I. Small energy and temperature demands II. Product composition under control	I. Catalyst is expensive II. The product has a low yield
CVD	I. Rapid expansion II. Product yield is high III Better option for developing epitaxial thin films	I. Difficult process II. Demand for extreme temperatures
Pyrolysis deposition CVD	I. Simple II. Plastic re-use with less solvent III. Highly porous with order	I. Structure damage occurs during the template etching method
Thermochemical decay	I. Wide surface area, adjustable pore diameter, and resistivity	I. Increased temperature results in a large pore volume, a low energy density, and poor conductivity.
Cross-linking in stages	I. An effective method for producing a cross-linked structure that is thermally stable.	I. Heating causes some structural degradation II. Moderate porosity
Thermochemical transformation	I. Resulted in increasing demand for conversion II. Large surface area III. Environmentally friendly products created by completely converting organic components into carbon-based products.	I. Morphology without control
Activation method	I. Low-temperature one-pot method II. Structure that is well-organized and has a large surface area	I. Reasonably low purity II. Washing requires a big volume of water.

(Continued)

Table 11.2 Synthesis methods for carbon-based materials comparison. (*Continued*)

Synthetic method	Advantages	Disadvantages
Template-based technique	I. Well-organized Structure	I. Structure collapse or disorder II. Clogging of pores while etching
One-pot synthesis	I. The method is straightforward, affordable, and repeatable.	I. The amount of polymer and the type of reactor
Hydrothermal carbonization	I. Straightforward method for obtaining important compounds with an oxygen functionality	I. Costly and underdeveloped porosity

from solid plastic wastes in a variety of ways. In order to produce CNTs, for illustration, a two-step pyrolysis process employing nickel sheet catalysts and polyethylene as a source of energy is used, with the reactor's quartz tube being maintained at a moderate temperature. Carbon nanofibers and defect-rich CNTs with thicknesses of between 10 nm to 20 nm were created using this procedure. The low temperature and energy requirements of this method, together with the controlled product composition, are its key benefits. On the other hand, the high catalyst cost and low product yield are its main drawbacks [44].

11.4.2 Batch Pyrolysis-Catalysis

A combined, fixed bedding, batch decomposition, and catalysis processor method is a further option for generating CNTs through catalytic pyrolysis of disposed of polypropylene (PP) in the existence of nickel catalyst at temperatures exceeding 600°C. This technology simplifies the fabrication of MWCNTs with thicknesses of between 10 nm to 25 nm with lengths of about 100 m [45].

11.4.3 CVD Method

CNTs may also be created using the pyrolysis approach in conjunction with the chemical vapor deposition technique. Two successive phases make up the entire conversion of plastic waste into CNTs. Waste plastic is transformed into volatile vapors in the first stage in a moderately heated, non-oxygenated environment. Using the CVD process, these vapors are converted into CNTs in a second step at increased pressure (1 MPa) but instead at temperature with the aid of a catalyst made of nickel. The rapid growth rate, higher product yield, and better alternatives for epitaxial thin film development are benefits of CVD. The drawback of this technology is a challenging procedure that requires high temperatures [46].

Single-layer graphene was also executed utilizing combined plastic pollution-padded envelopes that contained PS and PE. The hydrocarbon-rich gases generated in the reactor at around 500 °C are fed into the CVD system running at about 1020 °C in an Ar/H_2 atmosphere on a copper layer. A SiO_2/Si substrate with a PMMA sheet on top of monolayer graphene produces crystals with hexagonal shapes [47].

11.4.4 Pyrolysis-Deposition Followed by CVD

Another technique for generating mesoporous carbon combines the CVD procedure with the pyrolysis coating approach. Waste polystyrene is produced during the *in-situ* pyrolysis of polystyrene when a $Fe(NO_3)_3$ catalyst and a tough SBA-15 template are present. In the following stage, a combination of Iron SBA-15 and PS was pyrolyzed in a tube furnace. The small organic gases generated during pyrolysis were able to be absorbed, coated, and transformed into carbon-silica composites owing to the capillary effect of the Fe-SBA-15 microfluidic devices. Following the peeling of Iron and silica, nanostructured carbons (OMCs) with the same shape as SBA-15 were created. This method is highly recommended since it yields an ordered arrangement with acceptable porous, is simple, quick to use, requires little solvent, and recycles waste plastic [48].

11.4.5 Thermal Decomposition

To generate graphene from PET derived from used bottles, a simple, repeatable, and affordable thermal decomposition method was adopted. Using a conventional sieve shaker, raw PET was broken into the desired size particles, which ranged from 1 to 3 mm, in this process. The used PET was then transferred to a stainless steel sterilizer and pyrolyzed in a tubular furnace at 800°C for one hour. After that, graphene was synthesized by overnight cooling the product. This approach is suggested when a material's surface area, electrical properties, and pore diameter must all be controlled. Its employment under high-temperature conditions is nonetheless constrained by the high porosity volume at high temperatures that result in low energy density and conductivity. To generate graphene from PET derived from used bottles, a simple, repeatable, and affordable thermal decomposition method was adopted. Using a conventional sieve shaker, raw PET was broken into the desired size particles, which ranged from 1 to 3 mm, in this process [49].

11.4.6 Activation Techniques

Industrial raw materials are transformed into active chemicals through the activation process. Industrial waste is frequently converted into carbon-based compounds by the method of pyrolysis, which takes place at extreme temperatures in the absence of oxygen. Carbonaceous materials are also created through the activation process, which incorporates pyrolysis followed by material activation. This method is also well-known for being a one-pot operation with minimal temperature requirements, an arranged structure with adequate porosity, and a sizable surface area, but the product quality is poor and a lot of water is needed for cleaning [50].

11.4.6.1 Physical Activation Technique

After the pyrolysis procedure, which is performed at extremely high temperatures of around 1200°C in the presence of either air, vapor, carbon dioxide, or a mixture of those three, physical activation frequently occurs right away. The activating agents used in the aforementioned process include air, CO_2, and steam. Because it is cheap and doesn't require further processing to get rid of waste materials, steam is the most used activation agent for the conversion process. During the activation procedure, steam is often mixed with pyrolysis, and the oxygen-functionality-rich surface produced during the reaction enhances the hydrophilicity and electronic conductivity of the carbonaceous materials produced. The two phases that make up the physical activation process are as follows: (a) The first phase involves heating the carbon source to a temperature between 400 and 900°C in an inert environment to remove bulky volatile components; (b) Partially gasifying material at a temperature of between 400 and 900 °C while utilizing an oxidant constitutes the second stage. The oxidizing agents' active oxygen groups burn away the pyrolysis byproducts that have become lodged in the pores, which causes the clogged pores to open. Additionally, the microporous structure is formed as a result of the burning away of the carbon application's more reactive portions, which produce CO and CO_2. The activation temperature and gas utilized affect how much annealing occurs [51–53].

11.4.6.2 Chemical Activation Technique

When employing the one-pot chemical activation method, an activator such as a strong base (KOH, NaOH), an acid (H_3PO_4, H_2SO_4), or a salt (K_2CO_3, $ZnCl_2$, $FeCl_3$) is injected into the pores of the carbon precursor prior to pyrolysis and then heated to between 450 and 900 °C during annealing. At the laboratory level, it provides good pore size distribution control. However, handling contaminated water and the requirement for a lot of water to wash away the pollutants created during the activation process are downsides of chemical activation procedures. Due to its quick activation, low activation temperature requirement, high capacity, distinctive mesopore size distribution, and significant specific surface area of the resulting porous carbon structure with exceptional charge storage capacity, KOH is the alkali that is most frequently used to activate factory waste products. The molar ratio of the precursor material to the potassium hydroxide, the heating rate (2-10°C/min), the activation temperature, and the activation time are the four key regulating variables for the KOH activation process (1-4 hours). Since excessive activation creates big pores, which are to blame for the AC's high resistivity, low energy consumption, and reduced power capacity, time is the most crucial consideration. To create high-performance supercapacitors, porous carbons generated by the KOH activation process must have a superbly tailored nanoporous shape and surface quality chemistry. Researchers have worked hard to create customized nanostructured carbon materials for use in energy storage systems. For instance, the results of a thorough analysis of the research on the KOH activation of carbonaceous materials show that (1) large micropores are produced by activated carbon at high temperatures and a higher carbon/KOH ratio, (2) porosity and linked porous networks with distinctive textural properties are amplified by the incorporation of micropores and the presence of tiny porous structure in the carbon nanostructured framework, and (3) the dimension of the pores is amplified by the incorporation of micropores and the Additionally, due to the greater catalytic impact, the inclusion of heteroatoms via precursors rich in oxygen and nitrogen boosts the specific capacitance [54–58].

11.5 Use of Waste Materials for the Development of Carbon Allotropes

11.5.1 Synthesis of CNTs Using Waste Materials

CNTs were created using waste polymers in a variety of reactor systems, including quartz tubes, autoclaves, crucibles, and muffle furnaces [59].

Pyrolysis will be used to transform the large solid polymer into little carbon precursors, which is their common objective [60]. The bulk of waste plastics is composed of olefin monomer (C_nH_{2n}), which is the source of polyolefin, a kind of polymer with around 85.7 weight percent carbon. Polypropylene (PP) and polyethylene (PE), two waste polyolefin prototypes, have frequently been utilized in green synthesis processes for carbon materials. CNTs were effectively synthesized using plastics that contained elements other than C and H, including polyvinyl alcohol (PVA) and polyethylene terephthalate (PET). Gong *et al.* developed a unique synthesis method that involved beginning with waste PP and using activated carbon (AC) and Ni_2O_3 as catalysts in a quartz tube reactor. This synthesis method was given the moniker "layer-by-layer assembly". It was discovered that AC helped convert PP into light hydrocarbons and that it might work in synergy with Ni to catalyze the dehydrogenation and aromatization necessary to produce aromatic molecules [61]. In an open reaction system, Tang *et al.* showed how to create MWNTs from PP. The use of organically modified clay (OMC) caused the top charred ceramic layer to produce a winding path as a result of combustion. To prevent the leakage of carbon-gaseous precursors, a mimic micro-autoclave was built to give pyrolytic byproducts and catalytic Ni particles a place to interact. MWNTs were created from the pyrolytic products following their dehydrogenation and aromatization. When 10 wt % OMC is added, the maximum output of MWNTs (around 41.16 wt %) may be produced. It was established that this technique worked with different polymer materials [62].

Vegetable oils from plants, such as those with a high carbon content like coconut, neem, turpentine, eucalyptus, and palm, can be used to make extremely pure (90%) CNTs. Palm oil ($C_{55}H_{100}O_6$) stands out among them in delivering CNT with a notable characteristic. Due to the reduced production of amorphous carbon due to its low H concentration [63]. Vertically aligned CNTs (VACNTs) were created by Suriani *et al.* using used palm oil for frying. Two alumina tube furnaces were used to create the float-catalytic CVD apparatus, and the starting ingredients were filtered palm oil that had been thoroughly combined with ferrocene (5.33 wt%). One tube furnace's center was filled with the cleared silicon wafer, while another's center was used to deposit the composition into an alumina boat. The former tube was preheated to 450°C for half an hour after the latter tube had been preheated to 750°C in an Argon atmosphere in order to sufficiently break down waste palm oil into hydrocarbon molecules. In addition to promoting CNT development, ferrocene catalyzed the breakdown of palm oil [64]. The source material for Suriani *et al.*'s synthesis of quasi-aligned CNTs was waste engine oil (WEO), which has a substantially greater carbon content than other lubricants. The CVD setup was still employed in this circumstance, but the preparation and synthesis temperature changed from 500 to 750°C, respectively, and the ferrocene content was 17.99 wt%. It has been found that the oxygen concentration of WEO has a positive impact on enhancing the crystallinity of CNTs due to the ability of the oxygen content to supply hydroxyl radicals, which form the basis of the stable sp^2 structure of CNTs [65]. Mishra *et al.* developed CNTs from recycled PP using a CVD system (Figure 11.2) [45].

Chen *et al.* investigate the chemical recyclable materials of polyurethane (PU) as well as ethylene-vinyl acetate (EVA), two kinds of shoe waste polymers, using pyrolysis-chemical vapor deposition over a Fe/MgO catalyst. The results showed that although EVA tended to make CNTs with multiple walls, PU allowed for the highly selective manufacturing of few-walled CNTs with small diameters and a constrained size range [66]. Using a pyrolysis-catalysis technique, Yao *et al.* developed carbon nanotubes from post-consumer mixed waste plastics. Ni-Fe bimetals supported by four separate porous materials were used as catalysts. The Ni-Fe/MCM41 catalyst

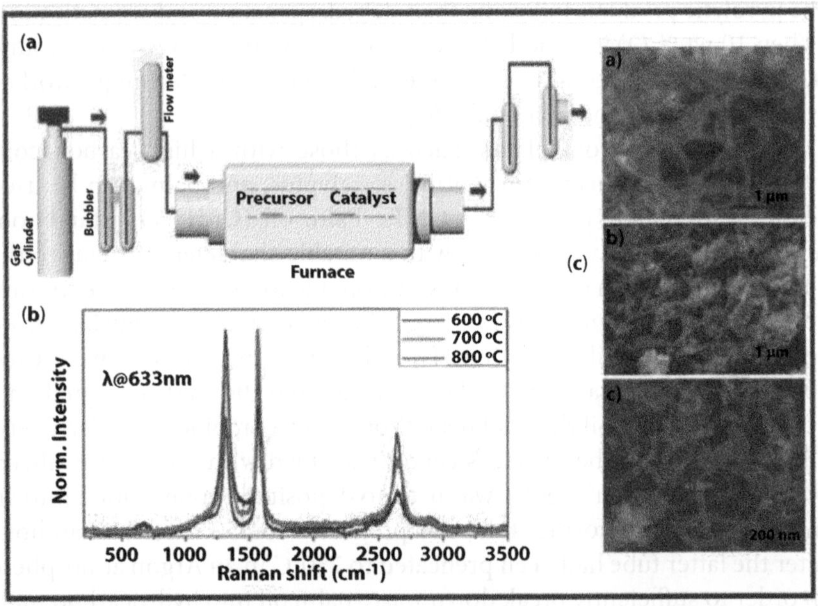

Figure 11.2 (a) Diagram of the unit used for CVD CNT synthesis. (b) Raman spectrum of MWCNTs synthesized. (c) SEM image of the MWCNTs synthesized [45] (reproduced with permission).

showed the highest catalytic performance for the pyrolysis-catalysis of plastic products, with a maximum output of carbon material of 55.60 wt% [67].

11.5.2 Synthesis of Graphene Using Waste Materials

It has been demonstrated that the degradation of plastics can result in the production of light hydrocarbons and aromatics. Different techniques were used to create graphene from different types of scrap plastic. Graphene flakes (GFs) with a high yield were produced by Gong *et al.* using waste PP as the catalyst and organically modified montmorillonite (OMMT) as the catalyzer (Figure 11.3). According to the original article's instructions, a crucible was filled with a homogenous combination of PP (about 89 wt%), talcum (about 11 wt%), and OMMT. The carbonized char was created by heating the combination to 750°C for 20 minutes, cooling it, and then immersing it in HF and HNO_3. HNO_3 oxidized amorphous carbon, whereas HF dissolved impurities (like MMT and powder additives). GFs were formed after successive centrifugation and isolation from the mixture [68].

Sharma *et al.* used solid waste polymers that were around 86% PE and 14% PS to undertake another CVD-based graphene production experiment (Figure 11.4). There are two furnaces in the CVD the authors utilized. During the first furnace, which had a temperature of around 500°C, plastic fragments (3 mg) were put in a ceramic boat, and in the second, which had a heat of roughly 1020°C, the copper foil was used as a substrate. Following pyrolysis, the carbonaceous byproducts were transported into the following furnace by a gas mixture (Ar/H_2: 98/2.5 sccm), where they interacted with the copper substrate to promote the formation of graphene. According to the authors, a key factor in the creation of graphene crystals

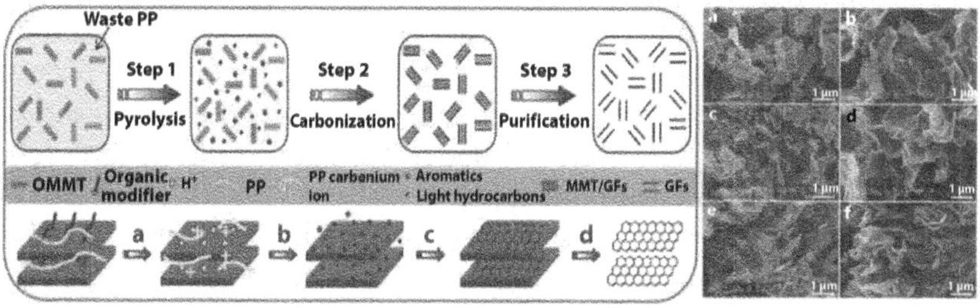

Figure 11.3 Procedures for converting waste PP into GFs and typical TEM images of GFs-B [68] (reproduced with permission).

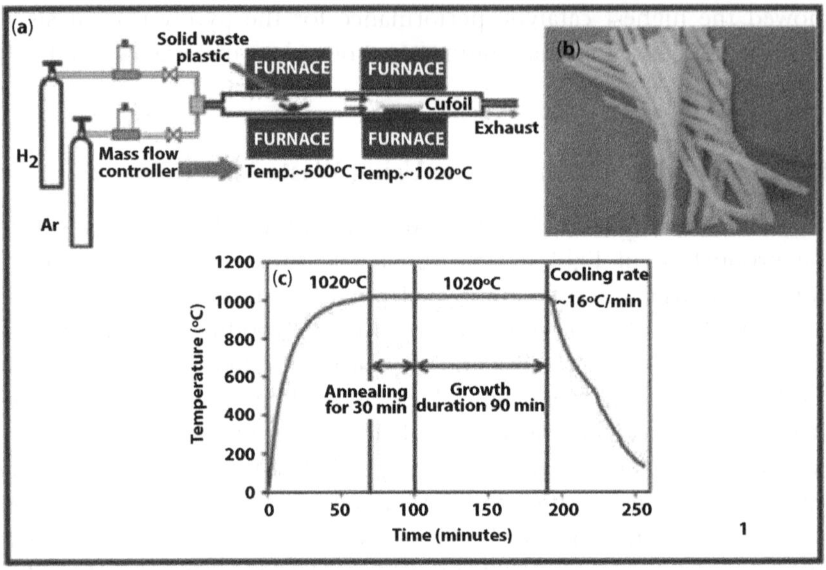

Figure 11.4 Two furnace-CVDs for big graphene crystal synthesis; (b) solid PE-PS wastes [69] (reproduced with permission).

is the pace at which degraded products are injected. Higher injection rates led to the creation of bilayer or even multilayer graphene, however, massive hexagonal-shaped single-layered graphene may develop at lower pyrolysis and injection rates as well [69].

Biomass wastes, which are by-products or residues of plant photosynthesis, might include wastewater deposits, ligneous plants, and agricultural waste. Biomass wastes have promise for usage as carbon sources for the production of lucrative CNMs, in addition to being valuable for renewable energy sources like biofuels. For example, Sun *et al.* used the waste biomass of coconut husks to make porous graphene nanosheets. In their process, coconut husk (3g), $ZnCl_2$, and the Ferric chloride solution (50mL, 3M, catalytic precursor) were mixed (9g, activating agent). After the mixed solution had been dried at about 100°C and evaporated at about 80°C for two hours, carbon precursor was collected. To produce graphene nanosheets, the carbon precursor was heated to about 900°C for one hour in a tube furnace with N_2 as the environment [70, 71].

To create GO and rGO, Akhavan *et al.* substituted affordable carbonaceous waste for expensive HPG (highly pyrolytic graphite) [72]. The processes described involved burning raw resources wrapped in aluminum foil at 350 to 500°C for a few days in an environment with little oxygen, followed by grinding them into fine powders and heating them at 450°C for a further day before dissolving them in an aqueous solution of Ferric chloride solution and Hydrochloride. The modified Hummers method was utilized to synthesize GO from the final graphitized product. Adolfsson *et al.* have presented a method for synthesizing graphene oxide quantum dots using scrap paper. In their research, H_2SO_4 was used as a catalytic agent to hydrothermally break down the high concentration of cellulose obtained from the waste paper at 160°C in microwave apparatus. The carbon nanospheres were heated and sonicated in a 70% Ammonium hydroxide solution to create graphene oxide quantum dots [73].

With the usage of sugarcane jute waste, Fehling's solution, and a novel one-pot greener method, ternary nanocomposites (TNCs) made of Au-Cu_2O attached on graphene oxide (Au-Cu_2O/RGO) were created. Toxic substances were omitted from the reagents and surfactants used in the process. In addition to the moderate reduction capacity of the polysaccharides in sugarcane jute waste, it also promotes the functionalization of RGO, limiting aggregation of the decreased layers and allowing *in situ* Cu_2O stabilization. The most recent study emphasizes a novel, environmentally friendly, and safe method for creating nanocomposites (NCs), as well as the study of decorative metal oxide as a nanocatalyst with excellent electrocatalytic potential; in comparison to unanchored metal oxide nanoparticles (NPs), the anchored NCs shown superior improvement and flexibility of recuperation [74].

11.6 Applications for Carbon-Based Materials

The three different carbon-based compounds that are commonly used include carbon nanotubes (CNTs), graphene, and activated carbon are presented in Table 11.3.

11.6.1 CNTs

Researchers have extensively studied carbon-based materials called carbon nanotubes, which already have outstanding electrical conductivity and low device resistance. The two key elements that can alter the material's specific capacitance are its shape and the purity of CNTs. Based on their structural alignment, CNTs may be split into two groups: (a) multi-walled carbon nanotubes (MWCNTs), and (b) single-walled carbon nanotubes (SWCNTs). The second is made up of out-of-phase sp^2 hybridized carbon atoms and cylindrical graphene sheets that are in phase. In accordance with

Table 11.3 Carbon-based compounds that are commonly used include carbon nanotubes (CNTs), grapheme, and activated carbon.

Carbon-based material	Advantages	Disadvantages	Appearance
Carbon nanotubes	I. Large surface area (400>1600 m^2/g) II. Good electrical and thermal conductivity and low ESR III. Fast charge-discharge time IV. Good ion mobility V. Low mass per unit volume	I. Comparatively low surface area	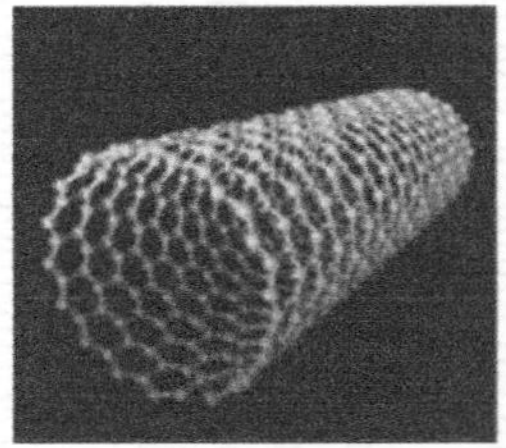
Graphene	I. High surface area (>3600m^2/g) II. Good mechanical strength III. Superior electrical and thermal conductivity	I. Surface area is relatively small	
Activated Carbon	I. High specific surface area (>3600m^2/g) II. Large pore volume III. Competitively priced IV. Capacitance values are high in both organic and aqueous electrolytes	I. Less Porous pathways II. Organic electrolytes with few electrolyte ions	

conductivity, structural patterns, and chirality, there are other SWCNT types that include metallic, semimetallic, and semiconducting SWCNTs. Due to the codimension arrangement of the SWCNTs, MWCNTs have a significantly larger diameter than SWCNTs. SWCNTs' electrochemical performance was significantly influenced by the electrolyte's ionic size and accessibility, whereas MWCNTs' mesoporous structure accelerated ion transport across the interface region [75]. The stronger carbon–carbon bond in CNTs is what gives them their tensile strength and molecular stiffness. However, their comparatively small surface area, which hampered their progress in this application, may be removed by either chemical or electrochemical processes, or by creating arrays or sponge-like foams containing porous holes that are larger than those seen in CNT films and result in increased capacitance [76–78].

11.6.2 Graphene

Graphene is the single-layer form of graphite and the thinnest form of carbon. It possesses the highest specific surface area, thermal conductivity, electrical conductivity, specific surface area, greatest thermal conductivity, greatest electrical conductivity, and best tensile strength because of a carbon-carbon bond linkage, highest high stability, and maximum carrier mobility. A graphene is a possible option for usage in a variety of applications due to all these characteristics. Graphene is a better option for an electrode material for supercapacitors than CNTs and ACs since the distribution of pores in the solid state does not significantly affect the electrodes' function. However, to employ graphene for technological applications, graphene sheets must be molecularly functionalized into a variety of electrical components [79, 80].

11.6.3 Activated Carbon

Activated carbon (AC) is the most commonly utilized carbon-based content used as an electrode in capacitances due to its large area of the specific surface (>1000 m^2/g), high porosity volume, relatively inexpensive, and adequate electrical characteristics with surface charge density in organic and water-based electrolytes of 120 F/g and >200 F/g, in both (range is 115–340 F/g). The vast variety of physicochemical characteristics of activated carbon governs its surface area, high porosity, and distribution of pore sizes. These characteristics depend on the carbon precursors used and the activation processes. A possible method for increasing the material's capacitance is to increase the specific surface area, but only up to a certain point. According to Wang *et al.*, the specific surface area expanded from 621 to 2685 m^2/g, and the capacitance jumped from 17.68 to 171.2 F/g. Small nanopores (0.5 nm) and a lack of permeable channels in the ACs created by the chemical activation of waste material (carbon precursors) limit the ion transport of the electrolyte, especially in organic electrolytes, and reduce capacitance. Well-ordered nanoporous carbons, a class of carbon-based materials having highly porous structure (2–50 nm in diameter), are developed as a solution to this problem. They demonstrate outstanding electrochemical activity during high power density, but their small physical properties decrease susceptibility [81, 82].

The following are some of the tactics used to improve the electrochemical responses of ACs: (i) oxidation of the activated carbon surface, (ii) inclusion of heterocyclic rings (S, O, and N) in the porous structure of activated carbon, (iii) a change in the ACs' Fermi level location caused by ultrasonic radiation, (iv) AC composition synthesis using additional carbon-based nanostructures (such as CNTs, CNF, etc.), and (v) using polymers to create a composite electrode [83–86].

11.7 Conclusions

CNMs can be employed not just in industrialization, but also in the remediation of environmental pollutants. As the use of CNMs increased, people began to investigate their toxicity. Researchers have obtained a better understanding of CNM toxicity in recent years. It can harm the ecosystem and endanger both animals and people. CNMs can combine with other substances in the environment, and investigation of the toxicity of conjugates in CNM toxicology studies has been restricted. More investigation and assessment of CNM migration in species and the environment is needed.

One important discovery is that practically all synthetic carbon molecules, whether large or nanoscale, are created by catalytic pyrolysis from organic sources. It is challenging to reduce the amount of energy used during pyrolysis, but it is possible with the application of specialized nanoscale catalysts, which might result in total cost reductions. One issue is the need for post-processing to eliminate the nanoparticles with a good dividend, which calls for a more focused investigation. In general, catalysts can aid in increasing process efficiency.

In spite of the considerable advancements achieved in the manufacture, composited treatment, and applications of low-dimensional nanostructured materials, such as carbon nanotubes, carbon nanofibers, and their derivatives in the removal of pollutants, issues, and opportunities in practical implementation continue to be created and addressed. Several of the technologies covered in this chapter have started in situ tests or are getting ready for commercialization, even if many of them are still in the lab. Adsorption or a system that combines adsorption and catalysis would be the most advantageous option for application on a large scale based on the cost, reliability, and maneuverability of nanostructured materials.

References

1. Adams, W.W. and Baughman, R.H., Richard E. Smalley (1943–2005). *Science*, 310, 5756, 1916–1916, 2005.
2. Nasrollahzadeh, M. and Sajadi, S.M., Green synthesis of copper nanoparticles using Ginkgo biloba L. leaf extract and their catalytic activity for the huisgen [3+ 2] cycloaddition of azides and alkynes at room temperature. *J. Colloid Interface Sci.*, 457, 141–147, 2015.
3. Carrola, J., Bastos, V., de Oliveira, J.M.P.F., Oliveira, H., Santos, C., Gil, A.M., Duarte, I.F., Insights into the impact of silver nanoparticles on human keratinocytes metabolism through NMR metabolomics. *Arch. Biochem. Biophys.*, 589, 53–61, 2016.
4. Lee, C.J., Lee, T.J., Park, J., Carbon nanofibers grown on sodalime glass at 500 C using thermal chemical vapor deposition. *Chem. Phys. Lett.*, 340, 5-6, 413–418, 2001.
5. Omoriyekomwan, J.E., Tahmasebi, A., Zhang, J., Yu, J., Formation of hollow carbon nanofibers on bio-char during microwave pyrolysis of palm kernel shell. *Energ. Convers. Manage.*, 148, 583–592, 2017.
6. De Volder, M.F., Tawfick, S.H., Baughman, R.H., Hart, A.J., Carbon nanotubes: Present and future commercial applications. *Science*, 339, 6119, 535–539, 2013.
7. He, H., Pham-Huy, L.A., Dramou, P., Xiao, D., Zuo, P., Pham-Huy, C., Carbon nanotubes: Applications in pharmacy and medicine. *BioMed. Res. Int.*, 2013, 1–12, 2013.
8. Serp, P., Corrias, M., Kalck, P., Carbon nanotubes and nanofibers in catalysis. *Appl. Catal. A Gen.*, 253, 2, 337–358, 2003.
9. Zhang, Y., Bai, Y., Yan, B., Functionalized carbon nanotubes for potential medicinal applications. *Drug Discovery Today*, 15, 11-12, 428–435, 2010.
10. Lee, H.M., Kwac, L.K., An, K.H., Park, S.J., Kim, B.J., Electrochemical behavior of pitch-based activated carbon fibers for electrochemical capacitors. *Energ. Convers. Manage.*, 125, 347–352, 2016.
11. Dai, H., Hafner, J.H., Rinzler, A.G., Colbert, D.T., Smalley, R.E., Nanotubes as nanoprobes in scanning probe microscopy. *Nature*, 384, 6605, 147–150, 1996.

12. Ahn, C.C., Ye, Y., Ratnakumar, B.V., Witham, C., Bowman Jr., R.C., Fultz, B., Hydrogen desorption and adsorption measurements on graphite nanofibers. *Appl. Phys. Lett.*, 73, 23, 3378–3380, 1998.
13. Bera, R.K., Mhaisalkar, S.G., Mandler, D., Magdassi, S., Formation and performance of highly absorbing solar thermal coating based on carbon nanotubes and boehmite. *Energ. Convers. Manage.*, 120, 287–293, 2016.
14. Zainoodin, A.M., Kamarudin, S.K., Masdar, M.S., Daud, W.R.W., Mohamad, A.B., Sahari, J., Optimization of a porous carbon nanofiber layer for the membrane electrode assembly in DMFC. *Energ. Convers. Manage.*, 101, 525–531, 2015.
15. Amiri, A., Sadri, R., Shanbedi, M., Ahmadi, G., Kazi, S.N., Chew, B.T., Zubir, M.N.M., Synthesis of ethylene glycol-treated graphene nanoplatelets with one-pot, microwave-assisted functionalization for use as a high performance engine coolant. *Energ. Convers. Manage.*, 101, 767–777, 2015.
16. Karaipekli, A., Biçer, A., Sarı, A., Tyagi, V.V., Thermal characteristics of expanded perlite/paraffin composite phase change material with enhanced thermal conductivity using carbon nanotubes. *Energ. Convers. Manage.*, 134, 373–381, 2017.
17. Jambeck, J.R., Geyer, R., Wilcox, C., Siegler, T.R., Perryman, M., Andrady, A., Narayan, R., Law, K.L., Plastic waste inputs from land into the ocean. *Science*, 347, 6223, 768–771, 2015.
18. Cózar, A., Echevarría, F., González-Gordillo, J.I., Irigoien, X., Úbeda, B., Hernández-León, S., Palma, Á.T., Navarro, S., García-de-Lomas, J., Ruiz, A., Fernández-de-Puelles, M.L., Plastic debris in the open ocean. *Proc. Natl. Acad. Sci. U. S. A.*, 111, 28, 10239–10244, 2014.
19. Lusher, A.L., Mchugh, M., Thompson, R.C., Occurrence of microplastics in the gastrointestinal tract of pelagic and demersal fish from the english channel. *Mar. Pollut. Bull.*, 67, 1-2, 94–99, 2013.
20. Lithner, D., Larsson, Å., Dave, G., Environmental and health hazard ranking and assessment of plastic polymers based on chemical composition. *Sci. Total Environ.*, 409, 18, 3309–3324, 2011.
21. Valavanidis, A., Iliopoulos, N., Gotsis, G., Fiotakis, K., Persistent free radicals, heavy metals and PAHs generated in particulate soot emissions and residue ash from controlled combustion of common types of plastic. *J. Hazard. Mater.*, 156, 1–3, 277–284, 2008.
22. Okan, M., Aydin, H.M., Barsbay, M., Current approaches to waste polymer utilization and minimization: A review. *J. Chem. Technol. Biotechnol.*, 94, 1, 8–21, 2019.
23. Hopewell, J., Dvorak, R., Kosior, E., Plastics recycling: Challenges and opportunities. *Philos. Trans. R. Soc. B Biol. Sci.*, 364, 1526, 2115–2126, 2009.
24. Brems, A., Baeyens, J., Dewil, R., Recycling and recovery of post-consumer plastic solid waste in a European context. *Therm. Sci.*, 16, 3, 669–685, 2012.
25. Munir, D., Irfan, M.F., Usman, M.R., Hydrocracking of virgin and waste plastics: A detailed review. *Renew. Sust. Energ. Rev.*, 90, 490–515, 2018.

26. Patni, N., Shah, P., Agarwal, S., Singhal, P., Alternate strategies for conversion of waste plastic to fuels. *Int. Sch. Res. Notices*, 2013, 1–7, 2013.
27. Karayıldırım, T., Yanik, J., Ucar, S., Sağlam, M., Yüksel, M., Conversion of plastics/HVGO mixtures to fuels by two-step processing. *Fuel Process. Technol.*, 73, 1, 23–35, 2001.
28. Butkutė, R. and Miknius, L., Properties of residual marine fuel produced by thermolysis from polypropylene waste. *Mater. Sci.*, 21, 2, 249–254, 2015.
29. Yu, J., Sun, L., Ma, C., Qiao, Y., Yao, H., Thermal degradation of PVC: A review. *Waste Manage.*, 48, 300–314, 2016.
30. Miskolczi, N., Angyal, A., Bartha, L., Valkai, I., Fuels by pyrolysis of waste plastics from agricultural and packaging sectors in a pilot scale reactor. *Fuel Process. Technol.*, 90, 7–8, 1032–1040, 2009.
31. Donate, P.M., Green synthesis from biomass. *Chem. Biol. Technol. Agric.*, 1, 1, 1–8, 2014.
32. Czajczyńska, D., Anguilano, L., Ghazal, H., Krzyżyńska, R., Reynolds, A.J., Spencer, N., Jouhara, H., Potential of pyrolysis processes in the waste management sector. *Therm. Sci. Eng. Prog.*, 3, 171–197, 2017.
33. Devi, M., Rawat, S., Sharma, S., A comprehensive review of the pyrolysis process: From carbon nanomaterial synthesis to waste treatment. *Oxford Open Mater. Sci.*, 1, 1, itab014, 1–30, 2021.
34. Walker, P.L., *Chemistry & Physics of Carbon: Volume 16*, CRC Press, Boca Raton, 2021.
35. Bacon, R., Growth, structure, and properties of graphite whiskers. *J. Appl. Phys.*, 31, 2, 283–290, 1960.
36. Leinweber, P. and Schulten, H.R., Advances in analytical pyrolysis of soil organic matter. *J. Anal. Appl. Pyrolysis*, 49, 1–2, 359–383, 1999.
37. Guéret, C., Daroux, M., Billaud, F., Methane pyrolysis: Thermodynamics. *Chem. Eng. Sci.*, 52, 5, 815–827, 1997.
38. Pierson, H.O., *Handbook of Carbon, Graphite, Diamonds and Fullerenes: Processing, Properties and Applications*, William Andrew, Noyes Publications, USA, 2012.
39. Zhang, Y., Zhu, H., Zhang, R., Yu, L., Liu, Z., Shearing, P.R., Brett, D.J., Williams, P.T., Study of tire pyrolysis oil model compound structure on carbon nanomaterial production. *ACS Sustain. Chem. Eng.*, 10, 2, 800–809, 2022.
40. De Volder, M.F., Tawfick, S.H., Baughman, R.H., Hart, A.J., Carbon nanotubes: Present and future commercial applications. *Science*, 339, 6119, 535–539, 2013.
41. Dai, L., Chang, D.W., Baek, J.B., Lu, W., Carbon nanomaterials for advanced energy conversion and storage. *Small*, 8, 8, 1130–1166, 2012.
42. Kumar, A., Sharma, K., Dixit, A.R., Carbon nanotube- and graphene-reinforced multiphase polymeric composites: Review on their properties and applications. *J. Mater. Sci.*, 55, 7, 2682–2724, 2020.

43. Bassyouni, M., Mansi, A.E., Elgabry, A., Ibrahim, B.A., Kassem, O.A., Alhebeshy, R., Utilization of carbon nanotubes in removal of heavy metals from wastewater: A review of the CNTs' potential and current challenges. *Appl. Phys. A*, 126, 1, 1–33, 2020.
44. Dresselhaus, M.S. (Ed.), *Carbon Nanotubes: Advanced Topics in the Synthesis, Structure, Properties and Applications, Vol. 111*, Springer, Berlin, 2008.
45. Mishra, N., Das, G., Ansaldo, A., Genovese, A., Malerba, M., Povia, M., Ricci, D., Di Fabrizio, E., Di Zitti, E., Sharon, M., Sharon, M., Pyrolysis of waste polypropylene for the synthesis of carbon nanotubes. *J. Anal. Appl. Pyrolysis*, 94, 91–98, 2012.
46. Wang, J., Shen, B., Lan, M., Kang, D., Wu, C., Carbon nanotubes (CNTs) production from catalytic pyrolysis of waste plastics: The influence of catalyst and reaction pressure. *Catal. Today*, 351, 50–57, 2020.
47. Jung, I., Pelton, M., Piner, R., Dikin, D.A., Stankovich, S., Watcharotone, S., Hausner, M., Ruoff, R.S., Simple approach for high-contrast optical imaging and characterization of graphene-based sheets. *Nano Lett.*, 7, 12, 3569–3575, 2007.
48. Liang, K., Liu, L., Wang, W., Yu, Y., Wang, Y., Zhang, L., Ma, C., Chen, A., Conversion of waste plastic into ordered mesoporous carbon for electrochemical applications. *J. Mater. Res.*, 34, 6, 941–949, 2019.
49. El Essawy, N.A., Ali, S.M., Farag, H.A., Konsowa, A.H., Elnouby, M., Hamad, H.A., Green synthesis of graphene from recycled PET bottle wastes for use in the adsorption of dyes in aqueous solution. *Ecotoxicol. Environ. Saf.*, 145, 57–68, 2017.
50. Wang, Z., Wei, F., Sui, Y., Qi, J., He, Y., Meng, Q., A novel core-shell polyhedron Co3O4/MnCo2O4. 5 as electrode materials for supercapacitors. *Ceram. Int.*, 45, 9, 12558–12562, 2019.
51. Jiang, H., Lee, P.S., Li, C., 3D carbon based nanostructures for advanced supercapacitors. *Energy Environ. Sci.*, 6, 1, 41–53, 2013.
52. Sevilla, M. and Mokaya, R., Energy storage applications of activated carbons: Supercapacitors and hydrogen storage. *Energy Environ. Sci.*, 7, 4, 1250–1280, 2014.
53. Rajagopal, R.R., Rajarao, R., Cholake, S.T., Sahajwalla, V., Sustainable composite panels from non-metallic waste printed circuit boards and automotive plastics. *J. Clean. Prod.*, 144, 470–481, 2017.
54. Sun, L., Wang, C., Zhou, Y., Zhang, X., Qiu, J., KOH-activated depleted fullerene soot for electrochemical double-layer capacitors. *J. Appl. Electrochem.*, 44, 2, 309–316, 2014.
55. Wang, J. and Kaskel, S., KOH activation of carbon-based materials for energy storage. *J. Mater. Chem.*, 22, 45, 23710–23725, 2012.
56. Yan, J., Miao, L., Duan, H., Zhu, D., Lv, Y., Xiong, W., Li, L., Gan, L., Liu, M., Core-shell hierarchical porous carbon spheres with N/O doping for efficient energy storage. *Electrochim. Acta*, 358, 136899, 2020.

57. Adibfar, M., Kaghazchi, T., Asasian, N., Soleimani, M., Conversion of poly (ethylene terephthalate) waste into activated carbon: Chemical activation and characterization. *Chem. Eng. Technol.*, 37, 6, 979–986, 2014.
58. Djahed, B., Shahsavani, E., Naji, F.K., Mahvi, A.H., A novel and inexpensive method for producing activated carbon from waste polyethylene terephthalate bottles and using it to remove methylene blue dye from aqueous solution. *Desalination Water Treat.*, 57, 21, 9871–9880, 2016.
59. Bazargan, A. and McKay, G., A review–synthesis of carbon nanotubes from plastic wastes. *Chem. Eng. J.*, 195, 377–391, 2012.
60. Zhuo, C. and Levendis, Y.A., Upcycling waste plastics into carbon nanomaterials: A review. *J. Appl. Polym. Sci.*, 131, 4, 39931 (1 of 14), 2014.
61. Gong, J., Liu, J., Wan, D., Chen, X., Wen, X., Mijowska, E., Jiang, Z., Wang, Y., Tang, T., Catalytic carbonization of polypropylene by the combined catalysis of activated carbon with Ni2O3 into carbon nanotubes and its mechanism. *Appl. Catal. A Gen.*, 449, 112–120, 2012.
62. Tang, T., Chen, X., Meng, X., Chen, H., Ding, Y., Synthesis of multiwalled carbon nanotubes by catalytic combustion of polypropylene. *Angew. Chem. Int. Edit.*, 44, 10, 1517–1520, 2005.
63. Suriani, A.B., Dalila, A.R., Mohamed, A., Mamat, M.H., Salina, M., Rosmi, M.S., Rosly, J., Nor, R.M., Rusop, M., Vertically aligned carbon nanotubes synthesized from waste chicken fat. *Mater. Lett.*, 101, 61–64, 2013.
64. Suriani, A.B., Nor, R.M., Rusop, M., Vertically aligned carbon nanotubes synthesized from waste cooking palm oil. *J. Ceram. Soc. Jpn.*, 118, 1382, 963–968, 2010.
65. Suriani, A.B., Alfarisa, S., Mohamed, A., Isa, I.M., Kamari, A., Hashim, N., Mamat, M.H., Mohamed, A.R., Rusop, M., Quasi-aligned carbon nanotubes synthesised from waste engine oil. *Mater. Lett.*, 139, 220–223, 2015.
66. Chen, W.Q., Fu, X., Veksha, A., Lipik, V., Lisak, G., Few-walled carbon nanotubes derived from shoe waste plastics: Effect of feedstock composition on synthesis, properties and application as CO2 reduction electrodes. *J. Clean. Prod.*, 356, 131868, 2022.
67. Yao, D., Yang, H., Hu, Q., Chen, Y., Chen, H., Williams, P.T., Carbon nanotubes from post-consumer waste plastics: Investigations into catalyst metal and support material characteristics. *Appl. Catal. B Environ.*, 280, 119413, 2021.
68. Gong, J., Liu, J., Wen, X., Jiang, Z., Chen, X., Mijowska, E., Tang, T., Upcycling waste polypropylene into graphene flakes on organically modified montmorillonite. *Ind. Eng. Chem. Res.*, 53, 11, 4173–4181, 2014.
69. Sharma, S., Kalita, G., Hirano, R., Shinde, S.M., Papon, R., Ohtani, H., Tanemura, M., Synthesis of graphene crystals from solid waste plastic by chemical vapor deposition. *Carbon*, 72, 66–73, 2014.
70. Letcher, T.M. (Ed.), *Future Energy: Improved, Sustainable and Clean Options for Our Planet*, Elsevier, Amsterdam, 2020.

71. Sun, L., Tian, C., Li, M., Meng, X., Wang, L., Wang, R., Yin, J., Fu, H., From coconut shell to porous graphene-like nanosheets for high-power supercapacitors. *J. Mater. Chem. A*, 1, 21, 6462–6470, 2013.
72. Akhavan, O., Bijanzad, K., Mirsepah, A., Synthesis of graphene from natural and industrial carbonaceous wastes. *RSC Adv.*, 4, 39, 20441–20448, 2014.
73. Adolfsson, K.H., Hassanzadeh, S., Hakkarainen, M., Valorization of cellulose and waste paper to graphene oxide quantum dots. *RSC Adv.*, 5, 34, 26550–26558, 2015.
74. Yadav, S., Chauhan, M., Jacob, M., Malhotra, P., Distinguished performance of biogenically synthesized reduced-graphene-oxide-based mesoporous Au–Cu2O/RGO ternary nanocomposites for sonocatalytic reduction of nitrophenols in water. *New J. Chem.*, 46, 20, pp.9685–9694, 2022.
75. Forouzandeh, P., Kumaravel, V., Pillai, S.C., Electrode materials for supercapacitors: A review of recent advances. *Catalysts*, 10, 9, 969, 2020.
76. Futaba, D.N., Hata, K., Yamada, T., Hiraoka, T., Hayamizu, Y., Kakudate, Y., Tanaike, O., Hatori, H., Yumura, M., Iijima, S., Shape-engineerable and highly densely packed single-walled carbon nanotubes and their application as super-capacitor electrodes. *Nat. Mater.*, 5, 12, 987–994, 2006.
77. Hsia, B., Marschewski, J., Wang, S., In, J.B., Carraro, C., Poulikakos, D., Grigoropoulos, C.P., Maboudian, R., Highly flexible, all solid-state micro-supercapacitors from vertically aligned carbon nanotubes. *Nanotechnology*, 25, 5, 055401, 2014.
78. Ye, J.S., Cui, H.F., Liu, X., Lim, T.M., Zhang, W.D., Sheu, F.S., Preparation and characterization of aligned carbon nanotube–ruthenium oxide nanocomposites for supercapacitors. *Small*, 1, 5, 560–565, 2005.
79. Gohardani, O., Elola, M.C., Elizetxea, C., Potential and prospective implementation of carbon nanotubes on next generation aircraft and space vehicles: A review of current and expected applications in aerospace sciences. *Prog. Aerosp. Sci.*, 70, 42–68, 2014.
80. Natsuki, T., Tsuchiya, T., Ni, Q.Q., Endo, M., Torsional elastic instability of double-walled carbon nanotubes. *Carbon*, 48, 15, 4362–4368, 2010.
81. Wang, C.M., Wen, C.Y., Che, Y.C., Chang, J.Y., Ho, C.H., Kao, K.S., Shih, W.C., Chiu, C.M., Shen, Y.A., The influence of specific surface area on the capacitance of the carbon electrodes supercapacitor, in: *The Proceedings of the Second International Conference on Industrial Application Engineering*, pp. 439–442, March 2015.
82. Simon, P. and Burke, A., Nanostructured carbons: Double-layer capacitance and more. *Electrochem. Soc. Interface*, 17, 1, 38, 2008.
83. Grishchenko, L.M., Tsapyuk, G.G., Ricco, M., Diyuk, V.E., Boldyrieva, O.Y., Mariychuk, R., Matushko, I.P., Pontiroli, D., Lisnyak, V.V., Scaravonati, S., Enhancing the performance of supercapacitor activated carbon electrodes by oxidation, in: *2020 IEEE 40th International Conference on Electronics and Nanotechnology (ELNANO)*, IEEE, pp. 173–177, April 2020.

84. Raj, F.R.M.S., Jaya, N.V., Boopathi, G., Kalpana, D., Pandurangan, A., S-doped activated mesoporous carbon derived from the Borassus flabellifer flower as active electrodes for supercapacitors. *Mater. Chem. Phys.*, 240, 122151, 2020.
85. Cheng, F., Yang, X., Zhang, S., Lu, W., Boosting the supercapacitor performances of activated carbon with carbon nanomaterials. *J. Power Sources*, 450, 227678, 2020.
86. Inal, I.I.G., Gokce, Y., Aktas, Z., Waste tea derived activated carbon/polyaniline composites as supercapacitor electrodes, in: *2016 IEEE International Conference on Renewable Energy Research and Applications (ICRERA)*, IEEE, pp. 458–462, November 2016.

12

Carbon Allotropes in a Sustainable Environment

Farhat A. Ansari

Department of Pharmacy, Faculty of Pharmaceutical Chemistry, Hygia Institute of Pharmaceutical Education and Research, Faizullaganj, Lucknow, India

Abstract

Several fields of chemicals, nanomaterials, and technology are extremely interested in carbon chemistry due to its distinctive and inherent features. Various physiochemical processes are used to develop sustainable carbon composite materials. Advanced carbon structures, which are components with desirable qualities, are also produced using design techniques. Developed carbon composite materials like fullerene, graphene, and carbon nanotubes (CNTs) have a tunable high porosity, large surface, good electrical conductivity, high thermal stability, as well as high corrosion resistance. They can be used in storing energy as electrocatalytic, electro-conductive admixtures, complexation hosts, and perfect substrates for nanomaterials. To improve solubilities, chemical biocompatibility covalent and noncovalent techniques are frequently performed. As further methods to improve their solubility, shear phase separation, ball grinding, electromagnetic stirring, and ultrasound mixing are frequently employed. The variety of synthetic carbon allotropes with interesting and aesthetically beautiful designs with exceptional material properties is expanding. The proposed article gives several manufacturing methods for the sustainability of carbon materials and emphasizes many applications.

***Keywords*:** Graphene, carbon nanotube, nanocomposites, nanotechnology, polymer matrixes

Email: farhataisha@gmail.com

Chandrabhan Verma and Chaudhery Mustansar Hussain (eds.) Carbon Allotropes and Composites: Materials for Environment Protection and Remediation, (257–302) © 2023 Scrivener Publishing LLC

12.1 Introduction

Carbon has been recognized and used for a variety of applications. Since ancient times, the main source of carbon has been hardwood, which has been burned as energy for centuries. In the 1980s, only three carbon polymorphs are known to occur: graphite, diamond, and unstructured carbon [1]. Amorphous carbon, which lacks a crystalline structure, is utilized in a variety of products. Almost all life depends on carbon owing to its unique capacity to connect with many other substances. There are many innovative substances available today, and the twentieth century unquestionably produced a lot more substances than other eras. Society is greatly affected by many substances. Due to its outstanding strength and lightweight, carbon has emerged as a significant material for revolutionizing the 21st century [2]. Several fields of materials, nanotechnology, and technology are extremely interested in carbon chemistry due to its distinctive and intrinsic features [3].

Two natural substances contain atomic carbon, i.e., graphite and diamonds are examples of allotropic forms. It includes a vast network of sp^3 and sp^2 hybridized carbon atoms respectively. The physical features of these forms are distinct, including toughness, heat capacity, lubricant behavior, and electrical properties. By changing the periodic binding domain in systems made up of sp^3, sp^2, and sp-hybridized carbon atoms. There are numerous other ways to create carbon allotropes that are theoretically possible. It has long been desirable to develop ideas for their synthesis on a macroscale because of the anticipated outstanding physical attributes of these obscure carbon allotropes. Carbon occurs in a variety of allotropes known as nanoallotropes of carbon, including graphene, diamonds, graphite, fullerene, and carbon nanotubes (CNTs), among many others (Figure 12.1). Covalent and noncovalent complexation techniques can be used to functionalize carbon allotropes [4–7]. Noncovalent complexation comprises weaker intermolecular force interactions between the molecules, whereas covalent formation incorporates covalent bonded carbon allotropic forms and polar substituents [4, 5, 8–10].

12.2 Functionalization of Carbon Allotropes

12.2.1 Covalent Functionalization

The restricted dispersion of the carbon materials prevents their usage in a broad range of commercial and biotechnological applications [11].

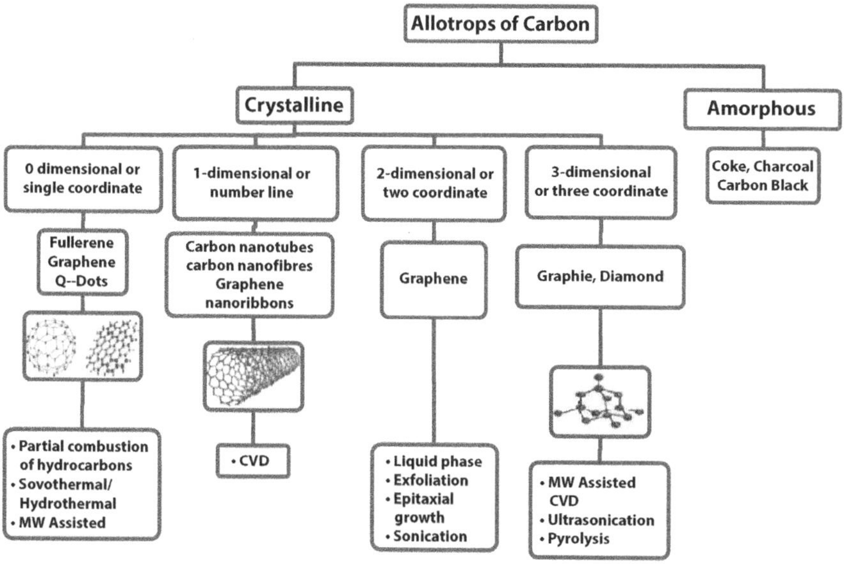

Figure 12.1 Carbon allotropes according to their respective dimensions and their synthesis processes.

Appropriate cross-linking and the addition of hydrophilic groups to the structural features of the allotropes of carbon can improve their solubility [33, 34]. Both covalent and noncovalent interaction functionalization techniques can be used to improve the properties of the carbon allotropes [35–38]. Noncovalent functionalization includes moderate intermolecular (van der Waals forces) interactions between the molecules, whereas covalent functionalization involves covalent bonds formed between the carbon allotropic forms and the polar substituents. Typically, carbon polymorphs have one or more locations where the covalent bonding can be used to promote cross-linking. For instance, graphene analogues can be chemically designed and synthesized to obtain -OH, -COOH, and R-O-R› (epoxide) rings utilizing condensing and esterification processes [12–14]. Although the covalent formation of > C = C (double bond) is challenging, it can be done by utilizing a free radical method, such as a dienophile approach. Graphene analogues are often covalently functionalized through processes like cyclic-addition, esterification, isocyanate, acyl formation, diazotization, and ring-opening under the influence of nucleophilic attack [15, 16].

Carbon tubes known as CNTs often have nanometer-sized dimensions (nm). Single-walled nanomaterials (SWCNTs) and multi-walled nanostructures (MWCNTs) are two categories. Iijima, Ichihashi, and Bethune *et al.* invented CNTs in 1993 [17]. Two or more rings of carbon nanotubes bound together via Vander Waals interactions form MWCNTs. They possess high mechanical, tensile strength, dynamism and nanoscale structure.

Covalent functionalization of carbon nanotubes (single and multiwalled nanotubes) results in the various physicochemical modifications listed below:

(i) Rise in the diffusion.
(ii) Improvement of interface bond strength.
(iii) Decrease in desorption of water.
(iv) A decline in the capacity for aggregation.
(v) Decline in mechanical strength.
(vi) Decrease in electric conductivity.

12.2.2 Noncovalent Functionalization

Hydrophobic contacts, electrostatic forces, van der Waals inter-molecular force of attachment, and pi(π)-pi(π) layering are the main methods used to noncovalently functionalize carbon allotropes [4, 5, 18]. Naturally, noncovalent interactions offer an intriguing method for functionalizing carbon allotropic forms because it retains the majority of the allotropes' features, particularly their electrical and thermal capabilities. Under this form of interaction, the heterocyclic ring(s), such as pyrene and perylene, interact effectively with the carbon allotropes' π-electron clouds.

Aromatic hydrocarbons and carbon allotropes interact by a process known as "π-π layering," whereas contact between forms of carbon (such as graphene, graphene oxide, carbon nanotubes, etc.) and aggregates, surfactants, and ionic liquids are known as aquaphobic interactions. The dispersal of allotropes of carbon in both water and organic solvents is mainly accomplished through the use of these contacts. Polymer electrolytes and graphene oxide (and possibly other allotropes as well) interact primarily through electrostatic interactions because grapheme oxide contains a variety of electrostatic interactions, including hydroxyl and carboxylic ionic groups. In addition to π-π stacking, hydrophobic, and electrostatic connections, H-bonding also has the potential to noncovalently functionalize carbon allotropes.

In the production and preservation of energy, renewable power sources primarily use various allotropic forms of carbon molecules. The application areas of nanocomposites in biological, healthcare, technology, energy technologies, and medicine delivery are enormous [19, 20]. For potential uses, researchers are already diligently working to obtain the unmatched electrochemical, mechanical, and heat transfer capabilities of their carbon molecules.

As a result, attention to the creation and use of sustainable products is increasing. Applications in the future chemical and power sectors are increasingly recognized, in particular. This chapter will outline the various methodologies for sustainable carbon materials and illustrate their practical uses in key fields of energy and sustainable environment.

12.3 Developments of Carbon Allotropes and Their Applications

The foundation of material science research and technology development is said to be due to advanced carbon nanomaterials, which include graphene, fullerene, carbon fibers, activated carbon, carbon nanotubes etc. These nanoparticles are produced utilizing various physical and chemical methods to enable high-quality materials with exceptional characteristics. Enhanced carbon materials are also used in digital equipment, organic photovoltaic (PV) systems, energy-saving technologies, and medicine distribution, among other things. Theoretical simulations have indicated that there may be numerous different structural allotropic forms of the extraordinarily varied bonding abilities of carbon. These are expected to form under circumstances of higher temperature and pressure, possibly the majority of them. However, remarkably few morphological allotropes have been found in practical highly pressurized experiments.

With the potential to store energy on a vast level, such innovative substances are the solution to renewable power storage systems. In the coming years, battery-based power generation is anticipated to be replaced with fuel cell technology. Microelectronics applications for graphene layers could prove revolutionary. The demand for cutting-edge composites will continue to rise with rapid technological innovation and development, healthcare [21], aeronautical, defense parts, domestic and power applications [22], cars, etc. [23, 24] are just a few of the uses listed in Figure 12.2. Recently, there has been a lot of interest in the use of carbon-based nanostructured material in the worldwide energy sector. Almost all major industries have seen a considerable increase in the area of materials nanotechnology and science in recent decades.

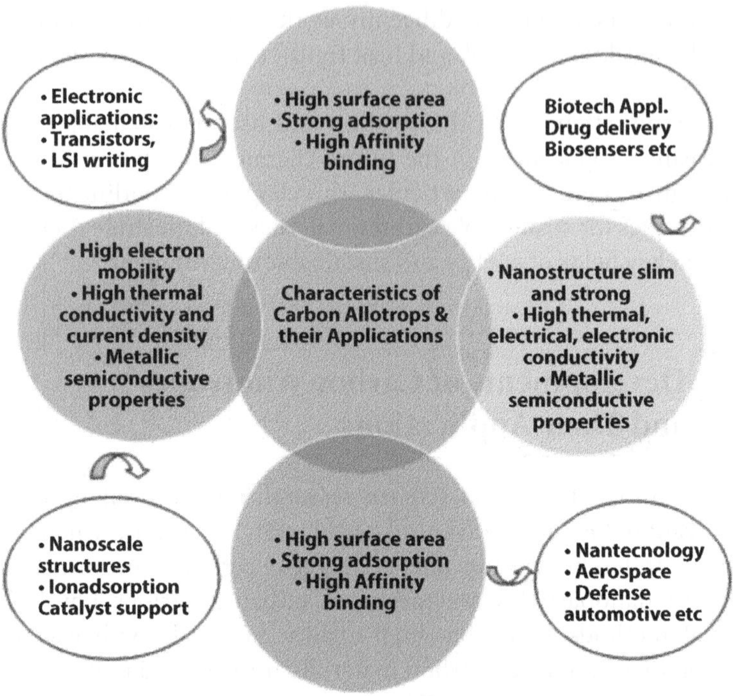

Figure 12.2 Various applications of advanced carbon composite material.

12.4 Carbon Allotropes in Sustainable Environment

Carbon nanotubes (CNTs), graphite and its derivatives, fullerenes, C-nanoparticles, etc. are examples of nanostructured materials. The use of CNMs can be an advantageous replacement for the standard method of treatment because it is affordable, practical, environmentally benign, and efficient. Worldwide, the presence of heavy metallic ions in the effluent is a severe environmental issue. Metals on the surface of organisms as well as inside their cells can significantly affect their metabolic processes.

The metal ion chromium is poisonous, human oncogenic, and has several harmful effects. They can slow down some plants' development and propagation, which raises the rate of reproduction and leads to death in living things. The majority of the heavy metal ions cannot be removed by traditional treatment techniques. The most popular water treatment techniques now use membranes and nanomaterials [24, 25]. Both single-wall (SWCNTs) and multi-wall carbon nanotubes (MWCNTs) can successfully capture gases like "ammonia (NH_3), nitric oxide (NO), nitrogen dioxide (NO_2), hydrogen (H_2), sulphur hexafluoride (SF_6), and chlorine (Cl_2)".

Since pharmaceuticals have been used for many years, it is surprising that there have not been many studies to look into the disposal of it in the environment. In addition, the medicines are not removed from wastewater. The hunt for water purification techniques takes on top significance in this situation. Technologies that show promise have been developed as a result of improvements in the treatment of water and sewage. Numerous studies have revealed that various prescription medications are removed during water treatment by this technique as the absorption method has been effectively used for the eradication of a broad range of organic contaminants in water.

Magnetized C-adsorbents, such as Activated carbon (AC), biochar (BC), nanomaterials, carbon nanotube, and other additional substances (such as "nitro-phenol, polyurethane, yeast alcohol dehydrogenase, low-density polyethene, acetate, acetylene, and oleates"), could be utilized to make a variety of carbon supports [26, 27]. Arsenic, drugs, toxic metals, pigments, fluorine, chemicals, and radioactive elements are a few of the important water pollutants that have been removed by adsorbent and the removal efficiency were examined.

12.5 Carbon Allotropes Purification Process in the Treatment of Wastewater

The need for clean water is one of the biggest issues the globe is now experiencing. Only 2.5% of the quantity of water on earth, which is believed to be 333 million cubic miles, is classified as freshwater; the remainder is saline and contaminated. Modern technology is essential for industrial wastewater because outdated techniques are harmful to get rid of some microorganisms and toxic metals. Carbon nanoparticles are regarded as the best necessary technology for treating wastewater in the coming years. Membrane separation approaches because of their many exceptional qualities and traits.

The special qualities of carbon nanomaterials, such as their structure, photonic, electrochemical, and catalyst properties, along with their large surface areas, have demonstrated considerable development opportunities for more effective treatment of wastewater. The functionalized molecules that aid in effective dispersal in solvent and matrix material is also produced by the synthesis process of nanotubes and graphene oxide (GO) sheets. The characteristics and capabilities of membranes have improved as a result of all this. Customized carbon nanocomposite membranes will emerge as the finest technique in treating wastewater and many other industries due to their excellent performance and continuous growth.

The superior quality of carbon-based adsorbents is rapidly replacing commonly used adsorbent materials. Carbon nanomaterial-based adsorbent has a larger surface area with more oxygen functional groups and improved corrosion protection than conventional adsorbents [28]. Functional groups promote surface properties, which can enhance contact with metallic ions and help eliminate heavy metals that prove harmful.

12.5.1 Fullerenes

Fullerenes are single-coordinate carbon allotropes that have a hollowed, cage-like shape. C60 allotropes of carbon are those that have been most frequently researched. Fullerenes are well suited for removing organisms from aqueous systems due to their high surface proportions and reactivity. Because of their extremely large surface area, fullerenes are frequently used as adsorbent materials in the purification of wastewater because they help eliminate pollutants such as heavy metals. Other nanostructures, such as zeolites and carbon black, are incorporated into fullerenes to serve as an affordable alternative absorbent. These adsorbents' increased porousness has also contributed to an improvement in adsorption capacities. In response, this enhances the surface properties. To create nanomaterials, fullerenes are combined with the polymeric membrane, which enhances the tensile and adsorption properties of the host material.

The polyvinyl pyrrolidone (PVP) grafted C60 exhibited excellent antimicrobial properties [29], suggesting potentiality for wastewater treatment. Proteins can cause biofouling. When in interaction with protein-containing effluent, membranes that had their surface modified by fullerenes and CNTs exhibit superior flux decreased recovery values. PPO-C60 membranes, which stand for poly-2, 6-dimethyl-1,4-phenylene oxide, contain larger pores than ordinary PPO membranes. This kind of film can be utilized to extract organic pollutants from wastewater [30]. Fullerenes exhibit a range of attributes that contribute to the treatment of wastewater pollutants; however, one of its main drawbacks is price.

12.5.2 Bucky Paper Membrane (BP)

Membranes made of buckypaper (BP) are thin, auto-supportive, and have remarkable structural and morphological characteristics. Owing to its unique hydrophobic qualities, it is employed in membrane filtration for the purification of seawater. The hydrophobicity and ability to make stable suspensions are increased by CNTs. With the use of vacuum filtering, a paper-like network of carbon nanotubes (CNTs) was created. Membrane technology is a widely used application in which the BP outcomes are connected using microscopes and a porosimeter.

Methods including scanning electron microscopy (SEM), TEM (transmission electron microscopy), and nitrogen sorption are used to characterize CNT BP's [31]. Two popular biomaterials, chitosan and carrageenan, are employed to speed up the creation of multi-walled carbon nanotubes (MWCNT) which shows easy dispersion. Since the elongated bio-based molecules are made up of flexible bridging between the tiny CNTs, they are particularly well suited for usage as bindings in CNTs in Bucky-paper membranes. Additionally, it improves the membrane's characteristics and meets the purpose of studies on salt, flowability, and the dismissal of contaminants [32]. The constructed membranes have eliminated of salt from simulated wastewater by about 99%, making it an ideal choice for desalinating [33].

12.5.3 Carbon Nanotubes (CNTs)

CNTs are cylindrical molecules made of single-layered carbon atom sheets that have been rolled up. CNTs can exist as single-walled carbon nanotubes (SWCNTs), or as many concentric cylinders referred to as multi-carbon nanotubes (MWCNTs). They stand out from other materials due to their exceptional qualities, which include their large surface area, excellent thermal conductivity, excellent optical and vibrational properties, high tensile strength, and young's modulus [34].

12.5.3.1 CNT Adsorption Mechanism

CNTs have been extensively used to extract varied organic contaminants, including methylene blue (MB) dye [35], phthalocyanine [36], direct blue 53 [37], and direct red 23 [38]. These CNTs have a distinctive submicron framework, relatively significant specific surface spaces, and powerful wettability. In adsorption systems, the pseudo-first order and pseudo-second order adsorption kinetics, respectively, are used. Sheets with a 3 mm diameter and 20–30 cm in length might be produced by extruding CNTs combined with an organic matrix [39]. The strips were capable of removing various organic materials from wastewater with efficiency. The strips also demonstrated good recycling capabilities [39]. Additionally, the phase transition approach can be used to produce CNTs hybrid polymer network which can successfully eliminate p-chlorophenol, [40].

According to Xu *et al.* findings, a fixed-bed columnar device may be used to create a fundamental structural CNT-based nanocomposite adsorbent material to adsorb 2-naphthol from contaminated water [41]. The outcomes demonstrated that the CNT-based nanocomposite adsorbent material has superior sorption qualities for removing 2-naphthol [41]. Additionally, sewage biochar enhanced by CNT nanocomposite (CNT-SBC) may be employed as an absorber to extract sulfamethoxazole (SMX) from water in low quantities. The capacity of CNT-SBC absorption was greatly boosted by the presence of CNTs. Physio-Chemical adsorption is the two traditional methods used to explain the interaction of organic pollutants with carbon nanotube surfaces [42–44]. Physisorption includes electronic forces, H-bonds, and pore filling. Chemisorption includes water-insoluble effects, electrostatic forces, and functional group complexation.

Additionally, magnetic CNT-based products were typically created by adding magnetic substances to CNTs, such as ferric oxide [40], ferric carbide [38, 44], ferrous oxide [44], and Fe metal-organic lattice (MIL-53(Fe)) to effectively segregate the adsorbed substances from the solution. In addition to having good adsorption capability, the magnetic nano-CNT-based composites were also simple to magnetically remove from an aqueous solution.

12.5.3.2 CNTs Ozone Method

According to the research, materials based on CNTs work well as catalysts for the catalysis ozonation of organic contaminants. According to reports, organic contaminants such as oxalic acid [45], sulphamethoxazole (SM) [46], and indigo [47] have been efficiently removed from effluent water using the CNTs-based nanomaterials catalysis ozonation technique. It is generally known that CNTs have several numbers of functional groups in their structure, which could have an impact on the ozonation process' catalytic properties. Ozone treatments, as discovered by Liu *et al.*, might alter the functional groups on the surface of MWCNT (multiwalled-CNT), changing the catalytic performance [48]. According to research by Zhang *et al.*, two processes were primarily implicated as CNTs stimulated the formation of ROS from ozonation:

By directly assaulting the active sites on the CNTs surface with O_3,

(i) by directly hitting the active surface site on the CNTs with O_3, which creates an interphase "HO zone" on the surface.

(ii) O_3 can change into $O_2^{\bullet-}$ when it comes into contact with an effective CNT initiator.

In the combined production of TiO_2-CNTs, UV, and O_3, Orge *et al.* evaluated the breakdown of oxamic acid (OMA), which demonstrated the best OMA removal efficiency among other process control [49]. On the other hand, electrochemical filtering might employ CNTs as an anode (ECF). The synergistic effects of the ECF procedure and O_3 on phenolic oxidation were shown, and nearly all of the breakdown products and phenol were mineralized as a result [50].

12.5.3.3 CNTs-Fenton-Like Systems

The superior capabilities, of CNTs-based materials, are now frequently employed in Fenton-like systems to eliminate organic contaminants. The breakdown of several organic pollutants, such as 17-methyl-testosterone [51], tetracycline [52], Orange II [53], bisphenol A [54], SMX [55], acetylsalicylic acid [56], and atrazine [57], was studied using CNTs/Fenton-like processes.

Based on the study, functionalized MWCNTs (FCNT-H) can significantly increase the removal rate of atrazine in the Fe (III)-mediated Fenton-like system [57]. To accomplish a quick Fe(III)/Fe(II) cycle, H_2O_2 might efficiently decrease the surface-bound Fe(III). According to the traditional processes in an aqueous system (Eqs. 3-7), 11% of the total Fe (II) was created [58]. Additionally, 9% of Fe (II) was produced by the direct reduction of Fe (III) on the interface of FCNT-H by carbonyl and quinone groups. Most significantly, 80% of the Fe(II) produced was due to the FCNT-H-Fe(III) complexes' quicker degradation by H_2O_2 [57].

Additionally, investigations on the degradation of organic pollutants using CNTs and a Fenton-like coupling mechanism that includes electrons, UV-radiation, and photoelectron are reported in [51–65]. Due to their great electrical conductance, CNTs can be utilized as a photoelectrode for the *in situ* creation of H_2O_2 in the electro-Fenton procedure. To effectively degrade SMX via the photo-Fenton process, Nawaz *et al.* produced highly photosensitive and magnetized recyclable MWCNTs including $NiFe_2O_4$ (NiFe-CNT) composite [55]. H_2O_2 and Fe^{3+} and Fe^{2+} may react to form $HO^{\bullet}$. Then, NiFe-CNT was exposed to UV-A light, producing an electron/hole (e-/h+) couple on the material's surface. The photogenerated electrons hVB+ can combine with H_2O or OH to make $HO^{\bullet}$, and the photogenerated electrons can immediately interact with H_2O_2 to generate $HO^{\bullet}$.

12.5.3.4 CNTs-Persulfates Systems

The potential use of *in situ* chemical oxidation (ISCO) and water/wastewater purification, peroxydisulfate (PDS) and peroxymonosulfate (PMS) derived (advanced oxidation processes) AOPs have subsequently received a significant amount of attention [66–68]. By cleaving the peroxide bonding in PDS and PMS, $SO_4^{\bullet}$ can be produced (persulfates). The advantages of $SO_4^{\bullet}$ over $HO^{\bullet}$ include a prolonged half-life, a strong oxidizing capacity, a larger pH range etc. [66].

Metal-based catalysts have become the subject of much research in recent decades [69, 70] because they can efficiently generate $SO_4^{\bullet}$ from persulfates. However, given to the possible risk of metal loss, it is important to carefully weigh the exchange between improving persulfates activation performance and the associated secondary pollution concern.

According to Sun *et al.* study, pure MWCNTs may efficiently catalyze the breakdown of persulfates to produce sulphate radicals that can be used to break down phenol [71]. It was proposed that the MWCNTs' sp^2 carbon and oxygen-containing units served as catalyst surfaces and were crucial for the stimulation of persulfates. Heteroatom doping has been presented and demonstrated as a viable method for enhancing the decomposition of organic pollutants in persulfate-based advanced oxidation processes (AOPs) to increase the activation effectiveness of CNTs.

Remarkably, the addition of the N atom to SWCNTs made NoCNTs exhibit exceptional stability since nonradical processes were present. That meant, graphitic N was crucial to this procedure. The PMS activation efficiency of N and sulphur (S) co-doped MWCNTs was also very good [72]. The breakdown of benzophenone-4 (BP-4) by PMS activation, demonstrated that the activity level of nitrogen and sulphur-doped CNT-COOH (NS-CNT-COOH) was about five times more than N and S-doped CNT and N-doped CNT-COOH.

An alternative nonradical process for CNTs/PDS was put out by Cheng *et al.* [73]. They proposed that this nonphotochemical mechanism produced $^1O_2$ and was crucial in the deterioration of 2, 4-dichlorophenol. In persulfates activation systems, materials based on CNTs operate as an accelerator to stimulate the breakdown of organic contaminants. Scientists have been constantly examining the possible mechanism. Researchers may need to make additional efforts in the future to develop a unified and comprehensive chemical reactivity.

12.5.3.5 CNTs-Ferrate/Permanganate Systems

CNTs exhibit reductive capabilities due to some reduction groups that might speed up the activity of ferrate/permanganate (Fe(VI)/Mn(VII)) for the breakdown of organic contaminants. According to Sun *et al.* research, CNTs could stimulate Fe(VI) and hasten the breakdown of bromophenols [74]. Fe(VI) may combine with the reducing groups present on the surface of CNTs to form high-valency metal-oxo precursors (Fe(V)/Fe(IV)), which had two to five times more reactive than Fe (VI). However, byproducts generated during the reaction mechanism might be taken up by the surface of CNTs, lowering the dangers of by-products in the wastewater.

CNTs might greatly speed up the breakdown of phenols by Mn, according to Zhao *et al.* (VII). The reducing capacity of CNTs in the CNTs/Mn(VII) combination might reduce Mn(VII) to produce MnO2, which might catalyse Mn(VII) oxidation of phenols.

According to Tian *et al.*, biochar's reducing ability may promote Mn(VII) to generate high-valent intermediate manganese molecules when compared to CNT, which significantly improved SMX removal [75]. In brief, using CNTs to activate Fe(VI)/Mn(VII) may have significant application potential for the treatment of wastewater.

12.5.4 Graphene

Graphene is a two-dimensional allotrope of carbon. Due to its molecular sieving property, nonporous graphene/graphene oxide (GO) is being designed for efficient and sophisticated membrane separation procedures. To acquire admirable characteristics that were not normally present in pristine graphene. Hummer's process uses chemical methods to produce graphene oxide sheets. In the lattices of graphene, this technique introduces functional groups like the -OH group and -COOH group. Graphene/Graphene Oxide sheets are employed for treating wastewater as they have high porosity and large surface area.

The outstanding anti-fouling capabilities in an oil and water emulsion of the newly designed GO/aminated polyacrylonitrile (APAN) membranes were related to their porous structure. They displayed a tremendous flow rate. The PA membrane's polyamide layer was covered with GO to improve its anti-fouling capabilities and retain durability [76].

The purpose of reducing the accumulation of heavy metals that seem to be present in high concentrations in the effluent, graphene can be utilized as an alternative to CNTs. Chemically oxidized graphene can be used to clean wastewater that contains too many Pb (II) ions. To treat heavy metals, graphene may also get rid of organic pollutants like phenolic compounds and pigments that are present in wastewater.

The term "carbon and graphene quantum dot" (CQD) designates carbon-based nanomaterials that have sp^3 and/or sp^2 hybridized atoms as their main constituents. High mechanical performance, electrical, and microscopic size effects, simple surface functionalization, and customizable chemistry have this special capacity to capture the energy of greater wavelength and have simple competence and toward the photo-reduction of many organic pollutants.

Unlike the micro-sized graphene oxide (GO) layers, GQDs are intact aromatic ring planar structures with a wealth of peripheral -COOH functionalized groups. Nitrogen pumping is primarily employed to alter and modify the structural functionalities, chemical properties, and electrical characteristics of graphene and GQDs [77].

12.6 Removal of Various Pollutants

Water pollution is a pertinent environmental hazard, and magnetic carbon-based materials are useful in helping to combat it. Consequently, the sorption of typical contaminants employing magnetic carbon-based adsorbent materials is investigated and discussed, comprising their elimination capabilities for contaminants like arsenic, medicine, toxic metals, pigments, fluorine, synthetic chemicals, and other pertinent substances.

12.6.1 Arsenic

Arsenic (As) is a surface and groundwater contaminant source of water which poses a significant public health risk and, as a consequence, is an ongoing environmental issue for the people of various Southeast Asian, Latin American, and European nations. The major sources of arsenic contamination are natural materials and human impacts (such as mines, the use of chemicals, and the burning of fossil fuels) [78].

The US Environmental Protection Agency (EPA) [79] and World Health Organization (WHO) [115] have therefore advised a maximum limit of 10 ppb for arsenic concentration in water in use for human consumption to safeguard the public provided by water supply [80–82]. This is because it exhibits serious toxic effects on living beings. But it's estimated over 100 million individuals could be subjected to contaminated waters with arsenic [83]. In infections, high blood pressure, diabetes, carcinoma, and peripheral vascular disorders can all result from prolonged contact with arsenic [84, 85]. This contaminant is generally found as arsenates or arsenites and also can exist in valence states of +5, +3, 0, and -3 [84]. As a result, the main composition of this toxin has a significant impact on its hazard. The most typical types of arsenic in water are arsenate $(AsO_4)_3$ with (+5) or arsenite $(AsO_3)_3$ with (+3), with arsenite approximately twenty times more hazardous than arsenate [86, 87]. Table 12.1 summarizes the carbon-based adsorbent materials to remove arsenic from aqueous solutions.

Fe_3O_4 based on MWCNTs was employed by Ntim and Mitra [88] as an arsenic absorber. MWCNTs were designed and synthesized using a microwave, and then they were ultrasonically disseminated in an aqueous medium of $FeCl_3$:$FeSO_4$. For As^{5+} and As^{3+}, this type of magnetic substance had adsorption capabilities of 189 and 1723 micrograms/g, respectively. The authors concluded that the complex formation with absorbent, oxyhydroxide domains had an important role in the arsenic adsorption on the magnetic sorbents.

On the other side, Meng *et al.* [89] investigated a new magnetic absorbent made from the orange peel as precursors. Utilizing a hydrothermal technique and iron solutions, this adsorbent was produced. On As5+ sorption, the effects of operational factors including liquid pH, temp, and concurrent inorganic sorbents were examined. On As^{5+} sorption, the effects of operational factors including solvent pH, temp, and concurrent inorganic adsorbents were studied. An iron/biomass ratio of 10 wt% led to an As^{5+} adsorption process of 81.3 mg/g. The arsenic adsorption mechanism was influenced by the pH value of the solution, and this elimination was highly exothermic. After three cycles of adsorption-regeneration, elimination effectiveness was determined to be 89%, and the adsorbent demonstrated superior mechanical durability and minimal iron release (i.e., 0.3 mg/L).

Ferrous salts and calcium-based magnetic BC with rice husk as a carbon source were used to study the multi-component sorption of As^{3+} and Cd^{2+} [78]. The findings showed that calcium improved Cd^{2+} uptake whereas iron favored As^{3+} adsorption. The adsorption of both contaminants in the studies with one adsorbed species was dependent on the solution pH. The pH 6 yielded the highest removal rate and maximum sorption capacity of 10.07 and 6.34 mg/g for Cd^{2+} and As^{3+}, accordingly. These adsorbents had both additive and antagonistic effects during multi-component-component adsorption.

For the As^{3+} adsorption from an aqueous solution, a magnetic GO was prepared using graphene and Fe salts [90]. With the use of FTIR, XRD,

Table 12.1 Magnetized Carbon-based adsorbent materials to remove arsenic from aqueous solutions: synthesis, extraction conditions, and adsorption details.

Synthesis of carbon-based adsorbent	Magnetic-material	Carbon source	Metal removal	Adsorption capability (mg/g)	Operating conditions		Ref.
					pH	Temp. (°C)	
Ca-based magnetic biochar (BC) produced through decomposition and chemical precipitation	Fe_3O_4	Rice straw	As^{5+}	6.34	5	25	[78]
Thermal penetration and hydrolysis of magnetized BC with microwave irradiation	Fe_3O_4	Fruit bunch of palm	As^{5+}	88	-	25	[80]
Using decomposition and thermal impregnation to create magnetic CNCs	Fe_3C	Pine resin	As^{5+}	264 357	7 4	-	[81]
Magnetic Activated-carbon (AC) by impregnation using Fe, Mn oxides	Fe_3O_4	Commercial AC	As^{5+}	19.35	3	25	[82]
Magnetized AC from co-precipitation, impregnation, and carbonization processes	Fe_3O_4	Cigarette soot	As^{3+} As^{5+}	81 108	7 3	25	[86]

(Continued)

Table 12.1 Magnetized Carbon-based adsorbent materials to remove arsenic from aqueous solutions: synthesis, extraction conditions, and adsorption details. (*Continued*)

Synthesis of carbon-based adsorbent	Magnetic-material	Carbon source	Metal removal	Adsorption capability (mg/g)	Operating conditions		Ref.
					pH	Temp. (°C)	
Magnetic Carbon nanotubes' from Catalytic CVD vapor deposition using an alkali-activated technique	Fe_3O_4 γ-Fe_2O_3	Ethanol	As^{3+} As^{5+}	8.1 9.73	5.5	25	[87]
Microwave-assisted, create magnetic Multi-walled CNTs	Fe_3O_4 γ-Fe_2O_3	Commercial MWCNTs	As^{3+} As^{5+}	1.72 0.18	7 4	40	[88]
Magnetized BC produced by hydrothermal impregnation and carbonization	Fe_3O_4	Orange peel	As^{5+}	81.3	6	2	[89]
Chitosan magnetic GO is produced by exfoliating and coprecipitating graphite,	Fe_3O_4	Graphite	As^{3+}	45	7.3	25	[90]

and field emission SEM/EDX tools, the properties of this magnetic GO were investigated. Response surface methodology was used to assess the effects of As^{3+} concentrations, contact time, pH, and adsorbent dosage on the As3+ sorption at batch conditions using a centralized composite design (CCD). At pH 7, 0.3 g/L of adsorbent dose, 77 min of treatment time, and a starting As^{3+} concentration of 100 mg/L, As^{3+} removal was 99.95%. After four adsorption–desorption cycles, the magnetic GO's removal performance dropped by up to 80%, and 0.1 M NaOH was employed to renew it.

12.6.2 Drugs and Pharmaceuticals

The propensity to contaminate water and pharmaceutical goods that have been used up or have expired is regarded as a specific kind of environmental waste and has been listed as a substance of increasing concern to the general population [84, 91, 92].

One million tonnes of pharmaceuticals are consumed worldwide each year, with a per-person consumption rate of 15 g [93]. Many pharmacological compounds are discharged into the surroundings because they cannot be fully absorbed and processed. Consequently, during its manufacturing, transportation, application, usage, and expulsion, a sizable quantity of medicine is released into the surroundings [84, 94].

Tetracycline (TC), for instance, is a significant antibacterial that is frequently utilized as an antibiotic in the cattle industry and human treatment. It is known that 50–80 per cent of TC used in agriculture can be eliminated through fecal matter as the unaltered constituent chemicals [95]. Pharmaceutical compounds are also present in underground water, effluents, and surface water sources in quantities varying from nanogram/liter to microgram/liter [94]. More than fifty-five medicines and one hundred and fifty everyday-used chemicals may be present in effluents in considerable concentrations [95–97].

The typical adsorbates examined in adsorption research include Carbamazepine (CBZ), atenolol, erythromycin, diclofenac, ibuprofen (IBP), and naproxen. A list of the magnetized carbon-based sorbent materials used for removing medicines Table 12.2. Magnetized nanomaterials were created by Baghdadi *et al.* [98] using commercialized CBZ adsorption testing. SEM, FTIR, N2 physical adsorption, VSM, XRD, and SEM were among the experimental methods used to analyze the surface characteristics, textural attributes, and magnetic behavior of this adsorbent. It was discovered that a high ferrite concentration in the adsorbent formulation resulted in a blocking of the composite pore structure and a decrease in the CBZ sorption capabilities. From 182.9 to 280 mg/g, the greatest CBZ sorption capabilities

Table 12.2 Preparation, removal parameters, and adsorption characteristics of magnetized carbon-based adsorbents for the elimination of drugs from aqueous systems.

Synthesis of carbon-based adsorbent	Magnetic-material	Carbon source	Drugs	Adsorption capability (mg/g)	Operating conditions		Ref.
					pH	Temp. (°C)	
Magnetic Activated-carbon by co-precipitation impregnation method	Fe_3O_4	Commercial Activated-carbon	Carbamazepine	182.9	-	25	[98]
Magnetic Graphene Oxide (GO) nanoplatelets through co-precipitation	Fe_3O_4	Commercial GO	Amoxicillin	106.38	5	25	[99]
Magnetic BC produced by Direct HTC and Fe-salt impregnation	Fe_3O_4	Sugarcane bagasse	Tetracycline	48.35	6.8	30	[101]
Magnetic AC by Direct hydrothermal technique through impregnation and functionalized with silane	Fe_3O_4	Commercial Activated-carbon	Carbamazepine Clofibric acid Ibuprofen	114 78.7 166.7	7	25	[102]

(*Continued*)

Table 12.2 Preparation, removal parameters, and adsorption characteristics of magnetized carbon-based adsorbents for the elimination of drugs from aqueous systems. (*Continued*)

Synthesis of carbon-based adsorbent	Magnetic-material	Carbon source	Drugs	Adsorption capability (mg/g)	Operating conditions		Ref.
					pH	Temp. (°C)	
Magnetic carbon microspheres (CMS) created by the decomposition and direct thermal impregnation	Fe_3O_4	Glucose	Sulphonamide	27.85	4	25	[103]
Magnetic BC by impregnation and decomposition method	Fe_3O_4	Astragalus membranaceus residue	Ciprofloxacin	68.9	6	25	[104]
Magnetic Activated-carbon by decomposition, impregnation and activation method	γ-Fe_2O_3	Corn stalks	Norfloxacin	7.69	10	25	[105]

(*Continued*)

Table 12.2 Preparation, removal parameters, and adsorption characteristics of magnetized carbon-based adsorbents for the elimination of drugs from aqueous systems. (*Continued*)

Synthesis of carbon-based adsorbent	Magnetic-material	Carbon source	Drugs	Adsorption capability (mg/g)	Operating conditions		Ref.
					pH	Temp. (°C)	
Magnetic Activated-carbon by co-precipitation impregnation method	Fe_3O_4	Commercial Activated-carbon	Carbamazepine	182.9	-	25	[98]
Magnetic Graphene Oxide (GO) nanoplatelets through co-precipitation	Fe_3O_4	Commercial GO	Amoxicillin	106.38	5	25	[99]
Magnetic BC produced by Direct HTC and Fe-salt impregnation	Fe_3O_4	Sugarcane bagasse	Tetracycline	48.35	6.8	30	[101]
Magnetic AC by Direct hydrothermal technique through impregnation and functionalized with silane	Fe_3O_4	Commercial Activated-carbon	Carbamazepine Clofibric acid Ibuprofen	114 78.7 166.7	7	25	[102]

(*Continued*)

Table 12.2 Preparation, removal parameters, and adsorption characteristics of magnetized carbon-based adsorbents for the elimination of drugs from aqueous systems. (*Continued*)

Synthesis of carbon-based adsorbent	Magnetic-material	Carbon source	Drugs	Adsorption capability (mg/g)	Operating conditions		Ref.
					pH	Temp. (°C)	
Magnetic carbon microspheres (CMS) created by the decomposition and direct thermal impregnation	Fe_3O_4	Glucose	Sulphonamide	27.85	4	25	[103]
Magnetic BC by impregnation and decomposition method	Fe_3O_4	Astragalus membranaceus residue	Ciprofloxacin	68.9	6	25	[104]
Magnetic Activated-carbon by decomposition, impregnation and activation method		Corn stalks	Norfloxacin	7.69	10	25	[105]

were measured. Additionally, as the total dissolved `solid (TDS) concentration increased, the CBZ adsorption marginally dropped. The researchers noted that the elimination of CBZ was an endothermic reaction and impulsive and that CBZ sorption was linked to -COOH groups.

According to Kerkez-Kuyumcu *et al.* [99], the antibacterial amoxicillin (AMX) was able to bind to magnetized graphene nanoplatelets (GNPs) in an aqueous solution. This absorbent was made using magnetite nanoparticles (NP's) and GNPs. According to the findings, magnetized GNPs outscored GNPs and magnetite as a whole. The highest adsorption capability of AMX, which was lowered by the rise in solution pH, was 14.10 mg/g. Magnetic GNPs and the AMX adsorption process were linked by electromagnetic interactions. The sudden and exothermic AMX sorption was well-fit by pseudo-second-order kinetics. Fluoroquinolones are another class of antibiotics that constitute a covert threat to human health (FQs). To investigate the adsorption ability of a humic acid-magnetic BC, ciprofloxacin (CIP), norfloxacin (NOR), and enrofloxacin (ENR) were used as FQs adsorbent surface [100].

The author developed a magnetic BC from the stems and leaves of potatoes coated with humic acid(HA). It displayed FQ absorption for these chemicals that were up to 1.8–3.4 times greater than any of those seen for crude and magnetic biochar (BC). Concerning the adsorbates ENR, NOR, and CIP, this magnetic material particularly displayed sorption capabilities of 6.0, 7.5, and 7.1 mg/g, accordingly, at 35 °C. Data from the experimental investigation were adjusted, and Langmuir isotherm and pseudo-second-order kinetic models were used to analyze them. The sorption of fluoroquinolones on this magnetized carbon-based absorbent was linked to electrostatic, hydrophobic, hydrogen-bond, and π-π-bond interactions.

12.6.3 Heavy Metals

A significant water contaminant that has high chemical resistance and poor biodegradability is heavy metals. At some dosage limits, heavy metals are harmful to humans and the environment. The toxic consequences of heavy metals include malignancy, high blood pressure, pulmonary disease, kidney problems, cognitive disorder, intestinal bleeding, and reproductive problems [106]. Heavy industrial activity, such as chemical treatment, oil refineries, electronics manufacture, and mineral processing, is linked to heavy metal pollution of the environment [107].

The eradication of Hg^{2+}, Pb^{2+}, Cd^{2+}, and Cu^{2+} using an RGO covalently functionalized with magnetized dithiocarbamate was examined by Fu and Huang [108]. The decreased and functionalized RGO composites were produced via GO bromination, polyethyleneimine substitution through a nucleophile, interaction with CS_2, and packing of ferric oxide NPs after the GO was prepared using a customized Hummers process. This magnetic adsorbent was characterized using XRD, TEM, FTIR, and XPS methods. The metal adsorption was enhanced by raising the solution's pH level from 2 to 4. The following pattern may be seen in this adsorbent's high adsorption capacities: 165 mg/g of Hg^{2+} > 135 mg/g of Pb^{2+} > 105 mg/g of Cd^{2+} > 90 mg/g of Cu^{2+}.

In affluent, heavy metals and medicines frequently interact, enhancing the toxic potential. To eliminate As^{5+}, Cd^{2+}, and TC, Huang *et al.* [109] investigated the use of magnetic GO and its reduced forms by chemical and annealing processes. These adsorbents' characteristics and the associated process for such contaminants were examined. The solution's pH impacted the sorption of these contaminants in aqueous systems with only a single adsorbate, and the sorption capacity displayed the following trend: magnetic annealed RGO < magnetic chemical RGO < magnetic GO. The magnetic GO in particularly had the best absorption capabilities, with values for TC, Cd^{2+}, and As^{5+} of 13.9, 234, and 252 mg/g.

The elimination of Hg^{2+}, Pb^{2+}, Cr^{3+}, and Cd^{2+}, Maneechakr and Mongkollertlop [110] produced and evaluated a magnetic BC modified with MnO_2. Fe_3O_4 and leftover palm kernel cake remnants were used to create this magnetized adsorbent. SEM-EDS, FTIR, Boehm titration, N2 physisorption, XRD, and VSM were some of the methods used to analyze the physicochemical characteristics of the adsorbent. Even though the process was endothermic, the adsorption temp. did not affect the removal of heavy metals. Cd^{2+}, Cr^{3+}, Pb^{2+}, and Hg^{2+} each had maximal observed sorption capacities of 19.90, 21.20, 49.64, and 13.69 mg/g, accordingly. Thermodynamic parameters as well as experimental measurements were modeled.

12.6.4 Pesticides and Other Pest Controllers

About 30% of the world's agricultural loss is due to pests and related diseases [111]. Insecticides are highly poisonous chemical substances used to eliminate bugs by spreading throughout their surroundings [112]. These substances are hard to bio-transformed, which results in their bioaccumulation over extended periods. Pesticide, herbicide, insecticide, and fungicide use have increased in crop cultivation, adding to the contamination of both water and soil [113, 114].

It can be noted that research findings conducted in the US found five or more pesticides across several specimens taken from various streams, while 400,000 tonnes of pesticides are being used yearly in China [115, 116]. This serves to highlight the severe pollution problems associated with these contaminants. Prolonged exposure to such compounds may have harmful effects on the body's hormonal, endocrine, mutagenesis, genotoxic, and breathing systems [112–114]. It may also cause carcinoma, and cardiovascular, and neurological illnesses. Particularly, the chemical glyphosate is utilized extensively in both agricultural and non-agricultural regions [113]. The preparation, removal parameters, and adsorption characteristics of magnetized carbon-based adsorbents for the elimination of some pesticides from aqueous systems are enlisted in Table 12.3.

Pesticides have been listed as one of the primary contaminants in sources of drinking water in Europe [117]. To create a magnetized adsorbent for the removal of organophosphates from soil and drinking water, MWCNTs and MOF ZIF-8 were used [118]. For this investigation, adsorbates included isazofos, sulfotep, ethoprop, methidathion, profenofos, phosalone, diazinon, and triazophos. According to the report's findings, coordinate polymers played a role in the integration of ZIF-8 molecules into the magnetized MWCNTs layer. The span of this magnetized adsorbent's sorption capabilities for the investigated contaminants was 2.18 to 3.89 mg/g, with ethoprop and profenofos having the smallest and maximum removal rates. Solution with pH 4 being the ideal value, pH had a significant impact on pesticide sorption. Freundlich's solution excelled with the Langmuir isotherm model in correlating the experimental results. It was determined that the interactions between the binding sites of the magnetized adsorbent and the pesticide compounds were controlled by electron pairs, with electrons swapping or sharing.

Mashile *et al.* [119] described the development of a recycled magnetized adsorbent for the concurrent adsorption of "propylparaben (PPB) and methylparaben (MPB)" utilizing chitosan and scrap tyres AC. FTIR, SEM, TEM, and XRD were used to characterize the adsorbent. This compound has a surface area of 1281 m^2/g. PPB and MPB had maximal Langmuir single-layer adsorption capabilities of 90.0 mg/g and 85.9 mg/g.

The atrazine sorption in water was assessed by Agani *et al.* [120] that used a fixed-bed stack packed with a magnetized composite made of AC, chitosan, and magnetite. The composite surface properties were analyzed using XRD and FTIR methods. According to the research observations, a highly porous adsorbent with a total area of 188.50 m^2/g and an atrazine adsorption ability of 62.32 mg/g was constructed. The recyclable capacity of this magnetic substance was confirmed through research using NaOH and HCl on composites restructuring and atrazine desorption.

Table 12.3 Preparation, removal parameters, and adsorption characteristics of magnetized carbon-based adsorbents for the elimination of pesticides, and fluorides from aqueous systems.

Synthesis of carbon-based adsorbent	Magnetic-material	Carbon source	Pesticides	Adsorption capability (mg/g)	Operating conditions		Ref.
					pH	Temp. (°C)	
Pesticides Removal							
Magnetic Graphene oxide (GO) produced from Chemical abrasion and co-precipitation penetration	Mn ferrite	Graphite	Gylphosphate	182.9	-	25	[98]
Magnetic MWCNTs are enhanced with the Zn-Metal organic framework (MOF) by using a co-precipitation impregnation and chemical coordination technique.	Fe_3O_4	Commercial MWCNT's	Triazophos\Diazinon\ Phosalone\ Profenofos Methidathion\ Ethoprop\Sulfotep\ Isazofos	3.12\2.59\3.80 3.89\2.34\2.18\ 2.84\3.00	4	-	[118]

(*Continued*)

Table 12.3 Preparation, removal parameters, and adsorption characteristics of magnetized carbon-based adsorbents for the elimination of pesticides, and fluorides from aqueous systems.

Synthesis of carbon-based adsorbent	Magnetic-material	Carbon source	Pesticides	Adsorption capability (mg/g)	Operating conditions		Ref.
					pH	Temp. (°C)	
Magnetic Activated-carbon coated produced from microwave irradiation and co-precipitation penetration	Fe_3O_4	Waste tyre	Methyl paraben Ethyl paraben	85.9 90.0	6	25	[119]
Activated-carbon composite synthesized by chemical activation by Phosphoric acid and coprecipitation	Fe_3O_4	Peanut Shell	Atrazine	62.32	5	-	[120]
Fluorides Removal							
Impregnation method	γ-Fe_2O_3	Cotton cellulose	Fluoride	18.6	4-9	25	[126]

(Continued)

Table 12.3 Preparation, removal parameters, and adsorption characteristics of magnetized carbon-based adsorbents for the elimination of pesticides, and fluorides from aqueous systems.

Synthesis of carbon-based adsorbent	Magnetic-material	Carbon source	Pesticides	Adsorption capability (mg/g)	Operating conditions		Ref.
					pH	Temp. (°C)	
Impregnated with $CaCl_2$ and $FeCl_3$ and carbonization at 300–700 °C for 2 hours	γ-Fe_2O_3	Cattle dung	Fluoride	15	7.5	25-45	[127]
Oak shell carbonization at 800° C with Fe-salt impregnation	Fe_3O_4	Oak shell	Fluoride	454.54	-	-	[128]
Activated carbon prepared by decomposition and impregnation at 500 °C for 2 hours	Fe_3O_4	Pea Peel	Fluoride	4.71	2-4	25-55	[124]

12.6.5 Fluoride

Fluorine is a groundwater contaminant that is widely distributed in the surroundings [121]. Numerous types of research have demonstrated its innate presence in underground water, minerals, and topsoil [122]. As an additional environmental source of contamination, fertilizers, electrolytic work, metal production, concrete, electronic manufacture, and other industrial companies have been linked to fluoride contamination [122, 123].

According to studies, well over 25 nations have water supply fluoride levels, that are higher than the WHO standard (1.5 mg/L) [122]. As a result, there is a chance that medical issues are linked to persistent fluoride exposure for nearly 300 million people around the world. As a result, moderate and efficient solutions for reducing fluoride levels in drinking water are needed, especially in economically weak nations. The most widely employed commercial fluoridated water depollution technique is adsorption. Some fluorine adsorption process employing magnetized carbon-based adsorbent materials is summarized in Table 12.3.

At 25 °C and pH 2, the highest fluoride adsorption efficiency for BC without and with magnetization, resp., were 6.0 and 4.11 mg/g. Both adsorbents material showed exothermic fluoride sorption between 25 and 45 °C. The pseudo-first-order approach was used to modify the dynamics of fluoride adsorption. Fe_3O_4 and -Fe_2O_3 were used to create a Douglas fir biochar (BC) by Dewage *et al.* [124]. Nitrogen gas physisorption, XPS, Diffraction, transmission electron microscopy (TEM), and SEM/EDX techniques were used to analyze the magnetized BC and its precursors. The distribution of Fe_3O_4 and -Fe_2O_3 granules led to adsorbent magnetism and the creation of a new adsorbent surface. It was discovered that fluoride elimination is endothermic and pH-dependent. The observational isotherms were examined and compared with the Langmuir model. Magnetized BC showed fluorides and nitrates adsorption capabilities of 9 and 15 mg/g, resp. Both adsorbates, fluoride, and nitrate, involved protonated interface-OH group on the iron oxides which forms hydrogen bonds in their respective adsorption mechanisms.

Together with aid of a magnetic AC made from pea peel material Sahu *et al.* [125] assessed the fluoride adsorption that had been coated with $FeCl_3$ and pyrolyzed at 500 °C. Columnar and turbulent tank techniques were used in removal efficiency testing. In adsorption process settings, the process variables were initial fluoride content, solution pH, adsorbent dose, and interaction period. In both bulk and columnar operating condition, this magnetized adsorbent exhibited improved efficiency in the removal of fluorine from contaminated water.

Radioactive isotopes are substances that produce α, β, γ, or neutrons [129]. They can be manufactured or found naturally (earthly and cosmogonic), and examples of the former include nuclear power plants, weapons development grounds, coal burning, and the usage of pesticides, insecticides etc. Due to rock extraction and mishaps at nuclear facilities, some radioisotopes could be detected in water [129, 130].

The toxicity of these contaminants is linked to carcinoma, reproductive abnormalities, and mental retardation [129–131]. To eliminate radionuclides, several magnetized adsorbents have been developed and compared. For instance, caesium (Cs) adsorption uses a magnetized graphene oxide [131]. The co-precipitation action of iron salts was used to create this absorbent, which was then examined using FTIR, Mössbauer, XRD, and TEM methods. The data comparison used the Langmuir adsorption isotherm, which had a capability of 362 mg/g. Since pH favoured cesium elimination, this was determined that ion exchange was considered as the process of cesium absorption.

The elimination of Co (cobalt) and Eu (europium), Maksoud *et al.* [132] produced carbon-based zirconia/spinel ferrous nanomatrices. At various pH and temp. levels, the radionuclides' adsorption was evaluated. The endothermic sorption had adsorption capabilities of 137 and 83 mg/g for Eu and Co, accordingly. The fit of the equilibrium condition and kinematics followed Langmuir and pseudo-second-order models were used.

The most apparent sign of water contamination is dyed, which are intricate chemical compounds [133]. These substances have an impact on aquatic environments and impair sunlight penetration, which affects photosynthesis [133]. With the considerable discharge of dyes and associated substances into waterbodies, the textile and dyeing industries are regarded as the major perpetrators of this environmental contamination. Magnetized carbon adsorbents are effective for the sorption process.

Mubarak *et al.* [134] test the removal of orange-G and methylthioninium chloride (or methylene blue) utilizing a magnetized BC created from fruit waste. This feedstock underwent an initial FeCl3 treatment before pyrolysis in a microwave furnace. The examination of preparation time, swirling rate, adsorbent dose, and pH were used to optimize dye removal. At pH 10, 120 minutes of contact hours, 125 rpm swirling, and 1.0 g of absorbent ratio resulted in the highest amount of methylene blue adsorption. The optimal removing parameters for orange-G were pH 2, 125 rpm swirling, same adsorbent dose, and contact time. The maximal adsorption efficiency for orange-G and methylene blue dyes were 32.36 and 31.25 mg/g.

Endocrine-disturbing chemical estradiol has raised awareness. As a result, Yin *et al.* [135] analysis of the sorption of cuprous ion (Cu^{2+}) and 17β-estradiol (E2) on active and magnetized BC. These adsorbents were produced using a single-step process at three temperatures respectively at 300, 500, and 700 °C. The magnetized activated carbons had larger oxygen functional moieties, porosities, and specific surface area than BC without alteration. Additionally, the elimination of such contaminants from solo and dual aqueous systems was improved by the magnetized adsorbent. The maximum adsorption potential for Cu^{2+} was demonstrated by the adsorbent produced at 300 °C, and the best adsorption rate for E2 was shown by the adsorbent produced at 700 °C, which had an adsorption capability of 153.2 mg/g for only one aqueous system. When compared to E2 adsorption suggested π-π interactions and H- bonds, it seemed that cuprous ion (Cu^{2+}) adsorption proceeded through a chemical complex formation with the oxygen functional groups.

To create a magnetized hybrid composite to capture the dyes reactive red 2 (RR2), acid red 1 (AR1), and Congo red (CR), two biomasses were used: graphene oxide (GO)-carboxymethyl cellulose (CMC) and ferric oxide (Fe_3O_4) [136]. Interface functionalization, packing, H-bonds, and electrostatic interactions have all been involved in the binding of these dyes on the magnetized composite. While the matrices and dye components reacted via ionic interactions with -OH and -COOH groups, there was a complex building specifically with Fe3O4. Up to chemisorption repetitions, this absorbent was recovered using a brine solution (0.1 Molar) with 65% efficiency.

12.7 Carbon Dioxide (CO_2) Adsorption

The increased amounts of CO_2 caused by human activity in the environment are causing climate change, which has been linked to serious repercussions like global warming, severe weather, rising ocean temperatures, rising sea level, and the loss of Arctic sea ice. Global climate change is one of the world›s top challenges in the twenty-first century. This promoted the development of technology that collects carbon dioxide emissions before they join the climate system or stores them from the environment. Due to its benefits over the absorbance method, 2 collection via porous materials has been introduced as an extremely promising way that has attracted considerable attention from researchers and scientists.

In addition to the more common graphite, diamond, and buckyball nanocomposite compositions, a vast array of unique nanocomposite formations also occur. Graphitic nanomaterials structures (CNM) are rising, extremely stable (geomorphic durability comparable to graphite), and cutting-edge substances can be desired CO_2 transformation products. Graphitic carbons and nongraphitic carbons are two different types of allotropes of carbons that can be distinguished based on the level of crystalline order in the third coordinate. Additionally, graphitized and nongraphitized carbons have just been isolated from nongraphitic carbons. From nanotechnology to the micro level, carbons exhibit incredibly unique shapes. The active carbons in granular and powdered are common carbon sorbent materials. Additionally, new forms of carbon are used as promising sorbents, such as textiles, active carbon fibers, and wool fibers, which can be made from a variety of substrates, such as viscose, viscose, hydrocarbon pitches, or coke.

Carbonization and physiochemical stimulation of a variety of organic resources (organic matter, sugars, plant-based, etc.) and synthesized precursors are the processes used to create active carbons for CO_2 adsorption (triazine-based POPs) [137, 138], petroleum pitch [139], etc. Active carbon-based materials are divided into powder-activated carbon, granular activated carbon, sphere-shaped activated carbon, infused carbon, and polymeric carbon based on their surface chemistry and method of production.

Sevilla *et al.* [140] made use of functional activated carbon known as AS-2-600, which was produced by thermally carbonizing sawdust and carbohydrates (cellulose and starch) at 600°C. At 25°C, it produced greater CO2 adsorption capacities of 4.8 mmol/g. The presence of 0.8 nm tiny small pores is primarily responsible for this exceptional capacity to absorb CO_2, with the BET (Brunauer-Emmett-Teller surface area analysis) surface area of 1260 m^2/g playing a less significant impact.

Wahby *et al.* [141] employed C-1012, which has a contact area of 2000 m^2/g and a CO_2 adsorption capacity of 232, 132, and 79.3 mg/g at temps. 0, 25, and 50 °C respectively, at 0.98 atm pressure-. They proposed that, in addition to a large number of small pores, the presence of a finely tuned microporosity considerably enhances the rate of Carbon dioxide capture. Moreover, these minute pores seem to be the primary factor in reaching complete CO_2 adsorption capability even when doing so at a temp. comparable to the temp of anthropogenic CO_2 emission.

Zeolites and MOFs are more polar than carbons, which results in weaker CO_2 adsorption affinity for carbon. By adding external moiety or starting substituent (often nitrogen, sometimes other elements) into the carbon networks, their typical drawbacks can be overcome. By incorporating straightforward nitrogen-functional groups into highly porous structures, it was hoped to increase the ability of CO_2 absorption and attraction toward its choice. The activated carbon's potential for adsorption will rise with its increased N concentration. Nandi *et al.* [142] employed mesoporous polyacrylonitrile (PAN) monoliths that had been carbonized and physically processed to generate high-quality Nitrogen-doped active microporous monoliths (ACMs) with a surface of 2501 m^2/g and 1.8 wt% N for CO_2 capture. At 25°C and 0°C, respectively, these ACMs demonstrated improved CO_2 reduction of 5.14 mmol/ g and 11.51 mmol/g under suitable pressure.

In comparison to other porous C-based adsorbents, Ma *et al.* [143] employed SA-2N-P with BET 1740 m_2/g and its N-content 3.38% for Carbon capture and found that it had a greater capacity of 4.57 mmol gl in ambient conditions and 8.99 mmol gl at 273 K . This can be attributed to the fact that pyrrolic N typically has a far greater impact on the adsorption of CO_2 than pyridinic nitrogen. Similar results were obtained by Dassanayake *et al.* [144] using ACS-4-6-2, which has a surface of 1079 m_2/g and a N-content of 3.7 wt%. These results showed CO_2 absorption of 5.42 mmol/g at 25 °C and 7.73 mmol/g at 0 °C at 1 bar pressure. This improved CO2 absorption was made possible by carefully controlling the activity of polypyrrole carbon microspheres while maintaining their sphere form right away and greatly increasing their ultra-microporosity. The active carbons employed by Waralee *et al.* [145] were made by incorporating urea, oxidizing the air, and activating KOH. They were N- and O-enriched. The samples with a high N content showed better pore absorption at ambient pressures (i.e., 1 bar) and adherence at low pressures at 25°C. It still exhibited superior affinity at 0 °C, although its adsorption rate was marginally reduced. Other nanocomposite C-forms with different surface patterns, such as CNTs, graphene, and graphene oxide (GO) have attracted attention as CO_2 adsorbent materials in recent times. For CO_2 capture, Wang *et al.* [146] employed multilayer N-doped C-nanotubes (NCTs) with carefully controlled aspect ratios. The external layer of Si-nanotubes is coated with various loadings of 3-aminophenol/formaldehyde resin (APF), which is subsequently carbonized in N_2 at 973 K.

The Si (silicon) and Ni (Nickel) templates are then removed by hydrofluoric (HF) acid etching and K_2CO_3 activation. The attained ANCTs demonstrated pore capacity and micro-surface area of 0.45 cm^3/g and 1195 m^2/g, resp. The ensuing absorption increased by 50% to 4.50 mmol/g at 273 K and 1 bar. Mishra *et al.* [147] developed large-scale, inexpensive graphene nanosheets with improved CO_2 removal of 21.6 mmol/g observed at 25 °C and 10.85 atm pressures, as compared to other nanocomposite forms. It was prepared by hydrogen to exfoliate, demonstrating its promise as a CO_2 absorbent variety of industrial usage.

12.8 Conclusion and Future Perspective

The abundance of articles outlining the unique features of carbon nanomaterials, especially graphene and its derived products, which may be directed toward specific types of applications, can be used to justify the special attention paid to them during recent decades. The prestigious Nobel prizes given for the discoveries of fullerene and graphene, as well as the Kavli Prize is given for contributing significantly to progressing awareness and insight of nanomaterials and the invention of carbon nanotubes, were additional factors in determining the distinctiveness of the carbon nanostructured materials.

It has certainly demonstrated the distinction between these carbon-based materials and other traditional noncarbon substances. A few difficulties and restrictions were also thought up in the synthesis protocol of certain such substances, with all the obvious potentials displayed by these carbon nanomaterials. This review contains references and guidelines for the creation of more environmentally friendly, effective, and durable carbon-based products in addition to summarizing in-depth views into the properties and processes of carbon allotropes-based substances used in treating wastewater.

Future studies should focus on the following significant issues that should be effective in the development of better adsorbent material:

1. The identification and selection of alternative green and affordable feedstocks for the creation of adsorbent.
2. Creating precise, affordable, and environmentally friendly preparation methods to increase the aim contaminant specificity of adsorbents (s).
3. The use of leftover feedstock and trash as biomasses to prevent pollution problems, lessen the effects of adsorption manufacturing on the environment, and cost-effectiveness of adsorption products.
4. The assessment of the adsorbent efficiency in actual water sources and functioning under dynamic conditions.

A lot of work has been done into developing efficient and unique magnetic adsorbents that have enriched the research data about the uses of these substances. It is clear from the small number of magnetic adsorbents developed at the industrial scale and the few patented articles on these materials. Overall, these signs show how better magnetic carbon-based materials must be further developed and researched to produce industrial application products and establish their use in practical applications. To achieve successful commercialization, it is crucial to overcome the technological obstacles that arise during the magnitude and industrial output of innovative magnetized adsorbent materials. Despite the multiple voids and present difficulties, there has been substantial and continuing progress in this important area of water purification and processing.

References

1. Hirsch, A., The era of carbon allotropes. *Nat. Mater.*, 9, 11, 868–871, 2010.
2. Haick, H., *Nanotechnology and Nanosensors: Introduction to Nanotechnology*, vol. Ebook Version 12, Technion Israel Institute of Technology, Israel, 2013.
3. Nasir, S., Hussein, M.Z., Zainal, Z., Yusof, N.A., Carbon-based nanomaterials/allotropes: A glimpse of their synthesis, properties and some applications. *Materials*, 11, 295, 2018.
4. Georgakilas, V., Otyepka, M., Bourlinos, A.B., Chandra, V., Kim, N., Kemp, K., Hobza, C.P., Zboril, R., Kim, K.S., Functionalization of graphene: Covalent and non-covalent approaches, derivatives and applications. *Chem. Rev.*, 112, 6156, 2012.
5. Marcia, F.M., Hirsch, A., Hauke, F., Perylene-based non-covalent functionalization of 2D materials. *Flat Chem.*, 1, 89, 2017.
6. Hirsch, A. and Hauke, F., Post-graphene 2D chemistry: The emerging field of molybdenum disulfide and black phosphorus functionalization. *Angew. Chem. Int. Ed.*, 57, 4338, 2018.
7. Punetha, V.D., Rana, S., Yoo, H.J., Chaurasia, J.A., McLeskey Jr., T., Ramasamy, M.S., Sahoo, N.G., Cho, J.W., Functionalization of carbon nanomaterials for advanced polymer nanocomposites: A comparison study between CNT and graphene. *Prog. Polym. Sci.*, 67, 1, 2017.
8. Englert, J.M., Dotzer, C., Yang, G., Schmid, M., Papp, C., Gottfried, J.M., Steinrück, H.P., Spiecker, E., Hauke, F., Hirsch, A., Covalent bulk functionalization of graphene. *Nat. Chem.*, 3, 279, 2011.
9. He, M. and Swager, T.M., Covalent functionalization of carbon nanomaterials with iodonium salts. *Chem. Mater.*, 28, 8542, 2016.
10. Jeon, I.-Y., Chang, D.W., Kumar, N.A., Baek, J.-B., *Carbon Nanotubes-Polymer Nanocomposites*, p. 91, InTech Janeza Trdine, Rijeka, Croatia, 2011.
11. Nakashima, N., Soluble carbon nanotubes: Fundamentals and applications. *Int. J. Nanosci.*, 4, 119, 2005.
12. Kharisov, B., II, Kharissova, O.V., Gutierrez, H.L., Méndez, U.O., Recent advances on the soluble carbon nanotubes. *Ind. Eng. Chem. Res.*, 48, 572, 2009.

13. Xu, Z., Wang, S., Li, Y., Wang., M., Shi, P., Huang, X., Covalent functionalization of graphene oxide with biocompatible poly (ethylene glycol) for delivery of paclitaxel. *ACS Appl. Mater. Interfaces*, 6, 17268, 2014.
14. Pham, T.A., Kumar, N.A., Jeong, Y.T., Covalent functionalization of graphene oxide with polyglycerol and their use as templates for anchoring magnetic nanoparticles. *Synth. Met.*, 160, 2028, 2010.
15. Yang, Q., Pan, X., Clarke, K., Li, K., Covalent functionalization of graphene with polysaccharides. *Ind. Eng. Chem. Res.*, 51, 310, 2012.
16. Gong, X., Liu, G., Li, Y., Yu, D.Y.W., Teoh, W.Y., Functionalized-graphene composites: Fabrication and applications in sustainable energy and environment. *Chem. Mater.*, 28, 8082, 2016.
17. Stergiou, A., Cantón-Vitoria, R., Psarrou, M.N., Economopoulos, S.P., Tagmatarchis, N., Functionalized graphene and targeted applications—Highlighting the road from chemistry to applications. *Prog. Mater. Sci.*, 114, 100683, 2020.
18. Sedaghat, S., Synthesis of clay-CNTs nanocomposite. *J. Nanostructure Chem.*, 3, 24, 2013.
19. Zhang, L. and Webster, T.J., Nanotechnology and nanomaterials: Promises for improved tissue regeneration. *Nano Today*, 4, 66–80, 2009.
20. Wang, H., Yuan, X., Wu, Y., Huang, H., Peng, X., Zeng, G., Zhong, H., Liang, J., Ren, M.-M., Graphene-based materials: Fabrication, characterizationand application for the decontamination of wastewater andwastegas and the hydrogen storage/generation. *Adv. Colloid Interface Sci.*, 195-196, 19–40, 2013.
21. Andre, V.P., Han, D., Shih, W.M., Yan, H., Challenges and Opportunities for structural DNA nanotechnology. *Nat. Nanotechnol.*, 6, 763–772, 2011.
22. Luther, W., *Application of Nano-Technologies in the Energy Sector*, p. 88, Hessen-Nanotech of the Hessian Ministry of Economy, Transport, Urban and Regional Development, Wiesbaden, Germany, 2008.
23. Park, B., Current and future applications of nanotechnology. Issues. *Environ. Sci. Technol.*, 24, 1–19, 2007.
24. Presting, H. and Ko, U., Future nanotechnology developments for automotive applications. *Mater. Sci. Eng. C*, 23, 737–741, 2003.
25. Chadha, U., A review of the function of using carbon nanomaterials in membrane filtration for contaminant removal from wastewater. *Mater. Res. Express*, 9, 012003, 2022.
26. Hao, L., Meng, X., Wang, C., Wu, Q., Wang, Z., Preparation of nickel-doped nanoporous carbon microspheres from metal-organic framework as a recyclable magnetic adsorbent for phthalate esters. *J. Chromatogr. A*, 1605, 460364, 2019.
27. Siddiqui, M.T., Nizamuddin, S., Baloch, H. *et al.*, Fabrication of advance magnetic carbon nano-materials and their potential applications: A review. *J. Environ. Chem. Eng.*, 7, 102812, 2019.

28. Zhang, Y.-N., Niu, Q., Xia, T.G., Nianjun, Y., Guohua, Z., Recent progress on carbon nanomaterials for the electrochemical detection and removal of environmental pollutants. *Nanoscale*, 11, 11979–2398, 2019.
29. Penkova, A., Polotskaya, G., Toikka, A., Trchová, M., Slouf, M., Urbanova, M., Brus, J., Brozová, L., Pientka, Z., Structure and pervaporation properties of poly (phenylene-iso-phthalamide) membranes modified by fullerene C60. *Macromol. Mater. Eng.*, 294, 432–40, 2009.
30. Sterescu, D., Stamatialis, D., Mendes, E., Wübbenhorst, M., Wessling, M., Fullerene-modified poly (2, 6-dimethyl-1, 4-phenylene oxide) gas separation membranes: Why binding is better than dispersing. *Macromolecules*, 39, 9234–42, 2012.
31. Dumée, L.F., Sears, K., Schütz, J., Finn, N., Huynh, C., Hawkins, S., Gray, S., Characterization and evaluation of carbon nanotube Bucky-Paper membranes for direct contact membrane distillation. *J. Membr. Sci.*, 351, 36–43, 2010.
32. Alshahrani, A., Alharbi, A., Alnasser, S., Almihdar, M., Alsuhybani, M., Al Otaibi, B., Enhanced heavy metals removal by a novel carbon nanotubes buckypaper membrane containing a mixture of two biopolymers: Chitosan and i-carrageenan. *Sep. Purif. Technol.*, 276, 119300, 2021.
33. Khan, F.S.A., Mubarak, N.M., Tan, Y.H., Khalid., M., Karri, R.R., Walvekar, R., Mazari, S.A., A comprehensive review on magnetic carbon nanotubes and carbon nanotube-based buckypaper-heavy metal and dyes removal. *J. Hazard. Mater.*, 413, 125375, 2021.
34. Popov, V., Carbon nanotubes: Properties and application. *Mater. Sci. Eng. R Rep.*, 43, 61–102, 2004.
35. Maazinejad, B., Mohammadnia, O., Ali, G.A.M. *et al.*, Taguchi L9 (34) orthogonal array study based on methylene blue removal by single-walled carbon nanotubes-amine: Adsorption optimization using the experimental design method, kinetics, equilibrium and thermodynamics. *J. Mol. Liq.*, 298, 112001, 2020.
36. Lawal, I.A., Lawal, M.M., Azeez, M.A., Ndungu, P., Theoretical and experimental adsorption studies of phenol and crystal violet dye on carbon nanotube functionalized with deep eutectic solvent. *J. Mol. Liq.*, 288, 110895, 2019.
37. Prola, L.D., Machado, F.M., Bergmann, C.P. *et al.*, Adsorption of direct Blue 53 dye from aqueous solutions by multi-walled carbon nanotubes and activated carbon. *J. Environ. Manage.*, 130, 166–175, 2013.
38. Konicki, W., Pełech, I., Mijowska, E., Jasinska, I., Adsorption of anionic dye Direct Red 23 onto magnetic multi-walled carbon nanotubes-Fe_3C nanocomposite: Kinetics, equilibrium and thermodynamics. *Chem. Eng. J.*, 210, 87–95, 2012.
39. Yin, Z., Duoni, Chen, H. *et al.*, Resilient, mesoporous carbon nanotube-based strips as adsorbents of dilute organics in water. *Carbon*, 132, 329–334, 2018.

40. Xu, L., Wang, Z., Ye, S., Sui, X., Removal of *p*-chlorophenol from aqueous solutions by carbon nanotube hybrid polymer adsorbents. *Chem. Eng. Res. Des.*, 123, 76–83, 2017.
41. Xu, L., Wang, S., Zhou, J., Deng, H., Frost, R.L., Column adsorption of 2-naphthol from aqueous solution using carbon nanotubebased composite adsorbent. *Chem. Eng. J.*, 335, 450–457, 2018.
42. Yang, K. and Xing, B., Adsorption of organic compounds by carbon nanomaterials in aqueous phase: Polanyi theory and its application. *Chem. Rev.*, 110, 5989–6008, 2010.
43. Ma, Y., Yang, L., Wu, L. *et al.*, Carbon nanotube supported sludge biochar as an efficient adsorbent for low concentrations of sulfamethoxazole removal. *Sci. Total Environ.*, 718, 137299, 2020.
44. Gong, J.L., Wang, B., Zeng, G.M. *et al.*, Removal of cationic dyes from aqueous solution using magnetic multi-wall carbon nanotube nanocomposite as adsorbent. *J. Hazard. Mater.*, 164, 1517–1522, 2009.
45. Liu, Z.Q., Ma, J., Cui, Y.H., Zhao, L., Zhang, B.P., Catalytic ozonation of diethyl phthalate in aqueous solution using graphite supported zinc oxide. *Sep. Purif. Technol.*, 78, 147–153, 2011.
46. Gonçalves, A.G., Órfão, J.J.M., Pereira, M.F.R., Catalytic ozonation of sulphamethoxazole in the presence of carbon materials: Catalytic performance and reaction pathways. *J. Hazard. Mater.*, 239-240, 167–174, 2012.
47. Qu, R., Xu, B., Meng, L., Wang, L., Wang, Z., Ozonation of indigo enhanced by carboxylated carbon nanotubes: Performance optimization, degradation products, reaction mechanism and toxicity evaluation. *Water Res.*, 68, 316–327, 2015.
48. Liu, Z.Q., Ma, J., Cui, Y.H., Zhang, B.P., Effect of ozonation pretreatment on the surface properties and catalytic activity of multi-walled carbon nanotube. *Appl. Catal. B*, 92, 301–306, 2009.
49. Orge, C.A., Soares, O.S.G.P., Faria, J.L., Pereira, M.F.R., Catalytic advanced oxidation processes for sulfamethoxazole degradation. *J. Environ. Chem. Eng.*, 5, 5599–5607, 2017.
50. Souza-Chaves, B.M., Dezotti, M., Vecitis, C.D., Synergism of ozonation and electrochemical filtration during advanced organic oxidation. *J. Hazard. Mater.*, 382, 121085, 2020.
51. Hu, X., Liu, B., Deng, Y. *et al.*, Adsorption and heterogeneous Fenton degradation of 17α-methyltestosterone on nano Fe_3O_4/MWCNTs in aqueous solution. *Appl. Catal. B*, 107, 274–283, 2011.
52. Li, Z., Shen, C., Liu, Y. *et al.*, Carbon nanotube filter functionalized with iron oxychloride for flow-through electro-fenton. *Appl. Catal. B*, 260, 118204, 2011.
53. Liu, Y., Liu, X., Zhao, Y., Dionysiou, D.D., Aligned α-FeOOH nanorods anchored on a graphene oxide-carbon nanotubes aerogel can serve as an effective Fenton-like oxidation catalyst. *Appl. Catal. B*, 213, 74–86, 2011.

54. Cleveland, V., Bingham, J.P., Kan, E., Heterogeneous fenton degradation of bisphenol A by carbon nanotube-supported Fe_3O_4. *Sep. Purif. Technol.*, 133, 388–395, 2014.
55. Nawaz, M., Shahzad, A., Tahir, K. *et al.*, Photo-fenton reaction for the degradation of sulfamethoxazole using a multi-walled carbon nanotube-$NiFe_2O_4$ composite. *Chem. Eng. J.*, 382, 123053, 2020.
56. Yang, H., Zhou, M., Yang, W., Ren, G., Ma, L., Rolling-made gas diffusion electrode with carbon nanotube for electro-Fenton degradation of acetylsalicylic acid. *Chemosphere*, 206, 439–446, 2018.
57. Yang, Z., Yu, A., Shan, C., Gao, G., Pan, B., Enhanced Fe(III)-mediated Fenton oxidation of atrazine in the presence of functionalized multi-walled carbon nanotubes. *Water Res.*, 137, 37–46, 2018.
58. Liao, Q., Sun, J., Gao, L., Degradation of phenol by heterogeneous fenton reaction using multi-walled carbon nanotube supported Fe_2O_3 catalysts. *Colloids Surf. A Physicochem. Eng. Asp.*, 345, 95–100, 2009.
59. Zhu, R., Zhu, Y., Xian, H. *et al.*, CNTs/ferrihydrite as a highly efficient heterogeneous fenton catalyst for the degradation of bisphenol A: The important role of CNTs in accelerating Fe (III)/Fe (II) cycling. *Appl. Catal. B*, 270, 118891, 2020.
60. Arshadi, M., Abdolmaleki, M.K., Mousavinia, F. *et al.*, Degradation of methyl orange by heterogeneous fenton-like oxidation on a nano-organometallic compound in the presence of multi-walled carbon nanotubes. *Chem. Eng. Res. Des.*, 112, 113–121, 2016.
61. Peng, J., Xue, J., Li, J. *et al.*, Catalytic effect of low concentration carboxylated multi-walled carbon nanotubes on the oxidation of disinfectants with Cl-substituted structure by a Fenton-like system. *Chem. Eng. J.*, 321, 325–334, 2017.
62. Liu, F., Liu, Y., Yao, Q. *et al.*, Supported atomically-precise gold nanoclusters for enhanced flow-through electro-fenton. *Environ. Sci. Technol.*, 54, 5913–5921, 2020.
63. Wei, L., Chen, S. *et al.*, Synthesis of nitrogen-doped carbon nanotubes-$FePO_4$ composite from phosphate residue and its application as effective Fenton-like catalyst for dye degradation. *J. Environ. Sci. (China)*, 76, 188–198, 2019.
64. Raizada, P., Aslam Parwaz Khan, A., Singh, P., Construction of carbon nanotube mediated Fe doped graphitic carbon nitride and Ag3VO4 based Z-scheme heterojunction for H_2O_2 assisted 2,4 dimethyl phenol photodegradation. *Sep. Purif. Technol.*, 247, 116957, 2020.
65. Khataee, A.R., Zarei, M., Moradkhannejhad, L., Application of response surface methodology for optimization of azo dye removal by oxalate catalyzed photoelectro-fenton process using carbon nanotube-PTFE cathode. *Desalination*, 258, 112–119, 2010.
66. Ghanbari, F. and Moradi, M., Application of peroxymonosulfate and its activation methods for degradation of environmental organic pollutants: Review. *Chem. Eng. J.*, 310, 41–62, 2017.

67. Peng, J., Lu, X., Jiang, X. *et al.*, Degradation of atrazine by persulfate activation with copper sulfide (CuS): Kinetics study, degradation pathways and mechanism. *Chem. Eng. J.*, 354, 740–752, 2018.
68. Peng, J., Zhou, H., Liu, W. *et al.*, Synergistic activation of peroxymonosulfate between Co and MnO for bisphenol A degradation with enhanced activity and stability. *Chem. Eng. J.*, 623, 775–786, 2022.
69. Liu, H., Bruton, T.A., Doyle, F.M., Sedlak, D.L., *In situ* chemical oxidation of contaminated groundwater by persulfate: Decomposition by Fe(III)- and Mn(IV)-containing oxides and aquifer materials. *Environ. Sci. Technol.*, 48, 10330–10336, 2014.
70. Jawad, A., Zhan, K., Wang, H. *et al.*, Tuning of persulfate activation from a free radical to a nonradical pathway through the incorporation of non-redox magnesium oxide. *Environ. Sci. Technol.*, 54, 2476–2488, 2020.
71. Sun, H., Kwan, C., Suvorova, A. *et al.*, Catalytic oxidation of organic pollutants on pristine and surface nitrogen-modified carbon nanotubes with sulfate radicals. *Appl. Catal. B*, 154-155, 134–141, 2014.
72. Liu, H., Sun, P., Feng, M. *et al.*, Nitrogen and sulfur co-doped CNT-COOH as an efficient metal-free catalyst for the degradation of UV filter BP-4 based on sulfate radicals. *Appl. Catal. B*, 187, 1–10, 2016.
73. Cheng, X., Guo, H., Zhang, Y., Wu, X., Liu, Y., Non-photochemical production of singlet oxygen via activation of persulfate by carbon nanotubes. *Water Res.*, 113, 80–88, 2017.
74. Sun, S., Jiang, J., Qiu, L. *et al.*, Activation of ferrate by carbon nanotube for enhanced degradation of bromophenols: Kinetics, products, and involvement of Fe(V)/Fe(IV). *Water Res.*, 156, 1–8, 2019.
75. Tian, S.Q., Wang, L., Liu, Y.L. *et al.*, Enhanced permanganate oxidation of sulfamethoxazole and removal of dissolved organics with biochar: Formation of highly oxidative manganese intermediate species and *in situ* activation of biochar. *Environ. Sci. Technol.*, 53, 5282–5291, 2019.
76. Chin, S.S., Chiang, K., Fane, A.G., The stability of polymeric membranes in a TiO_2 photocatalysis process. *J. Membr. Sci.*, 275, 202–211, 2006.
77. Nasrollahzadeh, M., Sajjadi, M., Iravani, S., Varma, R.S., Carbon-based sustainable nanomaterials for water treatment: State-of-art and future perspectives. *Chemosphere*, 263, 128005, 2020.
78. Wu, J., Dan, H., Xingmei, L., Jun, M., Caixian, T., Jianming, X., Remediation of As(III) and Cd(II) cocontamination and its mechanism in aqueous systems by a novel calcium-based magnetic biochar. *J. Hazard. Mater.*, 348, 10–19, 2018.
79. EPA, *National Primary Drinking Water Regulations*, 2021, https://www.epa.gov/ground-water-and-drinking-water/national-primary-drinking-water-regulations#Organic.
80. Mubarak, N., Sahu, J., Abdullah, E., Synthesis of novel magnetic biochar using microwave heating for removal of arsenic from waste water. *Iran. J. Chem. Chem. Eng.*, 37, 111–115, 2018.

81. Petala, E., Yiannis, G., Vasilis, K. *et al.*, Magnetic carbon nanocages: An advanced architecture with surface- and morphology-enhanced removal capacity for arsenites. *ACS Sustain. Chem. Eng.*, 5, 5782–5792, 2017.
82. Gallios, G., Athanasia, K., Ioannis, A., Katarina, S., Miroslava, V., Eleni, A., Adsorption of arsenate by nano scaled activated carbon modified by iron and manganese oxides. *Sustainability*, 9, 1684, 2017.
83. Xu, S., Meina, L., Lihao, Z., Shen, T., Packed bed column investigation on As (V) adsorption using magnetic iron oxide/bagasse biomass carbon composite adsorbent. *IOP Conf. Ser. Mater. Sci. Eng.*, 490, 032032, 2019.
84. WHO, *Guidelines For Drinking-Water Quality*, 4th Edition, World Health Organization, 2006.
85. Brahman, K., Kazi, T., Afridi, H. *et al.*, Simultaneously evaluate the toxic levels of fluoride and arsenic species in underground water of Tharparkar and possible contaminant sources: A multivariate study. *Ecotoxicol. Environ. Saf.*, 89, 95–107, 2013.
86. Sahu, U., Sumanta, S., Siba, S., Raj, K., Cigarette soot activated carbon modified with Fe_3O_4 nanoparticles as an effective adsorbent for As(III) and As(V): Material preparation, characterization and adsorption mechanism study. *J. Mol. Liq.*, 243, 395–405, 2017.
87. Ma, J., Zhiliang, Z., Bo, C. *et al.*, One-pot, large-scale synthesis of magnetic activated carbon nanotubes and their applications for arsenic removal. *J. Mater. Chem. A*, 1, 15, 4662–4666, 2013.
88. Ntim, S. and Mitra, S., Removal of trace arsenic to meet drinking water standards using iron oxide coated multiwall carbon nanotubes. *Chem. Eng. Data*, 56, 2077–2083, 2011.
89. Meng, F., Bowen, Y., Baodong, W., Shibo, D., Zhen, C., Wei, M., Novel dendrimerlike magnetic biosorbent based on modified orange peel waste: Adsorption-reduction behavior of arsenic. *ACS Sustain. Chem. Eng.*, 5, 11, 9692–9700, 2017.
90. Sherlala, A., Raman, A., Bello, M., Synthesis and characterization of magnetic graphene oxide for arsenic removal from aqueous solution. *Environ. Technol.*, 40, 1508–1516, 2019.
91. Sarici-Özdemir, C. and Önal, Y., Study to investigate the importance of mass transfer of naproxen sodium onto activated carbon. *Chem. Eng. Process. Process Intensif.*, 49, 1058–1065, 2010.
92. Bagheri, H., Roostaie, A., Baktash, M., A chitosanpolypyrrole magnetic nanocomposite as μ-sorbent for isolation of naproxen. *Anal. Chim. Acta*, 816, 1–7, 2014.
93. Wei, X., Wang, Y., Chen, J. *et al.*, Adsorption of pharmaceuticals and personal care products by deep eutectic solventsregulated magnetic metal-organic framework adsorbents: Performance and mechanism. *Chem. Eng. J.*, 392, 124808, 2020.
94. Yu, B., Yitong, B., Zhu, M. *et al.*, Adsorption behaviours of tetracycline on magnetic graphene oxide sponge. *Mater. Chem. Phys.*, 198, 283–290, 2017.

95. Sui, Q., Cao, X., Lu, S., Zhao, W., Qiu, Z., Yu, G., Occurrence, sources and fate of pharmaceuticals and personal care products in the groundwater: A review. *Emerg. Contam.*, 1, 14–24, 2015.
96. Huerta-Fontela, M., Galceran, M., Ventura, F., Occurrence and removal of pharmaceuticals and hormones through drinking water treatment. *Water Res.*, 45, 1432–1442, 2011.
97. Hounfodji, J., Kanhounnon, W., Kpotin, G. *et al.*, Molecular insights on the adsorption of some pharmaceutical residues from wastewater on kaolinite surfaces. *Chem. Eng. J.*, 407, 127176, 2021.
98. Baghdadi, M., Ghaffari, E., Aminzadeh, B., Removal of carbamazepine from municipal wastewater effluent using optimally synthesized magnetic activated carbon: Adsorption and sedimentation kinetic studies. *J. Environ. Chem. Eng.*, 4, 3309–3321, 2016.
99. Kerkez-Kuyumcu, O., Bayazit, S., Salam, M., Antibiotic amoxicillin removal from aqueous solution using magnetically modified graphene nanoplatelets. *J. Ind. Eng. Chem.*, 36, 198–205, 2016.
100. Zhao, J., Liang, G., Zhang, X. *et al.*, Coating magnetic biochar with humic acid for high efficient removal of fluoroquinolone antibiotics in water. *Sci. Total Environ.*, 688, 1205–1215, 2019.
101. Rattanachueskul, N., Saning, A., Kaowphong, S., Chumha, N., Chuenchom, L., Magnetic carbon composites with a hierarchical structure for adsorption of tetracycline, prepared from sugarcane bagasse _via_ hydrothermal carbonization coupled with simple heat treatment process. *Bioresour. Technol.*, 226, 164–172, 2017.
102. Wong, K.T., Yoon, Y., Snyder, S., Jang, M., Phenyl-functionalized magnetic palm-based powdered activated carbon for the effective removal of selected pharmaceutical and endocrine-disruptive compounds. *Chemosphere*, 152, 71–80, 2016.
103. Dai, K., Wang, F., Jiang, W., Chen, Y., Mao, J., Bao, J., Magnetic carbon microspheres as a reusable adsorbent for sulfonamide removal from water. *Nanoscale Res. Lett.*, 12, 528, 2017.
104. Kong, X., Liu, Y., Pi, J., Li, W., Liao, Q., Shang, J., Low-cost magnetic herbal biochar: Characterization and application for antibiotic removal. *Environ. Sci. Pollut. Res.*, 24, 6679–6687, 2017.
105. Wang, B., Yan, S., Fa, Y., Deng, Y., Preparation of biochar by simultaneous carbonization, magnetization and activation for norfloxacin removal in water. *Bioresour. Technol.*, 233, 159–165, 2017.
106. Zakir, H., Sharmin, S., Akter, A., Rahman, M., Assessment of health risk of heavy metals and water quality indices for irrigation and drinking suitability of waters: A case study of Jamalpur Sadar area, Bangladesh. *Environ. Adv.*, 2, 100005, 2020.
107. Zhu, F., Zheng, Y., Zhang, B., Dai, Y., A critical review on the electrospun nanofibrous membranes for the adsorption of heavy metals in water treatment. *J. Hazard. Mater.*, 401, 123608, 2021.

108. Fu, W. and Huang, Z., Magnetic dithiocarbamate functionalized reduced graphene oxide for the removal of Cu(II), Cd(II), Pb(II), and Hg(II) ions from aqueous solution: Synthesis, adsorption, and regeneration. *Chemosphere*, 209, 449–456, 2018.
109. Huang, D., Jizi, W., Lu, W. *et al.*, Novel Insight into adsorption and co-adsorption of heavy metal ions and an organic pollutant by magnetic graphene nanomaterials in water. *Chem. Eng. J.*, 358, 1399–1409, 2019.
110. Maneechakr, P. and Mongkollertlop, S., Investigation on adsorption behaviors of heavy metal ions (Cd2+, Cr3+, Hg2+ and Pb2+) through low-cost/active manganese dioxidemodified magnetic biochar derived from palm kernel cake residue. *J. Environ. Chem. Eng.*, 8, 104467, 2020.
111. Cosgrove, S., Jefferson, B., Jarvis, P., Pesticide removal from drinking water sources by adsorption: A review. *Environ. Technol. Rev.*, 8, 1–24, 2019.
112. Hernández, A., Parrón, T., Tsatsakis, A., Requena, M., Alarcón, R., López-Guarnido, O., Toxic effects of pesticide mixtures at a molecular level: Their relevance to human health. *Toxicology*, 307, 136–145, 2013.
113. Ueda, N., Bergamasco, R., Hamoudi, S., Magnetic $MnFe_2O_4$-graphene hybrid composite for efficient removal of glyphosate from water. *Chem. Eng. J.*, 295, 391–402, 2016.
114. Huang, T., Huang, Y., Huang, Y., Yang, Y., Zhao, Y., Toxicity assessment of the herbicide acetochlor in the human liver carcinoma (HepG2) cell line. *Chemosphere*, 24, 125345, 2020.
115. Zeliger, H., *Human Toxicology of Chemical Mixtures in Toxic Consequences Beyond the Impact of One-component Product and Environmental Exposures*, 2nd edition, Elsevier, Oxford, 2011.
116. Zheng, S., Chen, B., Qiu, X., Chen, M., Ma, Z., Yu, X., Distribution and risk assessment of 82 pesticides in Jiulong River and estuary in South China. *Chemosphere*, 144, 1177–1192, 2016.
117. Morton, P., Cassidy, R., Floyd, S., Doody, S., Colin, W., Jordan, P., Approaches to herbicide (MCPA) pollution mitigation in drinking water source catchments using enhanced space and time monitoring. *Sci. Total Environ.*, 755, 142827, 2021.
118. Liu, G., Li, L., Huang, X. *et al.*, Adsorption and removal of organophosphorus pesticides from environmental water and soil samples by using magnetic multi-walled carbon nanotubes @ organic framework ZIF-8. *J. Mater. Sci.*, 53, 10772–10783, 2018.
119. Mashile, G., Mpupa, A., Nqombolo, A., Dimpe, K., Nomngongo, P., Recyclable magnetic waste tyre activated carbon-chitosan composite as an effective adsorbent rapid and simultaneous removal of methylparaben and propylparaben from aqueous solution and wastewater. *J. Water Process Eng.*, 33, 101011, 2020.
120. Agani, I., Fatombi, J., Osseni, S. *et al.*, Removal of atrazine from aqueous solutions onto a magnetite/chitosan/activated carbon composite in a fixed-bed

column system: Optimization using response surface methodology. *RSC Adv.*, 10, 41588–41599, 2020.
121. Meng, X., Yao, Y., Ma, Y., Zhong, N., Alphonse, S., Pei, J., Effect of fluoride in drinking water on the level of 5-methylcytosine in human and rat blood. *Environ. Toxic. Pharma.*, 81, 103511, 2021.
122. Alhassan, A., He, Y., Huang, L., Wu, B., Yan, L., Deng, H., A review on fluoride adsorption using modified bauxite: Surface modification and sorption mechanisms perspectives. *J. Environ. Chem. Eng.*, 8, 104532, 2020.
123. He, J., Yang, Y., Wu, Z., Xie, C., Zhang, K., Kong, L., Review of fluoride removal from water environment by adsorption. *J. Environ. Chem. Eng.*, 8, 104516, 2020.
124. Dewage, N., Liyanage, A., Pittman, C., Mohan, D., Mlsna, T., Fast nitrate and fluoride adsorption and magnetic separation from water on α-Fe_2O_3 and Fe_3O_4 dispersed on Douglas fir biochar. *Bioresour. Technol.*, 263, 258–265, 2018.
125. Sahu, N., Bhan, C., Singh, J., Removal of fluoride from an aqueous solution by batch and column process using activated carbon derived from iron infused Pisum sativum peel: Characterization, isotherm, kinetics study. *Environ. Eng. Res.*, 26, 200241, 2021.
126. Zhao, Y., Li, X., Liu, L., Chen, F., Fluoride removal by Fe(III)-loaded ligand exchange cotton cellulose adsorbent from drinking water. *Carbohydr. Polym.*, 72, 144–150, 2008.
127. Rajkumar, S., Murugesh, S., Sivasankar, V., Darchen, A., Msagati, T., Chaabane, T., Low-cost fluoride adsorbents prepared from a renewable biowaste: Syntheses, characterization and modeling studies. *Arab. J. Chem.*, 12, 3004–3017, 2019.
128. Takmil, F., Esmaeili, H., Mousavi, S., Hashemi, S., Nanomagnetically modified activated carbon prepared by oak shell for treatment of wastewater containing fluoride ion. *Adv. Powder Technol.*, 31, 3236–3245, 2020.
129. Husnain, S.M., Um, W., Lee, W., Chang, Y.S., Magnetitebased adsorbents for sequestration of radionuclides: A review. *RSC Adv.*, 8, 2521–2540, 2018.
130. Choi, J.W., Cho, S., Choi, S.J., Ecofriendly, selective removal of radioactive strontium ions in aqueous solutions using magnetic banana peels. *Sci. Total Environ.*, 778, 146327, 2021.
131. Lujaniene, G., Semcuk, S., Lecinskyte, A. *et al.*, Magnetic graphene oxide based nano-composites for removal of radionuclides and metals from contaminated solutions. *J. Environ. Radioact.*, 166, 166–174, 2017.
132. Maksoud, M., II, Sami, N.M., Hassan, H.S., Bekhit, M., Ashour, A.H., Novel adsorbent based on carbon-modified zirconia/spinel ferrite nanostructures: Evaluation for the removal of cobalt and europium radionuclides from aqueous solutions. *J. Colloid Interface Sci.*, 607, 111–124, 2021.
133. Mall, I., Srivastava, I., Agarwal, N., Removal of orange-G and methyl violet dyes by adsorption onto bagasse fly ash–kinetic study and equilibrium isotherm analyses. *Dyes Pigments*, 69, 210–223, 2006.

134. Mubarak, N., Fo, Y., Al-Salim, H. *et al.*, Removal of methylene blue and orange-G from Waste Water using magnetic biochar. *Int. J. Nanosci.*, 14, 550009, 2015.
135. Yin, Z., Liu, Y., Liu, S. *et al.*, Activated magnetic biochar by one-step synthesis: Enhanced adsorption and coadsorption for 17β-estradiol and copper. *Sci. Total Environ.*, 639, 1530–1542, 2018.
136. Sirajudheen, P., Nikitha, M., Karthikeyan, P., Meenakshi, S., Perceptive removal of toxic azo dyes from water using magnetic Fe_3O_4 reinforced graphene oxide-carboxymethyl cellulose recyclable composite: Adsorption investigation of parametric studies and their mechanisms. *Surf. Interfaces*, 21, 100648, 2020.
137. Shao, L., Li, Y., Huang, J., Liu, Y.-N., Synthesis of triazine-based porous organic polymers derived N-enriched porous carbons for CO_2 capture. *Ind. Eng. Chem. Res.*, 57, 2856–2865, 2018.
138. Shao, L., Wang, S., Liu, M., Huang, J., Liu, Y.-N., Salt template synthesis of nitrogen and sulfur co-doped porous carbons as CO_2 adsorbents. *Chem. Eng. J.*, 339, 509–518, 2018.
139. Wahby, A., Ramos-Fernandez, J.M., Martinez-Escandell, M., Sepulveda-Escribano, A., Silvestre-Albero, J., Rodriguez-Reinoso, F., High-surface-area carbon molecular sieves for selective CO_2 adsorption. *ChemSusChem*, 3, 974–981, 2010.
140. Sevilla, M. and Fuertes, A.B., Sustainable porous carbons with a superior performance for CO_2 capture. *Energy Environ. Sci.*, 4, 1765, 2011.
141. Wahby, A., Silvestre-Albero, J., Sepulveda-Escribano, A., Rodriguez-Reinoso, F., CO_2 adsorption on carbon molecular sieves. *Microporous Mesoporous Mater.*, 164, 280–287, 2012.
142. Nandi, M., Okada, K., Dutta, A., Bhaumik, A., Maruyama, J., Derks, D., Uyama, H., Unprecedented CO_2 uptake over highly porous N-doped activated carbon monoliths prepared by physical activation. *Chem. Commun.*, 48, 10283, 2012.
143. Ma, X., Li, Y., Cao., M., Hu, C., A novel activating strategy to achieve highly porous carbon monoliths for CO_2 capture. *J. Mater. Chem. A*, 2, 4819–4826, 2014.
144. Zhou, J., Liu, Y., Han, Y., Jing, F., Chen, J., Bone derived biochar and magnetic biochar for effective removal of fluoride in groundwater: Effects of synthesis method and coexisting chromium. *Water Environ. Res.*, 91, 588–597, 2019.
145. Dilokekunakul, W., Teerachawanwong, P., Klomkliang, N., Supasitmongkol, S., Chaemchuen, S., Effects of nitrogen and oxygen functional groups and pore width of activated carbon on carbon dioxide capture: Temperature dependence. *Chem. Eng. J.*, 124413, 2020.
146. Wang, H., Xu, C., Zhou, Y., Zhao, W., Zhong, J., Huang, W., Chen, R., Fabrication of hierarchical N-doped carbon nanotubes for CO_2 adsorption. *Nano*, 14, 06, 1950072, 2019.
147. Mishra, A.K. and Ramaprabhu, S., Carbon dioxide adsorption in graphene sheets. *AIP Adv.*, 1, 3, 032152, 2011.

13

Carbonaceous Catalysts for Pollutant Degradation

Poonam Kaswan[1], Santimoy Khilari[2], Ankur Srivastava[3], Girijesh Kumar[2], Pratap K. Chhotaray[4], Mrituanjay D. Pandey[3*] and Kamalakanta Behera[2†]

[1]*Department of Applied Chemistry (CBFS - ASAS), Amity University, Gurugram, Manesar, Panchgaon, Haryana, India*
[2]*Department of Chemistry, Faculty of Science, University of Allahabad, Prayagraj, U.P., India*
[3]*Department of Chemistry, Institute of Science, Banaras Hindu University Varanasi, U.P., India*
[4]*Department of Chemistry, School of Applied Sciences, Centurian University, Ramachandrapur, Jatni, Odisha, India*

Abstract
To enhance the quality of the environment, a lot of studies have been done on carbonaceous materials to develop different architectures and qualities. The development and importance of carbonaceous photocatalysts is briefly highlighted in this chapter, which also includes different semiconductors based on activated carbon, carbon nanotubes, carbon dots, fullerene, graphene, aerogels and carbon sponges, etc. In the present chapter, we have considered two types of strategies to discuss the entitled topic, one is a carbon-based catalyst, and the second is carbonaceous material with other metal composites to enhance the activity. Various difficulties with a metal catalyst, such as limited light adsorption capacity, large band gap energy, and insufficient physicochemical stability, continue to severely restrict photocatalysts for practical uses. To get rid of this, some unique catalysts based on carbon materials have been designed and inserted into a photocatalytic system to enhance activity. They are found to be ideal for various types of applications ranging from the field of renewable energy to waste water management and treatment, thanks to the synergistic combination of metal nanoparticles and carbonaceous nanostructures. Carbon-based metal nanostructures are much essential

**Corresponding author*: mdpandey.chem@bhu.ac.in
†*Corresponding author*: kamala.iitd@gmail.com

Chandrabhan Verma and Chaudhery Mustansar Hussain (eds.) Carbon Allotropes and Composites: Materials for Environment Protection and Remediation, (303–336) © 2023 Scrivener Publishing LLC

to dissolve a variety of colorful and colorless hazardous contaminants, such as organic dye, NO_X elimination, water splitting, etc., when they are subjected to visible light.

Keywords: Carbonaceous catalysts, carbon allotropes, pollutants degradation, nanocomposites, photocatalysts

13.1 Introduction

Environmental pollution has been emerging as a major global crisis nowadays. Water and air pollutions are considered the as prominent challenges for the sustainable progress of human civilization [1–3]. It has a serious negative socioeconomic impact. The relevance of innovation to improve environmental quality has been studied vastly [4, 5]. The polluting agents cause many diseases, such as asthma [6] and other breath-related issues [7]. Carbon is available in plentiful amounts on our planet. In nature, it generally exists with hydrogen as hydrocarbon, which provides a source of renewable energy storage [8–11]. Carbon-based systems play critical role in the fruitful utilization of renewable and sustainable energy sources. This system is used in energy storage devices, biofuels, catalysis, CO_2 capture/conversion, and photocatalysis [12–18]. Carbon-based materials are desirable because they are inexpensive, environmentally friendly, have high chemical and thermal stability, are easy to manufacture, and have a low framework density. Carbon-based materials significantly broaden the potential use of metal nanoparticles in photocatalysis. Carbon-based nanomaterials are well known for their extraordinary functionality in photo-induced applications, which include visible light-assisted capacitor materials, hydrogen generation, CO_2 reduction/utilization, water photoelectrolysis, fine chemical synthesis, etc. [11, 16, 18–23]. Graphene [24, 25], carbon nanotubes [12, 26], carbon dots [27–31], and graphitic carbon nitride (g-C_3N_4) [32–34] have been largely explored as a way of environmental improvement (Figure 13.1) [16, 23].

Metal-organic frameworks (MOFs), in combination with carbonaceous materials (Figure 13.2), like graphene oxide (GO), carbon nanotubes (CNTs), improve their physicochemical properties. It includes chemical robustness, high mechanical strength, unique electronic property, and thermal stability [8]. In the recent years, several nanocatalysts were used to remediate organic contaminants from water through chemical reduction, with $NaBH_4$ being a well-established method [35]. The development of the Earth depends on carbon. The two naturally occurring carbon allotropes

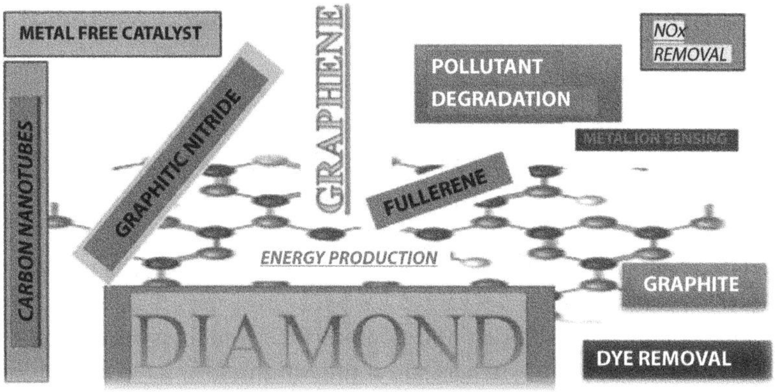

Figure 13.1 Pictorial representation of different carbon-based catalyst and their applications.

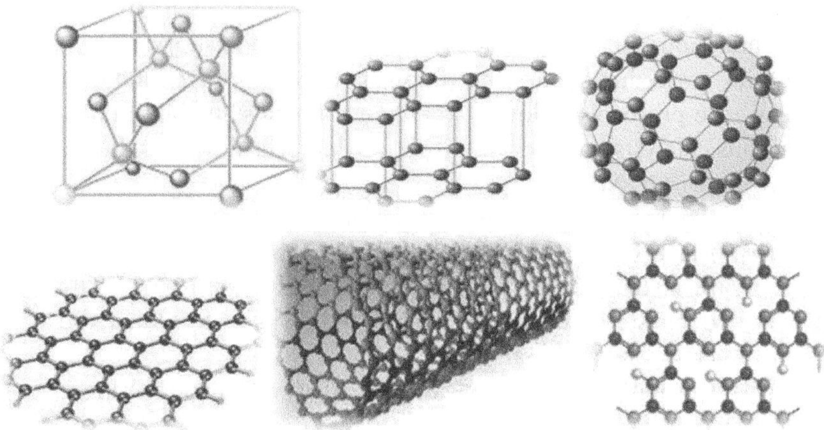

Figure 13.2 Important carbon catalysts/support materials for pollutant degradation.

are diamond and graphitic. Although graphite is a condensed lubricant due to a movable interlamellar link structure in graphite, the carbon-carbon sp^3 hybridization behaves diamond as the stiffest natural substance. Several nanostructured carbons, including graphene, CNTs, fullerene, porous carbon, and others, have been discovered over the past few decades. These carbonaceous nanostructured materials exhibit encouraging activity for the purification of wastewater for the destruction of dye and other pollutants [16, 36–38].

13.2 Strategies to Develop Carbon-Based Material

A range of various carbon materials, i.e., porous carbon, graphene, carbon nanotubes, activated carbon, nitrogen containing carbon-based substances like graphitic-carbon nitride are being utilized as catalyst owing to their noteworthy features, which are listed below (Figure 13.3) [39, 40].

(i) Noteworthy chemical stability both in the acidic, as well as basic medium.
(ii) Better dispersibility and large surface area than other catalysts supports, such as silica and alumina.
(iii) Possibility to tailor the surface chemistry and porosity with opportunity to synthesize/couple metal and metal-based materials on the carbon support.

Catalyst dissolution, sintering, and agglomeration are the major bottleneck of metal nanoparticle catalysts. Catalytic activity can be strengthened by metal nanoparticles supported on a suitable carbon support [41, 42].

Several procedures have been explored to improve photocatalytic efficacy. The protocols followed to enhance the sustainability of the carbonaceous catalysts include graphitization degree is desired to be high-

(i) Nonradical pathways are promoted
(ii) Metal doping can be a unique way to generate different binding sites and reduce the photocatalyst band gap. Easily defects could be incorporated, which helps to reduce the charge recombination.
(iii) Carbon-based materials with metal prevent agglomeration.

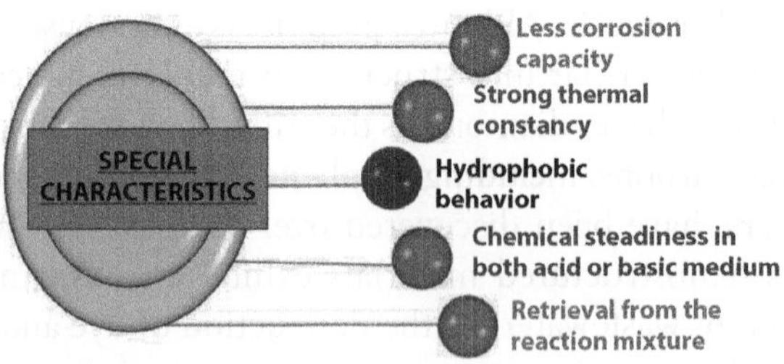

Figure 13.3 Special characteristic properties of a carbon-based catalyst.

(iv) Carbon-based materials are considered to possess greater capacity of absorption and outstanding photocatalytic activity under visible radiation [26].

Metal-free heterogeneous catalysts paid considerable attention to wastewater treatment by peroxymonosulfate (PMS) activation. They considered the good electronic conductivity of polyaniline (PANI), which facilitates the charge transfer during PMS activation. Hence, PANI was utilized as a PMS activator for the methyl orange (MO) catalytic degradation. Degradation of MO on PANI in the presence of PMS is reported to be proceeding through both radical and non-radical processes as depicted in Figure 13.4. Further, it is found that PANI outperformed both metal-based catalysts as well as common N-doped carbon catalysts [43].

Metal carbonaceous material is synthesized by a precursor of metal ions into the setting of carbonaceous material. The following strategies involve in designing this carbon-based catalyst.

- Metals, like Cu, Pd, W, Zr, Fe, etc., are doped to alter carbonaceous materials' electronic and optical properties.
- Metal doping can improve light absorption capacity, reduce band gap energy, and offers faster charge transport for better photocatalysis.
- The free electrons harvest the photonic energy in its conduction band, which is improved by the metal's surface plasmon resonance (SPR) activity. It significantly affects the photocatalytic activation of molecules and the rate of photocatalysis.

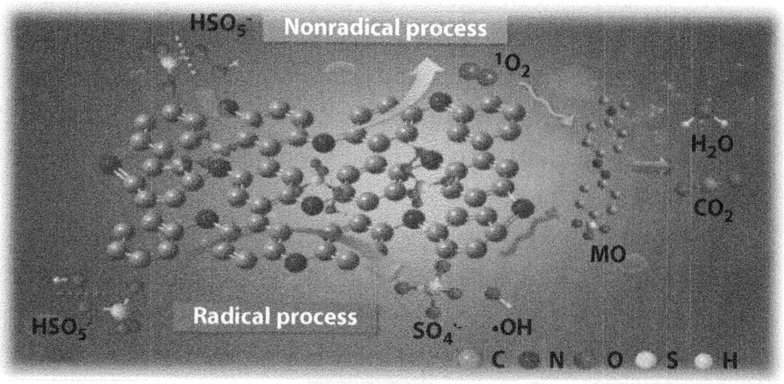

Figure 13.4 The radical and non-radical process by polyaniline [43] (reproduced with permission from ACS).

13.3 Advantages of Carbon-Based Metal Nanocomposites

Metal-free carbon nanocatalysts (zero-dimensional diamond, one-dimensional carbon nanofibers and carbon nanotubes, two dimensional sheets of graphene and graphene containing carbon nitride, as well as 3-dimensional carbon nanostructures) offer many advantages, ease of manufacturing, lower cost, high stability, and eco-friendly advantages. Metal-free carbon nanocatalysts were increasing focus on their applications as photocatalysts for pollutant degradation via persulfate activation and EOCs decontamination [8, 11]. Porous carbons are well known to remove various organic molecules due to their efficient adsorbent nature. Furthermore, porous carbon materials are also used as a catalyst in many chemical processes (Table 13.1):

- The large surface area of carbonaceous materials tends to make the hybrid a potential photocatalyst with a decent capacity for absorbing organic contaminants throughout the water and in the air. The intermediates produced during photocatalysis may be effectively absorbed by the carbon catalyst for further degradation in the next cycle, thereby boosting photocatalytic capacity.
- The photocatalytic activity of the nanocomposite system can be improved by the addition of metal nanoparticles either by reducing charge recombination of photoinduced electron-hole pairs.
- The outstanding stability in acidic and basic circumstances is one of the most significant characteristics.
- By integrating metal nanoparticles with carbon-based supports, it achieves two goals: it reduces photo-generated
- CNT's distinctive optical, electrical, and mechanical features could enhance the photocatalysts that operate on the basis of photo-response semiconductor. Electron-hole pair recombination provides an aqua-friendly environment on the photocatalyst's surface.
- Heterogeneous technology has the remarkable capacity to emerge as an alternate method for wastewater decontamination [46].
- Metal-free Fenton-like system is also popular, as shown in Figure 13.5. The synergetic effect of carbon nanocomposites is commonly implemented for enhanced applications [47].

Table 13.1 Carbon-based catalysts and their various applications for environmental purification.

S. no.	Catalyst	Used precursor	Applications	References
1.	C_3N_4/Fe_3O_4/BiOI nanocomposites	Melamine	Photodegradation of pollutant	[48]
2.	Alkalinized-g-C_3N_4-Fe	Melamine	Photodegradation of Trimethoprim	[49]
3.	$CaTiO_3$/g-C_3N_4/AgBr Z-scheme	Melamine	Photodegradation of dye	[50]
4.	Boron-doped diamond (BDD)	–	Oxidatively remove Rhodamine B (RhB)	[51]
5.	Nanodiamond (ND)	–	Photocatalytic H_2 production	[52]
6.	Au-graphene	Graphene sheet	Photodegradation/photoelectrochemical	[53]
7.	Self-assembled g-C_3N_4 nanoarchitectures	Cyanamide	Hydrogen production	[54]
8.	2D g-C_3N_4/TiN/N-doped graphene	–	Sunlight-assisted photoelectrochemical water reduction	[55]
9.	g-C_3N_4	Urea	Photocatalytic hydrogen evolution	[56]
10.	g-C_3N_4-graphene	Cyanamide	Photocatalytic H_2 evolution	[57]
11.	g-C_3N_4-graphene	Dicyandiamide	H_2 generation	[58]
12.	g-C_3N_4-ZIF	Melamine	CO_2 reduction	[59]
13.	Sulfur assisted g-C_3N_4	Trithiocyanuric acid	Water oxidation	[60]

(Continued)

Table 13.1 Carbon-based catalysts and their various applications for environmental purification. (*Continued*)

S. no.	Catalyst	Used precursor	Applications	References
14.	g-C_3N_4/Pd	Cyanamide	Photocatalytic performance in Suzuki coupling reactions	[61]
15.	S-doped g-C_3N_4	Thiourea and melamine	Photocatalytic carbon dioxide reduction	[62]
16.	g-C_3N_4nanosheet	Melamine	Photocatalytic degradation of target pollutants (rhodamine B, methylene blue, and methyl orange), and hydrogen evolution	[63]
17.	FMFN	Ferrocene	Adsorption of Ciprofloxacin antibiotic	[64]
18.	$BiVO_4$-APS-C_{60}	C_{60}	Photocatalytic degradation of pharmaceutical pollutants	[65]
19.	Activated Fullerene Soot	Fullerene	Electro–Fenton Degradation of Tetracycline	[66]
20.	Fullerene-cored star-shaped polyporphyrinZnCPP-Fullerol-incorporated TiO_2	–	Degradation of rhd B	[67]
21.	C_{60}/$CsPbBr_3$ composite	C_{60}	Photocatalytic CO_2 reduction	[68]
22.	Co/g-C_3N_4	Sm	Photocatalytic HER	[69]

(*Continued*)

Table 13.1 Carbon-based catalysts and their various applications for environmental purification. (*Continued*)

S. no.	Catalyst	Used precursor	Applications	References
23.	Fe-Co oxide/carbon-based magnetic composite	Prussian blue analogue and graphene oxide	Refractory drug degradation by activation peroxymonosulfate (PMS)	[13]
24.	$La_2(MoO_4)_3$/TPCN composite	–	TC degradation	[70]
25.	g-C_3N_4/Fe_3O_4	–	Rhodamine B degradation	[71]
26.	g-C_3N_4	Acetic acid	Rhodamine B degradation	[72]
27.	P, Se-codoped g-C_3N_4 (PSeCN) nanosheets	Melamine	Reduction of numerous nitroaromatic compounds	[73]
28.	Alkalinized g-C_3N_4	–	Photodegradation of volatile organic molecules (VOCs)	[45]
29.	Graphene oxide (GO) nanosheets	–		[74]
30.	TiO_2 and C_{60} 3DOM	–	Photocatalytic degradation of rhodamine B	[75]
31.	$CaFe_2O_4$/g-C_3N_4/CNT	–	Decontamination of polluted water	[76]
32.	Ti^{3+}/TiO_2/CNT	–	Degrade organic pollutants	[77]
33.	ZnO/PPy/CNTs	–	Degradation of 4-Chlorophenol (4-CP)	[78]
34.	Graphene/g-C_3N_4	–	Pollutant degradation	[79]
35.	Cu/g-C_3N_4	Melamine and pluronic P-123	Pollutant degradation	[80]

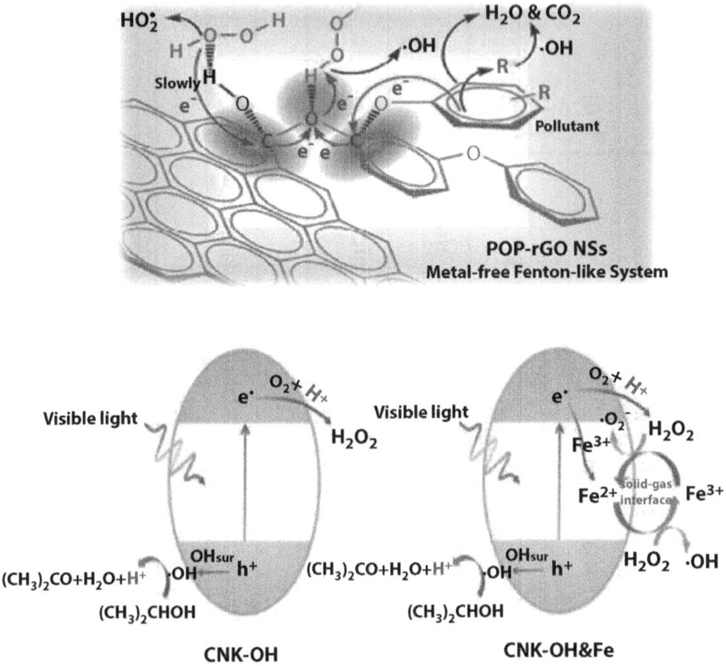

Figure 13.5 Metal-free Fenton-like system (top) [44] and proposed mechanism (bottom) of O-related radicals generation in photocatalytic systems without (CNK-OH) and with solid-gas interfacial Fenton reaction coupling (CNK-OH and Fe) [45] (reproduced with permission from ACS © 2019).

13.4 Methods for the Development of Carbon-Based Nanocomposites

Usual methods to synthesize carbon-based photocatalysts are shown in Figure 13.6. The manufacturing process may be carried out in either a controlled environment (e.g., N_2, Ar as the inert medium) or an air atmosphere without significant changes in the structure, but obtaining large product yield and the surface characteristics can be obtained. Cyanamide to dicyanamide and nitrogen-rich components are simple and effective synthetic techniques for producing somewhat imperfect products. The cyanamide precursor can be converted to g-C3N4 at 550 °C by polymerization [81].

- Phosphorus doped g-C_3N_4 (PCN) was modified with fullerene (C60), considered as conducting substances. A simple ultrasonic method was developed to synthesize C60/PCN

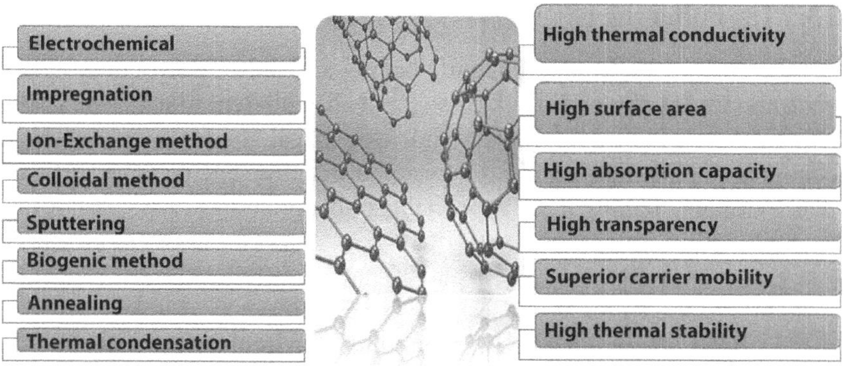

Figure 13.6 Summarized synthesis method and advantages of carbon-based material for the catalytic application.

nanocomposites to apply for the imidacloprid (IMI) pesticide degradation. The photocatalytic IMI degradation indicated a loading of optimal concentration around 0.04 wt.% of C60 [82].

- The electrostatic self-assembly approach was used to create S, N co-doped graphene quantum dots/$(BiO)_2CO_3$ hollow microspheres. Because of their superior electron transport and peroxidase mimetic capabilities, nanosized S, N: GQDs are produced by a bottom-up method for photocatalytic applications. The fundamental roles of S, N: GQDs include photo-induced charge carrier separation and an enzymatic-like catalytic center to accelerate H_2O_2 breakdown to create •OH [83].
- The copolymerization approach was used to synthesized the oxygen and carbon dual-doped g-C_3N_4 by taking ascorbic acid as the oxygen and carbon sources and urea as the g-C_3N_4 precursor that induces electronic structure reconfiguration [84].

13.5 Carbon-Based Photocatalyst

The highly porous carbon architecture promotes pollutant/oxidant adsorption and diffusion throughout the framework, facilitating electron transport inside the conjugated system. Here are a few examples of carbon-based nanostructures.

13.5.1 Fullerene (C_{60})

A fullerene is a molecule frequently having a caged ring structure, with a central core "bucky onion" like a soccer ball. Fullerenes have a very specific property, like comparatively much electrical conductivity with high chemical stability and a three-dimensional structure having large surfaces, that is unique, making them appreciable photocatalysts [23, 85, 86]. A few examples are as follows.

- P-doped g-C_3N_4 (PCN) has been utilized to couple with C60, where P doping was found to be helpful for the separation and transportation of charge where C60 captured electrons from PCN, which reduced the charge separation and increased the surface area almost double according to the Brunauer-Emmett-Teller (BET) isotherm with enhanced catalytic activity [82].
- TiO_2 and C60 3DOM structures may have a synergistic impact on light absorption, charge separation, and transfer efficiency in pollutant degradation catalysis. The 3DOM structure of TiO_2 may increase specific surface areas, exposing more active sites for RhB degradation. The best-performing OM photocatalytic system is reported to have a degradation rate of ~97 % per hour [75].

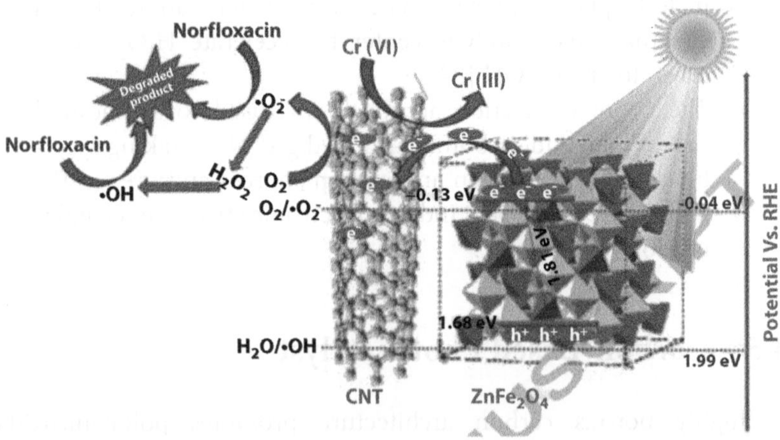

Figure 13.7 ZFO@carbon nanocomposites for the degradation of antibiotics (Norfloxacin) and hexavalent Chromium were investigated under open sunlight illumination. Reproduced by permission of elsevier copyright 2019 [88].

- The catalytic thermal decomposition of PET bottle wastes obtained as a precursor of magnetite nanoparticles and a feedstock and ferrocene catalyst that is FMFNc was primarily conducted via hydrogen bonding and electrostatic π-stacking interaction [87].
- ZnFe2O4 (ZFO)/carbon derivatives that act as photocatalytic degradation of antibiotics such as Norfloxacin and hexacoordinated chromium in the presence of sunlight, as shown in Figure 13.7. It is revealed that ZFO carbon derivatives nanoparticles improve the capacity of visible light absorption as well as excitation separation efficiency [88].

13.5.2 Carbon Nanotubes

- Carbon nanotubes (CNTs) are peculiarly one-dimensional carbon nanostructures, having the property of large surface area to design and preparation of innovative photocatalysts [16, 39, 89]. Carbon nanotubes are utilized for environmental improvement as catalysts [36], such as $CaFe_2O_4$/g-C_3N_4/CNT composite has been used for water purification treatment through visible light by the degradation of toxic pollutants [76]. Moreover, the Ti^{3+}/TiO_2/CNT, as well as Ti^{3+}/TiO_2/C nanocomposites were obtained to destroyed the organic contaminants by reducing band gap of TiO_2 [77].
- The nanocomposite ZnO/PPy/CNTs considered an efficient photocatalyst in respect of 4-Chlorophenol (4-CP) degradation excreted for industrial wastage through radiation of the simulated UV solar light where the optical band gaps of different catalysts were estimated, and then slightly decreasing optical band gap values obtained for the ZnO by the addition of CNTs and PPy in zinc oxide nanomaterials. However, 100% 4-CP degradation in water for ZnO/PPy, ZnO, and ZnO/PPy/CNTs was obtained in general after two to three hour exposure to UV light, and with the exposure to simulated solar radiation, after 150, 120, and 100 minutes, respectively [78].

13.5.3 Graphene

The predominance of 3D GBG photocatalysts for the degradation of environmental pollutants was developed with the particular attention of diverse researchers in promoting the sensible use of 3D GBGs for practical applications in a pollutant-free environment [90]. Finally, in terms of reaction rate and mineralization efficiency, the graphene-based composites were analyzed. It was also analyzed process efficiency and to reduced time for complete degradation [91].

- Optimized N-doped graphene showed a very, high H_2O_2 selectivity, and low electric energy consumption and high H_2O_2 generation rate. N-doped graphene is much more effective for degrading organic pollutants in alkaline and neutral solutions and is less affected by initial pH [92].
- LIB-rGO, which was made of pure graphite, showed intense activity for the organic pollutants by catalytic ozonation. This work investigated suitable catalytic ozonation functions for graphene-based materials and ROS generation and offered an approach for recovering abandoned LIB anodes [93].
- With a 3.9-fold quicker kinetic, the SA-Cu/rGO/PMS system performs well compared to the rGO/PMS system. Surprisingly, the SA-Cu/rGO/PMS system removed nearly all of the total organic carbon (TOC) in 120 minutes, revealing its superior processing capabilities. The improved catalytic efficiency might be attributed to a synergistic interaction between the rGO matrix and efficient Cu active sites, which allowed for forming a high number of reactive species [94].
- Graphene oxide (GO) nanosheets with melamine self-assembled is found more suitable for the 3D nitrogen-doped graphene aerogels (NGA) having porosity and in fused morphological macrostructures. NGA demonstrated outstanding catalytic activity in peroxymonosulfate (PMS) activation for ibuprofen oxidative degradation (IBP). The activity of NGA and surpassed graphene aerogel (GA) and reduced graphene oxide doped with nitrogen was found 44 and 8 times increased [74].
- Without metal, graphene-based catalysts were found to be an excellent precursor in modern oxidation processes (AOPs). The materials (NrGOb) were heated at 500 to 1000 °C, temperature with the NrGOb generated from 800 °C, demonstrating PMS activation at high performance [36, 95].
- A highly efficient with dual reaction centers metal-free Fenton-like catalyst was prepared as 4-phenoxy phenol-based functionalized rGO nanosheets (POP-rGO NSs) where dual

reaction centers are produced on the C-O-C bridge, the electron-rich around O shows an effective H2O2 reducer to •OH, whilst the electron-poor center around C absorbs electrons from adsorbed contaminants and redirects them to the electron-rich area through the C-O-C bridge [44].

13.5.4 Graphitic Carbon Nitride (g-C_3N_4)

Tris-triazine-functionalized motifs connect the polymeric substance graphitic carbon nitride (g-C_3N_4). Because of the existence of H and N atoms, it exhibits electron-rich features, and interconnecting themes. As a result, it is seen as a potential replacement for carbon-based use [96–99]. Except for it, the material's excellent thermal stability allows it to be widely used in heterogeneous catalysis [100]. The colored carbon nitride-based photoelectrodes were found to be satisfactory photocatalytic applications [101, 102] g-C_3N_4-based catalysts included in AOPs beyond photocatalysis. g-C_3N_4-based catalysts in three classical AOPs in the form of catalytic ozonation and previously discussed persulfates activation and Fenton-based processes [103]. A few examples are the following:

- Protonation pre-treatment of a g-C_3N_4 matrix is suitable to allocate growth site silver bromide grains and also for g-C_3N_4 matrix with increment in the porosity. The result of which the g-C_3N_4/AgBr nanocomposites photocatalyst rate over MO is 2.1 and 33.8 times greater when separately applied pure silver bromide and g-C_3N_4 samples, respectively [104].
- Magnetic Fe_3O_4/CeO_2/g-C_3N_4 (FCG) composites successfully synthesized to response under visible light like heterogeneous Fenton-like catalyst to degrade TCH. Photodegradation studies found that the best performance occurred when the composite's mass ratio was 1:0.75:0.75 (F1C0.75G0.75) [105].
- 3D, rGO/g-C_3N_4 quantum dot (CNQD) aerogel hybrids are developed for collecting visible light, created with the approach of hydrothermal. The CNQDs generated a photogenerated charge when exposed to visible light [106].
- g-C_3N_4 has sparked widespread applications in the existing issues that might be solved by carbon nitride nanotubes CNNTs which are electron rich, having a hydrogen evolution rate of 18.06 mmol•h-1•g-1 when exposed to visible light (> 400 nm) and an apparent quantum yield when

exposed to Pt as co-catalyst (420 nm). Approx 92% degradation rate of bisphenol A (BPA) was found simultaneously with the hydrogen generation of 13.63 μmol photocatalytic after irradiation for 5 h [107].

- The acid treatment of bulk graphitic carbon nitride (GCN) synthesized polymeric oxygen-rich exfoliated. In exposure to the short-time photocatalytic visible light ($\lambda > 420$ nm), high valance band hole transfer had been obtained for the photocatalytic mineralization of bisphenol A (BPA), Whereas more BPA removal was gained due to the formation of OH• radicals from the EGCN (99 %). A total pH range catalyst dose and pollutant concentration have been optimized to obtain more BPA degradation. A very slight rate decreases from 0.045 to 0.029 min^{-1} was found on repeated cycle degradation of BPA [108].
- display enhanced OER performance was found much enhanced with the Cl-doped g-C_3N_4 nanorings applying a small overpotential comparable to OER reactivity with the valuable ieon and ruthenium metal oxides. With the outstanding catalytic performance of Cl-doped g-C_3N_4 nanorings generated by the high oxidation ability, more active sites opened, that's why efficient charge transfer occurs [109].
- A 2D g-C_3N_4/graphene hybrid in-plane heterostructure (DCN-Cg) with 32.0 times greater photocatalytic activity, compared to bulk g-C_3N_4, for the unwanted particle degradation was created using a self-conversion approach. The carbon nitride part is considered to transfer photoexcited electrons to graphene by the use of a conjugated bond-seamed structure, which was trapped by an oxygen molecule to create [rad]O^{2-} for pollutant decomposition. Adsorbed impurities are eliminated by shifting electrons to holes in the DCN-Cg valence band [79].
- The protonated g-C_3N_4 and through hydrogen bonding and complexation, 2D Co-MOF molecules adsorb triethanolamine and Eosin Y molecules to form a complete photocatalytic system [110, 111].

13.5.5 Diamond

Diamond composites are becoming increasingly popular for their valuable properties and high-potential uses in various sectors. The performance of multiple types of diamonds like normal metal-diamond, either alloy-diamond and other more oxide-diamond or carbide-diamond or nitride-diamond and organic-diamond composites in either electrocatalytic sensing, water splitting, photoelectrochemistry, and supercapacitor are examined [104].

- Rhodamine B (RhB), which is known as a persistent organic pollutant, can be oxidatively removed by Boron doped diamond (BDD), where it acts as an anode, from the wastewater by electrolysis as well as photo-electrolysis, and having chemically oxidized phosphate or sulfate as supporting electrolytes. RhB is oxidized effectively by chemical oxidation and by electrolysis with the oxidants [51].
- A comparative study was carried out for the fabricated ND sample and pristine ND samples. The fabricated ND found remarkable H_2 gas production through photocatalytic activity with HER for about 400 μmol h^{-1} compared to pristine ND (197 μmol h^{-1}). The hydrogen generation extensively enhanced the quantum yield, a clear indication of hydrogen-terminated sites that work as a reservoir for electrons [52].

13.6 Applications

- The development of various metal and carbon-based heterogeneous catalysts to achieve a pollution-free environment is presented here. Metal nanoparticles and carbon-based nanostructures' synergistic properties have contributed to employment in several unique applications, from H_2O treatment procedures to the renewable energy industry.

13.6.1 Dye Degradation

Nanostructures based on carbon and metal having enhanced compatible quality are required to degrade a broad spectrum of hazardous contaminants, both colored and colorless. This updated review may help you better grasp the many properties of carbon-based metal nanostructures, notably those utilized in photocatalysis and water purification. Three-dimensional graphene-based materials (3D GBMs) have been extensively employed in various areas as developing materials. The application potential of 3D GBMs, in particular, their catalytic activity and degradation processes [22].

Nanostructures based on carbon and metal are essential to generate a wide range of hazardous pollutants, both colored and colorless. This chapter may aid in gaining a better understanding of numerous nanostructures based on carbon and metal used in photocatalysis and water purification applications [112].

- ✓ Doping and surface modification of g-C_3N_4 nanosheets increase photoactivity via thiophene group insertion, which enhances absorption into the g-C_3N_4 matrix, considerably extending the optical characteristic of g-C_3N_4. Furthermore, the thiophene group can enhance the spacing of photogenerated electron-hole pairs by enhancing electron delocalization [70].
- ✓ The Fe_3O_4 nanoparticles are equally distributed across the g-C_3N_4 sheets, avoiding aggregation of the Fe_3O_4 nanoparticles. Even after six cycles of visible-light irradiation, the outstanding behaviour of the g-$C_3N_4Fe_3O_4$ as photocatalysts for degradation of rhodamine B is because of the synergistic effect involving a wide surface area with high visible radiation absorption effectiveness [71].
- ✓ PANI has a higher activity than standard metal-based catalysts. Because of the conjugated chains' oxidation resistance and stable N sites. It produces both radical and non-radical species for breaking down pollutants with equal potential. PANI can increase its performance and catalytically activate PMS in the presence of specific anions [43].
- ✓ The rGO1-CoPc1/PMS system removes dye before the 9 minutes at a high degrading rate utilizing RhB (25 M) and rGO1-CoPc1 (0.5 mg/mL). PMS (0.1 mM).rGO1-CoPc1/PMS system had long-lasting catalytic activity even after 11 cycles, and the elimination efficiency was found to be as high as 100%, indicating that the. Quenching tests revealed that sulfate (SO4•) radicals dominate the breakdown of hydroxyl (OH•) radicals. Furthermore, the catalyst was successfully used to efficiently degrade pentachlorophenol (PCP) within 20 minutes [113].
- ✓ Calcinated acetic acid-assisted supramolecular assembly technique was used to construct a porous photocatalyst having a hexagonal-prismatic g-C_3N_4(ACNH) structure. When compared to bulk g-C_3N_4, an improved RhB degradation was observed using the ACNH sample by nearly 13-fold.

This discovery implies that acid-assisted supramolecular assembly might become a viable approach for producing photocatalysts based on g-C_3N_4 having extended surface area and greater photocatalytic activity [72].

13.6.2 Organic Transformation

The carbon nanocatalysts engineering for the removal of organic impurities (EOCs) through persulfate-activated demonstrates encouraging potential as compared to metal-based counterparts because of its ability to provide a unique advantage of low toxicity and high stability [8, 11].

- Carbon nitride g-C_3N_4 having exfoliated graphite with LED for degradation of phenol is utilized. The expected degrading efficiency is 88.62%, whereas the experimental efficiency will be 83.75% [114].
- The combination of a stable H_2O_2 reproducer with an alkalinized *in situ* g-C_3N_4-based photocatalysts and Fe^{3+} surface-decorator as an activator of H_2O_2 conversion produces incredibly effective and unique activity for volatile organic molecules photodegradation. Using isopropanol photooxidation as a model process, photoactivity increased by 2-3 orders compared to pristine g-C_3N_4; at about 420 nm, it shows a high quantum yield of approximately 50% [45].
- Carbon-based Fe-Co oxide as a magnetic composite was created from graphene oxide and Prussian blue analogue, which demonstrated good efficacy for numerous types of resistant drug breakdown via PMS activation. A minor performance loss in MCFC detected during reuse was recoverable via a simple thermal procedure [13].
- Hydrogen evolution reaction (HER) that functions as dye-sensitive photocatalytic activity was constructed sheet-like Co-MOF with protonated g-C_3N_4 combined to generate a 2D-2D heterojunction self-assembly that adsorbs triethanolamine (TEOA) molecules and EY through complexation and hydrogen bond for photocatalytic activity. The photocatalytic HER rate of g-C_3N_4 is found to be about 1.8 times greater as compared to original g-C_3N_4 whereas the doping of Sm^{3+} provided the photocatalytic enhanced rate of composite material obtained at 73.42 $\mu mol.h^{-1}$ in 5 h by simulated solar radiation.

13.6.3 NOx Removal

The primary air pollutants are NOx emissions from automotive exhaust and fossil fuel burning. It is desirable to employ photocatalytic technologies to reduce NOx levels in the atmosphere. Academics have become interested in the oxidation of NOx to nonvolatile nitrates on the surface of graphene and g-C_3N_4 photocatalysts in earlier periods because of their specific features such as thermal and chemical resilience, with larger specific surface area and increased solar energy (visible radiation) consumption. NOx reduction using graphene and g-C_3N_4 composites. For the photocatalytic NOx removal, oxygen radicals are considered the predominant species at the surface, having a great majority of g-C_3N_4 and graphene-based photocatalysts [115].

13.7 Factors Affecting Degradation

Many factors affect degradation, such as Ph, radiation type and time, temperature, and carbonaceous material, which are discussed here in brief:

13.7.1 Radiation

A comparison experiment revealed that single-wavelength LED immersion light performed better as a photocatalyst than using a full-spectrum xenon lamp. A full-spectrum xenon lamp and a specially constructed single-wavelength LED immersion light were used in photodegradation tests; however, the results demonstrated that the LED immersion light functioned better as a photocatalyst. Visible light has gained so much interest as it is an evident and high-energy source. For this reason, g-C_3N_4 is potential nonmetallic material. However, if not exposed to light, the catalytic oxidation of g-C_3N_4 suffers from lower efficiency due to its chemical inertness [84].

13.7.2 Exfoliation

When exfoliated and bulk g-C_3N_4 were compared, exfoliated g-C_3N_4 destroyed the pollutant in 90 minutes, but bulk g-C_3N_4 only decomposed 25% in 180 minutes. This is because of the increased active sites availability in exfoliated g-C_3N_4, which enhances phenol degradation.

13.7.3 pH

Experiments at various pH levels revealed that the degradation process enhanced at acidic pH. Exfoliated g-C_3N_4 has shown outstanding photocatalytic activity in the photodegradation of other materials [116]. The highest active nitrogen content and hydrogen peroxide selectivity, including the spontaneous [rad]OH production of the optimum N-GE at 400 °C, provided a more significant phenol degradation rate in the neutral pH at about 180 min. The RSM model demonstrates analyzed experimental values of photocatalytic phenol degradation.

13.7.4 Reaction Condition

The topographical morphology for the S-CN/CN can be altered through the void cocoon to 2D nanosheets by substituting the raw material ratio and annealing conditions. The Tafel slope for corresponding data obtained just 57.71 mV/dec, outperforming even the most recent and precious IrO_2 metal catalyst. Furthermore, once exposed to visible radiation, the rate of photodegradation for RhB is 2.38, which is observed to be 47.6 times more significant than bulk CN. The current method may create effective multifunctional metal-free catalysts [117]. The hydrothermally synthesized g-C_3N_4/BiOCl structure for photocatalytic removal of MB is larger than the solitary g-C_3N_4 and BiOCl. When the photocatalysis was prolonged to 240 minutes, the efficiency of degradation of almost g-C_3N_4/BiOCl was 91.53% [118, 119].

13.7.5 Carbonaceous Material

Numerous operation modes tested flat sheet ultrafiltration (UF) membranes for the pharmaceutical compound degradation and diphenhydramine (DP) in visible light [119]. The same experiment was done for the methyl orange, under visible and near UV-vis radiation. The highest photocatalytic activity was observed for the membrane which was prepared with GOT composite (M-GOT) because of the lower GOT band gap energy.

13.8 Challenges

Despite substantial advancements in manufacturing and application of carbon-based catalysts, few obstacles remain-

(i) A novel green approach for carbon-based catalysts must be developed that makes it sustainable.
(ii) Manufacturing processes need to be advanced for metal-carbonaceous material-based nanocomposites.
(iii) Carbon-based materials as metal-free heterogeneous catalysts are less durable.
(iv) Light is required for the activation of the carbon-based photocatalyst.
(v) Catalysis mechanisms are not so precise till now for further study of the factors affecting catalytic performance.
(vi) Production is complicated. Proper synthetic methods, production scaling, and economic potential are expected.

13.9 Conclusion and Future Aspects

We intend to evaluate and summarise contemporary literature and research and development in carbon-based catalytic reduction of different types of contaminants. In contrast, different nanocatalytic moieties are utilized for dye reduction. Moreover, the chemical reduction of nitrophenols and specific dyes in a carbon-based catalyst was also observed. Some points are given to develop new applications:

(i) Cost-effective and eco-friendly nature is essential for the fabricated carbon-based sheeted structures.
(ii) The synergistic effect of the potential application and utility of nanocomposites in catalysis based on carbon materials and biosensing analysis can better understand the synergistic effect.
(iii) The more accurate fabrication of metal nanocomposites based on carbon materials, with accuracy in crystallinity, size and shape with can enhance the photocatalytic utility.
(iv) The combination of other metals and its synergetic effects on the carbon-based metal nanostructures can effectively enhance the progress of the photocatalytic organic dye degradation of pollutants.
(v) Carbon-based metal nanostructures with unique characteristics and applications created for waste-water treatment and environmental issues, as well as energy issues.

Acknowledgments

MDP acknowledge the Institute of Eminence (IoE)-Banaras Hindu University for the funding.

Abbreviations

rGO	Reduced graphene oxide material
RhB	Rhodamine B
PCP	Pentachlorophenol
PMS	Peroxymonosulfate
AOP	Advanced oxidation processes
LIB	Lithium-ion battery
MO	Methyl orange
NGA	Nitrogen-doped graphene aerogels
CTAB	cetyltrimethylammonium bromide
PCN	P-doped g-C_3N_4
FMFNc	functionalized magnetic fullerene nanocomposite
PET	Poly ethylene terephthalate
HER	H_2 evolution reaction
rGO	CoPc Cobalt phthalocyanine-supported reduced graphene oxide material
HNCs	Co-based homobimetallic hollow nanocages
BDD	Boron doped diamond
ND	Nanodiamond
MOF	Metalorganic framework
ZnO/PPy/CNTs	ZnO/polypyrrole/carbon nanotube
4-CP	4-Chlorophenol
GN	Graphene
g-C3N4	Graphitic carbon nitride
M-GOT	Graphene oxide-based ultrafiltration membranes
BET	Brunauer-Emmett-Teller
PLAL	Laser ablation in liquid

References

1. Manisalidis, I., Stavropoulou, E., Stavropoulos, A., Bezirtzoglou, E., Environmental and health impacts of air pollution: A review. *Front. Public Health*, 8, 1–13, 2020. https://doi.org/10.3389/fpubh.2020.00014.

2. Landrigan, P.J. and Fuller, R., Global health and environmental pollution. *Int. J. Public Health*, 60, 7, 761–762, 2015. https://doi.org/10.1007/s00038-015-0706-7.
3. Zandalinas, S., II, Fritschi, F.B., Mittler, R., Global warming, climate change, and environmental pollution: Recipe for a multifactorial stress combination disaster. *Trends Plant Sci.*, 26, 6, 588–599, 2021. https://doi.org/10.1016/j.tplants.2021.02.011.
4. Ibrahim, M. and Vo, X.V., Exploring the relationships among innovation, financial sector development and environmental pollution in selected industrialized countries. *J. Environ. Manage.*, 284, 112057, January 2021. https://doi.org/10.1016/j.jenvman.2021.112057.
5. Wu, H., Xia, Y., Yang, X., Hao, Y., Ren, S., Does environmental pollution promote Çhina's crime rate? A new perspective through government official corruption. *Struct. Change Econ. Dyn.*, 57, 292–307, 2021.
6. Chatkin, J., Correa, L., Santos, U., External environmental pollution as a risk factor for asthma. *Clin. Rev. Allergy Immunol.*, 62, 1, 72–89, 2022.
7. Marcantonio, R., Javeline, D., Field, S., Fuentes, A., Global distribution and coincidence of pollution, climate impacts, and health risk in the anthropocene. *PloS One*, 16, 1–18, 2021. https://doi.org/10.1371/journal.pone.0254060.
8. Zhu, L., Meng, L., Shi, J., Li, J., Zhang, X., Feng, M., Metal-organic frameworks/carbon-based materials for environmental remediation: A state-of-the-art mini-review. *J. Environ. Manage.*, 232, 964–977, 2019. https://doi.org/10.1016/j.jenvman.2018.12.004.
9. Quílez-Bermejo, J., Morallón, E., Cazorla-Amorós, D., Metal-free heteroatom-doped carbon-based catalysts for ORR. A critical assessment about the role of heteroatoms. *Carbon N. Y.*, 165, 434–454, 2020. https://doi.org/10.1016/j.carbon.2020.04.068.
10. Basavaraj, N., Sekar, A., Yadav, R., Review on green carbon dot-based materials for the photocatalytic degradation of dyes: Fundamentals and future perspective. *Mater. Adv.*, 2, 7559–7582, 2021.
11. Guo, D., You, S., Li, F., Liu, Y., Engineering carbon nanocatalysts towards efficient degradation of emerging organic contaminants via persulfate activation: A review. *Chin. Chem. Lett.*, 33, 1–10, 2021. https://doi.org/10.1016/j.cclet.2021.06.027.
12. De, B. and Karak, N., Recent progress in carbon dot-metal based nanohybrids for photochemical and electrochemical applications. *J. Mater. Chem. A*, 5, 5, 1826–1859, 2017. https://doi.org/10.1039/C6TA10220D.
13. Wang, Y., Song, Z., Zhang, Y., Li, R., Liu, Y., Qi, F., Novel carbon based Fe-Co oxides derived from prussian blue analogues activating peroxymonosulfate: Refractory drugs degradation without metal leaching. *Chem. Eng. J.*, 379, 122274–122287, 2019. https://doi.org/10.1016/j.cej.2019.122274.
14. Hu, W., Zheng, M., Xu, B., Wei, Y., Zhu, W., Li, Q., Pang, H., Design of hollow carbon-based materials derived from metal-organic frameworks for

electrocatalysis and electrochemical energy storage. *J. Mater. Chem. A*, 9, 7, 3880–3917, 2021. https://doi.org/10.1039/d0ta10666f.
15. Online, V.A., Saputra, E., Muhammad, S., Sun, H., Wang, S., Activated carbons as green and effective catalysts for aqueous phenol. *RSC Adv.*, 3, 21905–21910, 2013. https://doi.org/10.1039/c3ra42455c.
16. Ge, J., Zhang, Y., Park, S.J., Recent advances in carbonaceous photocatalysts with enhanced photocatalytic performances: A mini review. *Mater. (Basel)*, 12, 12, 1916–1941, 2019. https://doi.org/10.3390/ma12121916.
17. Hu, C., Lin, Y., Connell, J.W., Cheng, H., Gogotsi, Y., Carbon-based metal-free catalysts for energy storage and environmental remediation. *Adv. Mater.*, 31, 1806128, 1–14, 2019. https://doi.org/10.1002/adma.201806128.
18. Phang, S.J. and Tan, L.L., Recent advances in carbon quantum dot (CQD)-based two dimensional materials for photocatalytic applications. *Catal. Sci. Technol.*, 9, 5882–5905, 2019. https://doi.org/10.1039/c9cy01452g.
19. Kuvarega, A.T. and Mamba, B.B., TiO2-based photocatalysis: Toward visible light-responsive photocatalysts through doping and fabrication of carbon-based nanocomposites. *Crit. Rev. Solid State Mater. Sci.*, 42, 1–52, 2016. https://doi.org/10.1080/10408436.2016.1211507.
20. Khan, M.E., State-of-the-art developments in carbon-based metal nanocomposites as a catalyst: Photocatalysis. *Nanoscale Adv.*, 3, 1887–1900, 2021. https://doi.org/10.1039/d1na00041a.
21. Wang, N., Ma, W., Ren, Z., Du, Y., Xu, P., Han, X., Prussian blue analogues derived porous nitrogen-doped carbon microspheres as high-performance metal-free peroxymonosulfate activators for non-radical-dominated degradation of organic pollutants. *J. Mater. Chem. A*, 6, 3, 884–895, 2018. https://doi.org/10.1039/C7TA08472B.
22. He, K., Chen, G., Zeng, G., Chen, A., Huang, Z., Shi, J., Huang, T., Peng, M., Hu, L., Three-dimensional graphene supported catalysts for organic dyes degradation. *Appl. Catal. B Environ.*, 228, 19–28, 2018. https://doi.org/10.1016/j.apcatb.2018.01.061.
23. Pan, Y., Liu, X., Zhang, W., Liu, Z., Zeng, G., Shao, B., Liang, Q., He, Q., Yuan, X., Huang, D. *et al.*, Advances in photocatalysis based on fullerene C60 and its derivatives: Properties, mechanism, synthesis, and applications. *Appl. Catal. B Environ.*, 265, 118579, 2020. https://doi.org/10.1016/j.apcatb.2019.118579.
24. Pedrosa, M., Drazic, G., Tavares, P.B., Figueiredo, J.L., Silva, A.M.T., Metal-free graphene-based catalytic membrane for degradation of organic contaminants by persulfate activation. *Chem. Eng. J.*, 369, 223–232, 2019. https://doi.org/10.1016/j.cej.2019.02.211.
25. Gandhi, M.R., Vasudevan, S., Shibayama, A., Yamada, M., Graphene and graphene-based composites: A rising star in water purification-a comprehensive overview. *ChemistrySelect*, 1, 15, 4358–4385, 2016. https://doi.org/10.1002/slct.201600693.
26. Regulska, E., Olejnik, P., Zubyk, H., Czyrko-Horczak, J., Chaur, M.N., Tomczykowa, M., Butsyk, O., Brzezinski, K., Echegoyen, L.,

Plonska-Brzezinska, M.E., Nanostructural catalyst: Metallophthalocyanine and carbon nano-onion with enhanced visible-light photocatalytic activity towards organic pollutants. *RSC Adv.*, 10, 18, 10910–10920, 2020. https://doi.org/10.1039/d0ra00896f.

27. Mehta, A., Mishra, A., Basu, S., Shetti, N.P., Raghava, K., Saleh, A., Aminabhavi, T.M., Band gap tuning and surface modification of carbon dots for sustainable environmental remediation and photocatalytic hydrogen production-a review. *J. Environ. Manage.*, 250, 2019. https://doi.org/10.1016/j.jenvman.2019.109486.
28. Zhu, B., Li, X., Wang, Y., Liu, N., Tian, Y., Yang, J., Visible-light-driven photocatalytic degradation of RhB by carbon-quantum-dot-modified g-C3N4 on carbon cloth. *CrystEngComm*, 23, 27, 4782–4790, 2021. https://doi.org/10.1039/D1CE00396H.
29. Chen, B.B., Liu, M.L., Huang, C.Z., Carbon dot-based composites for catalytic applications. *Green Chem.*, 22, 13, 4034–4054, 2020. https://doi.org/10.1039/d0gc01014f.
30. Yang, C., Liu, H., Zhong, J., Li, J., Huang, S., Fan, H., Carbon quantum dots modified BiOCl for highly efficient degradation of contaminants benefited from effective generation of O2-. *Mater. Sci. Semicond. Process.*, 136, 106165, 2021. https://doi.org/10.1016/j.mssp.2021.106165.
31. Long, C., Jiang, Z., Shangguan, J., Qing, T., Zhang, P., Feng, B., Applications of carbon dots in environmental pollution control: A review. *Chem. Eng. J.*, 406, 126848, 2021. https://doi.org/10.1016/j.cej.2020.126848.
32. Balakumar, V., Manivannan, R., Chuaicham, C., Karthikeyan, S., Sasaki, K., A simple tactic synthesis of hollow porous graphitic carbon nitride with significantly enhanced photocatalytic performance. *Chem. Commun.*, 57, 55, 6772–6775, 2021. https://doi.org/10.1039/d1cc02355a.
33. Chen, X., Wang, H., Meng, R., Chen, M., Porous graphitic carbon nitride synthesized via using carbon nanotube as a novel recyclable hard template for efficient visible light photocatalytic organic pollutant degradation. *ChemistrySelect*, 4, 20, 6123–6129, 2019. https://doi.org/10.1002/slct.201901227.
34. Wang, A., Wang, C., Fu, L. *et al.*, Recent advances of graphitic carbon nitride-based structures and applications in catalyst, sensing, imaging, and LEDs. *Nano-Micro Lett.*, 9, 1–21, 2017.
35. Naz, M., Rafiq, A., Ikram, M., Haider, A., Ahmad, S.O.A., Haider, J., Naz, S., Elimination of dyes by catalytic reduction in the absence of light: A review. *J. Mater. Sci.*, 56, 28, 15572–15608, 2021. https://doi.org/10.1007/s10853-021-06279-1.
36. Javaid, A., Latif, S., Imran, M., Hussain, N., Rajoka, M.S.R., Iqbal, H.M.N., Bilal, M., Nanohybrids-assisted photocatalytic removal of pharmaceutical pollutants to abate their toxicological effects-a review. *Chemosphere*, 291, 133056, 2022. https://doi.org/https://doi.org/10.1016/j.chemosphere.2021.133056.
37. Xin, L., Hu, J., Xiang, Y., Li, C., Fu, L., Li, Q., Wei, X., Carbon-based nanocomposites as fenton-like catalysts in wastewater treatment applications: A review. *Mater. (Basel)*, 14, 10, 1–21, 2021. https://doi.org/10.3390/ma14102643.

38. Rajendran, S., Inwati, G.K., Yadav, V.K., Choudhary, N., Solanki, M.B., Abdellattif, M.H., Yadav, K.K., Gupta, N., Islam, S., Jeon, B.-H., Enriched catalytic activity of TiO2 nanoparticles supported by activated carbon for noxious pollutant elimination. *Nanomaterials*, 11, 2808–2820, 2021.
39. Pérez-Mayoral, E., Matos, I., Bernardo, M., Ventura, M., Fonseca, I.M., Carbon-based materials for the development of highly dispersed metal catalysts: Towards highly performant catalysts for fine chemical synthesis. *Catalysts*, 10, 12, 1–24, 2020. https://doi.org/10.3390/catal10121407.
40. Zhang, P., Zhu, H., Dai, S., Porous carbon supports: Recent advances with various morphologies and compositions. *ChemCatChem*, 7, 18, 2788–2805, 2015. https://doi.org/10.1002/cctc.201500368.
41. Wang, X.X., Swihart, M.T., Wu, G., Achievements, challenges and perspectives on cathode catalysts in proton exchange membrane fuel cells for transportation. *Nat. Catal.*, 2, 7, 578–589, 2019.
42. Kalz, K.F., Kraehnert, R., Dvoyashkin, M., Dittmeyer, R., GlÃ¤ser, R., Krewer, U., Reuter, K., Grunwaldt, J.D., Future challenges in heterogeneous catalysis: Understanding catalysts under dynamic reaction conditions. *ChemCatChem*, 9, 1, 17–29, 2017. https://doi.org/10.1002/cctc.201600996.
43. Sun, B., Ma, W., Wang, N., Xu, P., Zhang, L., Zhao, H., Lin, K.A., Du, Y., Polyaniline: A new metal-free catalyst for peroxymonosulfate activation with highly efficient and durable removal of organic pollutants. *Environ. Sci. Technol.*, 53, 9771–9780, 2019.
44. Lyu, L., Yu, G., Zhang, L., Hu, C., Sun, Y., 4-phenoxyphenol-functionalized reduced graphene oxide nanosheets: A metal-free fenton-like catalyst for pollutant destruction. *Environ. Sci. Technol.*, 2018, 52, 2, 747–756, 2017. https://doi.org/10.1021/acs.est.7b04865.
45. Li, Y., Ouyang, S., Xu, H., Wang, X., Bi, Y., Zhang, Y., Ye, J., Constructing solid-gas-interfacial fenton reaction over alkalinized-C3N4 photocatalyst to achieve apparent quantum yield of 49% at 420 Nm. *J. Am. Chem. Soc.*, 138, 40, 13289–13297, 2016. https://doi.org/10.1021/jacs.6b07272.
46. Oh, W. and Lim, T., Design and application of heterogeneous catalysts as peroxydisulfate activator for organics removal: An overview. *Chem. Eng. J.*, 358, 110–133, 2018. https://doi.org/10.1016/j.cej.2018.09.203.
47. Wang, X., Vasileff, A., Jiao, Y., Zheng, Y., Qiao, S.Z., Electronic and structural engineering of carbon-based metal-free electrocatalysts for water splitting. *Adv. Mater.*, 31, 13, 1–8, 2019. https://doi.org/10.1002/adma.201803625.
48. Mousavi, M. and Habibi-Yangjeh, A., Magnetically separable ternary G-C3N4/Fe3O4/BiOI nanocomposites: Novel visible-light-driven photocatalysts based on graphitic carbon nitride. *J. Colloid Interface Sci.*, 465, 83–92, 2016. https://doi.org/10.1016/j.jcis.2015.11.057.
49. Li, R., Huang, J., Cai, M., Huang, J., Xie, Z., Zhang, Q., Liu, Y., Liu, H., Lv, W., Liu, G., Activation of peroxymonosulfate by Fe doped G-C3N4/graphene under visible light irradiation for trimethoprim degradation. *J. Hazard. Mater.*, 384, 121435, 2020. https://doi.org/10.1016/j.jhazmat.2019.121435.

50. Yan, Y., Yang, H., Yi, Z., Li, R., Xian, T., Design of ternary CaTiO3/g-C3N4/AgBr Z-scheme heterostructured photocatalysts and their application for dye photodegradation. *Solid State Sci.*, 100, 106102, 2020. https://doi.org/10.1016/j.solidstatesciences.2019.106102.
51. de Araújo, D.M., Sáez, C., Cañizares, P., Rodrigo, M.A., Martínez-Huitle, C.A., Improving the catalytic effect of peroxodisulfate and peroxodiphosphate electrochemically generated at diamond electrode by activation with light irradiation. *Chemosphere*, 207, 774–780, 2018. https://doi.org/10.1016/j.chemosphere.2018.05.121.
52. Khan, M., Hayat, A., Mane, S.K.B., Li, T., Shaishta, N., Alei, D., Zhao, T.K., Ullah, A., Zada, A., Rehman, A.U. *et al.*, Functionalized nano diamond composites for photocatalytic hydrogen evolution and effective pollutant degradation. *Int. J. Hydrogen Energy*, 45, 53, 29070–29081, 2020. https://doi.org/10.1016/j.ijhydene.2020.07.274.
53. Khan, M.E., Khan, M.M., Cho, M.H., Green synthesis, photocatalytic and photoelectrochemical performance of an Au-graphene nanocomposite. *RSC Adv.*, 5, 34, 26897–26904, 2015. https://doi.org/10.1039/c5ra01864a.
54. Jiang, Y., Qu, F., Tian, L., Yang, X., Zou, Z., Lin, Z., Self-assembled g-C3N4 nanoarchitectures with boosted photocatalytic solar-to-hydrogen efficiency. *Appl. Surf. Sci.*, 487, 59–67, 2019. https://doi.org/10.1016/j.apsusc.2019.05.056.
55. Shanker, G.S., Bhosale, R., Ogale, S., Nag, A., 2D nanocomposite of G-C3N4 and TiN embedded N-doped graphene for photoelectrochemical reduction of water using sunlight. *Adv. Mater. Interfaces*, 5, 24, 1–8, 2018. https://doi.org/10.1002/admi.201801488.
56. Xu, C.Q. and Zhang, W.D., Facile synthesis of nitrogen deficient G-C3N4 by copolymerization of urea and formamide for efficient photocatalytic hydrogen evolution. *Mol. Catal.*, 453, 85–92, 2018. https://doi.org/10.1016/j.mcat.2018.04.029.
57. Yu, Q., Guo, S., Li, X., Zhang, M., Template free fabrication of porous G-C3N4 graphene hybrid with enhanced photocatalytic capability under visible light. *Mater. Technol.*, 29, 3, 172–178, 2014. https://doi.org/10.1179/1753555714Y.0000000126.
58. Zou, J.P., Wang, L.C., Luo, J., Nie, Y.C., Xing, Q.J., Luo, X.B., Du, H.M., Luo, S.L., Suib, S.L., Synthesis and efficient visible light photocatalytic H2 evolution of a metal-free g-C3N4/graphene quantum dots hybrid photocatalyst. *Appl. Catal. B Environ.*, 193, 103–109, 2016. https://doi.org/10.1016/j.apcatb.2016.04.017.
59. Meng, Y., Zhang, L., Jiu, H., Zhang, Q., Zhang, H., Ren, W., Sun, Y., Li, D., Construction of G-C3N4/ZIF-67 photocatalyst with enhanced photocatalytic CO2 reduction activity. *Mater. Sci. Semicond. Process.*, 95, 35–41, 2019. https://doi.org/10.1016/j.mssp.2019.02.010.
60. Zhang, J., Sun, J., Maeda, K., Domen, K., Liu, P., Antonietti, M., Fu, X., Wang, X., Sulfur-mediated synthesis of carbon nitride: Band-gap engineering and

improved functions for photocatalysis. *Energy Environ. Sci.*, 4, 3, 675–678, 2011. https://doi.org/10.1039/c0ee00418a.
61. Wang, N., Ma, L., Wang, J., Zhang, Y., Jiang, R., Graphitic carbon nitride (g-C3N4) supported palladium species: An efficient heterogeneous photocatalyst surpassing homogeneous thermal heating systems for suzuki coupling. *ChemPlusChem*, 84, 8, 1164–1168, 2019. https://doi.org/10.1002/cplu.201900360.
62. Wang, K., Li, Q., Liu, B., Cheng, B., Ho, W., Yu, J., Sulfur-doped g-C3N4 with enhanced photocatalytic CO2-reduction performance. *Appl. Catal. B Environ.*, 176–177, 44–52, 2015. https://doi.org/https://doi.org/10.1016/j.apcatb.2015.03.045.
63. Singla, S., Sharma, S., Basu, S., Shetti, N.P., Aminabhavi, T.M., Photocatalytic water splitting hydrogen production via environmental benign carbon based nanomaterials. *Int. J. Hydrogen Energy*, 46, 68, 33696–33717, 2021. https://doi.org/10.1016/j.ijhydene.2021.07.187.
64. Elessawy, N.A., Elnouby, M., Gouda, M.H., Hamad, H.A., Taha, N.A., Gouda, M., Mohy Eldin, M.S., Ciprofloxacin removal using magnetic fullerene nanocomposite obtained from sustainable PET bottle wastes: Adsorption process optimization, kinetics, isotherm, regeneration and recycling studies. *Chemosphere*, 239, 124728, 2020. https://doi.org/10.1016/j.chemosphere.2019.124728.
65. Rokesh, K., Sakar, M., Do, T.-O., Integration of aminosilicate functionalized-fullerene (C60) QDs on bismuth vanadate (BiVO4) nanolayers for the photocatalytic degradation of pharmaceutical pollutant. *Catal. Today*, 407, 252-259, 2021. https://doi.org/https://doi.org/10.1016/j.cattod.2021.10.006.
66. Li, C., Lian, Y., Zhou, Q., Synergistic effect between Ni and Ce dual active centers initiated by activated fullerene soot for electro-fenton degradation of tetracycline. *Catalysts*, 2022. https://doi.org/10.3390/catal12050509.
67. Wu, Z.Y., Xu, Y.J., Huang, L.J., Zhang, Q.X., Tang, D.L., Fullerene-cored star-shaped polyporphyrin-incorporated TiO2 as photocatalysts for the enhanced degradation of rhodamine B. *J. Environ. Chem. Eng.*, 9, 5, 106142, 2021. https://doi.org/10.1016/j.jece.2021.106142.
68. Zhang, Z., Shu, M., Jiang, Y., Xu, J., Fullerene modified CsPbBr3 perovskite nanocrystals for efficient charge separation and photocatalytic CO2 reduction. *Chem. Eng. J.*, 414, 128889, 2021. https://doi.org/https://doi.org/10.1016/j.cej.2021.128889.
69. Su, P., Zhou, M., Song, G., Du, X., Lu, X., Efficient H2O2 generation and spontaneous [Rad]OH conversion for in-situ phenol degradation on nitrogen-doped graphene: Pyrolysis temperature regulation and catalyst regeneration mechanism. *J. Hazard. Mater.*, 397, 122681, 2020. https://doi.org/10.1016/j.jhazmat.2020.122681.
70. Zhang, J., Zhang, W., Yue, L., Hu, X., Lin, H., Zhao, L., He, Y., Thiophene insertion and lanthanum molybdate modification of G-C3N4 for enhanced

visible-light-driven photoactivity in tetracycline degradation. *Appl. Surf. Sci.*, 592, 153337, 2022. https://doi.org/https://doi.org/10.1016/j.apsusc.2022.153337.

71. Kumar, S., Kumar, B., Baruah, A., Shanker, V., Synthesis of magnetically separable and recyclable g-CN4-Fe3O4 hybrid nanocomposites with enhanced photocatalytic performance under visible-light irradiation. *J. Phys. Chem. C*, 117, 49, 26135–26143, 2013.
72. Chen, Y., Ding, F., Khaing, A., Yang, D., Jiang, Z., Acetic acid-assisted supramolecular assembly synthesis of porous g-C3N4 hexagonal prism with excellent photocatalytic activity. *Appl. Surf. Sci.*, 479, 757–764, 2019. https://doi.org/10.1016/j.apsusc.2019.02.176.
73. Piri, M., Heravi, M.M., Elhampour, A., Nemati, F., Silver nanoparticles supported on P, Se-codoped g-C3N4 nanosheet as a novel heterogeneous catalyst for reduction of nitroaromatics to their corresponding amines. *J. Mol. Struct.*, 1242, 130646, 2021. https://doi.org/10.1016/j.molstruc.2021.130646.
74. Wang, J., Duan, X., Dong, Q., Meng, F., Tan, X., Liu, S., Wang, S., Facile synthesis of N-doped 3D graphene aerogel and its excellent performance in catalytic degradation of antibiotic contaminants in water. *Carbon N. Y.*, 144, 781–790, 2019. https://doi.org/10.1016/j.carbon.2019.01.003.
75. Hou, J., Lan, X., Shi, J., Xin, L., Wang, L., The synergistic effect of fullerene and 3D ordered macroporous structure on promoting photocatalytic performance. *J. Photochem. Photobiol. A Chem.*, 400, 112680, 2020. https://doi.org/10.1016/j.jphotochem.2020.112680.
76. Synthesis of a well-dispersed CaFe2O4_g-C3N4_CNT composite towards the degradation of toxic water pollutants under visible light. *RSC Adv.*, 9, 25750–25761, 2019.
77. Akter, J., Hanif, Md.A., Islam, Md.A., Sapkota, K.P., Hahn, J.R., Selective growth of Ti^3+/TiO_2/CNT and Ti^3+/TiO_2/C nanocomposite for enhanced visible-light utilization to degrade organic pollutants by lowering TiO_2-bandgap. *Sci. Rep.*, 11, 1, 9490, 2021.
78. Xing, H., An, C., Wu, L., Xu, Y., Fabrication of ZnO/polypyrrole/carbon nanotube nanocomposite and its application for removal of 4-chlorophenol from wastewater. *Int. J. Electrochem. Sci.*, 17, 2022. https://doi.org/10.20964/2022.05.26.
79. Li, F., Tang, M., Li, T., Zhang, L., Hu, C., Two-dimensional graphene/g-C3N4 in-plane hybrid heterostructure for enhanced photocatalytic activity with surface-adsorbed pollutants assistant. *Appl. Catal. B Environ.*, 268, 118397, 2020. https://doi.org/10.1016/j.apcatb.2019.118397.
80. Shahzeydi, A., Ghiaci, M., Farrokhpour, H., Shahvar, A., Sun, M., Saraji, M., Facile and green synthesis of copper nanoparticles loaded on the amorphous carbon nitride for the oxidation of cyclohexane. *J. Chem. Eng.*, 370, 2019. https://doi.org/10.1016/j.cej.2019.03.227.
81. Tahir, M., Cao, C., Mahmood, N., Butt, F.K., Mahmood, A., Idrees, F., Hussain, S., Tanveer, M., Aslam, I., Multifunctional g-C(3)N(4) nanofibers:

A template-free fabrication and enhanced optical, electrochemical, and photocatalyst properties, *ACS Appl. Mater. Interfaces*, 6, 1258–1265, 2014.
82. Sudhaik, A., Raizada, P., Singh, P., Hosseini-Bandegharaei, A., Thakur, V.K., Nguyen, V.H., Highly effective degradation of imidacloprid by H2O2/fullerene decorated P-doped g-C3N4 photocatalyst. *J. Environ. Chem. Eng.*, 8, 6, 104483, 2020. https://doi.org/10.1016/j.jece.2020.104483.
83. Ding, P., Di, J., Chen, X., Ji, M., Gu, K., Yin, S., Liu, G., Zhang, F., Xia, J., Li, H., S, N codoped graphene quantum dots embedded in (BiO)2CO3: Incorporating enzymatic-like catalysis in photocatalysis. *ACS Sustain. Chem. Eng.*, 6, 8, 10229-10240, 2018. https://doi.org/10.1021/acssuschemeng.8b01552.
84. Zhu, Y., Chen, Z., Gao, Y., Hu, C., General synthesis of carbon and oxygen dual-doped graphitic carbon nitride via copolymerization for non-photochemical oxidation of organic pollutant. *J. Hazard. Mater.*, 394, 122578, 2020. https://doi.org/10.1016/j.jhazmat.2020.122578.
85. Chen, M., Guan, R., Yang, S., Hybrids of fullerenes and 2D nanomaterials. *Adv. Sci.*, 6, 1, 1800941-1800946, 2019. https://doi.org/10.1002/advs.201800941.
86. Kausar, A., Advances in polymer/fullerene nanocomposite: A review on essential features and applications. *Polym. Plast. Technol. Eng.*, 1999, 1–42, 2006.
87. Elessawy, N.A., El-Sayed, E.M., Ali, S., Elkady, M.F., Elnouby, M., Hamad, H.A., One-pot green synthesis of magnetic fullerene nanocomposite for adsorption characteristics. *J. Water Process Eng.*, 34, 101047, 2020. https://doi.org/10.1016/j.jwpe.2019.101047.
88. Behera, A., Mansingh, S., Das, K.K., Synergistic ZnFe2O4-carbon allotropes nanocomposite photocatalyst for norfloxacin degradation and Cr (VI) reduction. *J. Colloid Interface Sci.*, 544, 96–111, 2019. https://doi.org/10.1016/j.jcis.2019.02.056.
89. Baig, N., Kammakakam, I., Falath, W., Kammakakam, I., Nanomaterials: A review of synthesis methods, properties, recent progress, and challenges. *Mater. Adv.*, 2, 6, 1821–1871, 2021. https://doi.org/10.1039/d0ma00807a.
90. Zhang, F., Li, Y.H., Li, J.Y., Tang, Z.R., Xu, Y.J., 3D graphene-based gel photocatalysts for environmental pollutants degradation. *Environ. Pollut.*, 253, 365–376, 2019. https://doi.org/10.1016/j.envpol.2019.06.089.
91. Olatunde, O.C. and Onwudiwe, D.C., Graphene-based composites as catalysts for the degradation of pharmaceuticals. *Int. J. Environ. Res. Public Health*, 18, 1–36, 2021. https://doi.org/10.3390/ijerph18041529.
92. Su, P., Zhou, M., Lu, X., Yang, W., Ren, G., Cai, J., Electrochemical catalytic mechanism of N-doped graphene for enhanced H2O2 yield and *in-situ* degradation of organic pollutant. *Appl. Catal. B Environ.*, 245, 583–595, 2019. https://doi.org/10.1016/j.apcatb.2018.12.075.
93. Wang, Y., Cao, H., Chen, L., Chen, C., Duan, X., Xie, Y., Song, W., Sun, H., Wang, S., Tailored synthesis of active reduced graphene oxides from waste graphite: Structural defects and pollutant-dependent reactive radicals in

aqueous organics decontamination. *Appl. Catal. B Environ.*, 229, 71–80, 2018. https://doi.org/10.1016/j.apcatb.2018.02.010.
94. Chen, F., Wu, X.L., Yang, L., Chen, C., Lin, H., Chen, J., Efficient degradation and mineralization of antibiotics via heterogeneous activation of peroxymonosulfate by using graphene supported single-atom Cu catalyst. *Chem. Eng. J.*, 394, 124904, 2020. https://doi.org/10.1016/j.cej.2020.124904.
95. Hirani, R.A.K., Asif, A.H., Rafique, N., Wu, H., Shi, L., Zhang, S., Duan, X., Wang, S., Saunders, M., Sun, H., Three-dimensional nitrogen-doped graphene oxide beads for catalytic degradation of aqueous pollutants. *Chem. Eng. J.*, 446, 137042, 2022. https://doi.org/https://doi.org/10.1016/j.cej.2022.137042.
96. Wang, Y., Zhang, C., Zeng, Y., Cai, W., Wan, S., Li, Z., Zhang, S., Zhong, Q., Ag and MOFs-derived hollow Co3O4 decorated in the 3D g-C3N4 for creating dual transferring channels of electrons and holes to boost CO2 photoreduction performance. *J. Colloid Interface Sci.*, 609, 901–909, 2022. https://doi.org/https://doi.org/10.1016/j.jcis.2021.11.153.
97. Xiao, M., Jiao, Y., Luo, B., Wang, S., Chen, P., Lyu, M., Du, A., Wang, L., Understanding the roles of carbon in carbon/g-C3N4 based photocatalysts for H2 evolution. *Nano Res.*, 22, 48–54, 2021. https://doi.org/10.1007/s12274-021-3897-7.
98. Zhang, H., Bao, C., Hu, X., Wen, Y., Li, K., Zhang, H., Synthesis of tunnel structured G-C3N4 through a facile vapor deposition method using SBA-15 and KIT-6 as templates and their photocatalytic degradation of tetracycline hydrochloride and phenol. *J. Environ. Chem. Eng.*, 10, 3, 107871, 2022. https://doi.org/https://doi.org/10.1016/j.jece.2022.107871.
99. Tubular G-C3N4_carbon framework for high-efficiency photocatalytic degradation of methylene blue. *RSC Adv.*, 11, 18519–18524, 2021.
100. Li, F., Li, T., Zhang, L., Jin, Y., Hu, C., Enhancing photocatalytic performance by direct photo-excited electron transfer from organic pollutants to low-polymerized graphitic carbon nitride with more C-NH/NH2 exposure. *Appl. Catal. B Environ.*, 296, 120316, 2021. https://doi.org/10.1016/j.apcatb.2021.120316.
101. Xiong, W., Huang, M., Huang, F., Zhang, R.Q., Colorful carbon nitride based composite films. *Appl. Surf. Sci.*, 511, 145535, 2020. https://doi.org/10.1016/j.apsusc.2020.145535.
102. Ajiboye, T.O., Kuvarega, A.T., Onwudiwe, D.C., Graphitic carbon nitride-based catalysts and their applications: A review. *Nano-Struct. Nano-Objects*, 24, 100577, 2020. https://doi.org/10.1016/j.nanoso.2020.100577.
103. Yang, Y., Li, X., Zhou, C., Xiong, W., Zeng, G., Huang, D., Zhang, C., Wang, W., Song, B., Tang, X. *et al.*, Recent advances in application of graphitic carbon nitride-based catalysts for degrading organic contaminants in water through advanced oxidation processes beyond photocatalysis: A critical review. *Water Res.*, 184, 116200, 2020. https://doi.org/10.1016/j.watres.2020.116200.

104. Jian, Z., Xu, J., Yang, N., Han, S., Jiang, X., A perspective on diamond composites and their electrochemical applications. *Curr. Opin. Electrochem.*, 30, 100835, 2021. https://doi.org/10.1016/j.coelec.2021.100835.
105. Wang, S., Long, J., Jiang, T., Shao, L., Li, D., Xie, X., Xu, F., Magnetic Fe3O4/CeO2/g-C3N4 composites with a visible-light response as a high efficiency fenton photocatalyst to synergistically degrade tetracycline. *Sep. Purif. Technol.*, 278, 119609, 2022. https://doi.org/10.1016/j.seppur.2021.119609.
106. Zhao, P., Jin, B., Yan, J., Peng, R., Fabrication of recyclable reduced graphene oxide/graphitic carbon nitride quantum dot aerogel hybrids with enhanced photocatalytic activity. *RSC Adv.*, 11, 56, 35147–35155, 2021. https://doi.org/10.1039/d1ra06347b.
107. Xu, F., Mo, Z., Yan, J., Fu, J., Song, Y., El-Alamia, W., Wu, X., Li, H., Xu, H., Nitrogen-rich graphitic carbon nitride nanotubes for photocatalytic hydrogen evolution with simultaneous contaminant degradation. *J. Colloid Interface Sci.*, 560, 555–564, 2020. https://doi.org/10.1016/j.jcis.2019.10.089.
108. Sahu, R.S., Shih, Y.H., Chen, W.L., New insights of metal free 2D graphitic carbon nitride for photocatalytic degradation of bisphenol A. *J. Hazard. Mater.*, 402, 123509, 2021. https://doi.org/10.1016/j.jhazmat.2020.123509.
109. Han, E.X., Li, Y.Y., Wang, Q.H., Huang, W.Q., Luo, L., Hu, W., Huang, G.F., Chlorine doped graphitic carbon nitride nanorings as an efficient photoresponsive catalyst for water oxidation and organic decomposition. *J. Mater. Sci. Technol.*, 35, 10, 2288–2296, 2019. https://doi.org/10.1016/j.jmst.2019.05.057.
110. Vasseghian, Y., Le, V.T., Joo, S.-W., Dragoi, E.-N., Kamyab, H., Chelliapan, S., KlemeÅ¡, J.J., Spotlighting graphene-based catalysts for the mitigation of environmentally hazardous pollutants to cleaner production: A review. *J. Clean. Prod.*, 365, 132702, 2022. https://doi.org/https://doi.org/10.1016/j.jclepro.2022.132702.
111. Li, X., Yu, J., Wageh, S., Al-Ghamdi, A.A., Xie, J., Graphene in photocatalysis: A review. *Small*, 12, 48, 6640–6696, 2016. https://doi.org/10.1002/smll.201600382.
112. Shinde, S.S., Sami, A., Lee, J.H., Sulfur mediated graphitic carbon nitride/S-Se-graphene as a metal-free hybrid photocatalyst for pollutant degradation and water splitting. *Carbon N. Y.*, 96, 929–936, 2016. https://doi.org/10.1016/j.carbon.2015.10.050.
113. Marinescu, C., Ben Ali, M., Hamdi, A., Cherifi, Y., Barras, A., Coffinier, Y., Somacescu, S., Raditoiu, V., Szunerits, S., Boukherroub, R., Cobalt phthalocyanine-supported reduced graphene oxide: A highly efficient catalyst for heterogeneous activation of peroxymonosulfate for rhodamine B and pentachlorophenol degradation. *Chem. Eng. J.*, 336, 465–475, 2018. https://doi.org/10.1016/j.cej.2017.12.009.
114. Rana, A.G. and Minceva, M., Analysis of photocatalytic degradation of phenol with exfoliated graphitic carbon nitride and light-emitting diodes using response surface methodology. *Catalysts*, 11, 898–912, 2021.

115. Nikokavoura, A. and Trapalis, C., Graphene and G-C 3 N 4 based photocatalysts for NO x removal: A review. *Appl. Surf. Sci.*, 430, 18–52, 2018. https://doi.org/10.1016/j.apsusc.2017.08.192.
116. Rana, A.G., Tasbihi, M., Schwarze, M., Minceva, M., Efficient advanced oxidation process (AOP) for photocatalytic contaminant degradation using exfoliated metal-free graphitic carbon nitride and visible light-emitting diodes. *Catalysts*, 11, 662–678, 2021.
117. Zhang, H.W., Lu, Y.X., Li, B., Huang, G.F., Zeng, F., Li, Y.Y., Pan, A., Chai, Y.F., Huang, W.Q., Acid-induced topological morphology modulation of graphitic carbon nitride homojunctions as advanced metal-free catalysts for OER and pollutant degradation. *J. Mater. Sci. Technol.*, 86, 210–218, 2021. https://doi.org/10.1016/j.jmst.2021.01.030.
118. Qi, S., Liu, X., Zhang, R., Zhang, Y., Xu, H., Preparation and photocatalytic properties of G-C3N4/BiOCl heterojunction. *Inorg. Chem. Commun.*, 133, 108907, 2021. https://doi.org/10.1016/j.inoche.2021.108907.
119. Pastrana-MartÃnez, L.M., Morales-Torres, S., Figueiredo, J.L., Faria, J.L., Silva, A.M.T., Graphene oxide based ultrafiltration membranes for photocatalytic degradation of organic pollutants in salty water. *Water Res.*, 77, 179–190, 2015. https://doi.org/10.1016/j.watres.2015.03.014.

14

Importance and Contribution of Carbon Allotropes in a Green and Sustainable Environment

Ajay K. Singh

Department of Applied Science & Engineering, Indian Institute of Technology, Roorkee/Saharanpur, India

Abstract

Increased industrialization, construction activity, etc. are leading to deforestation, pollution, the rise of the earth's temperature due to the depletion of the ozone layer and greenhouse effect causing snow to melt and rising water levels in seas, thus enhancing chances of the flood, the disappearance of flora and fauna on our planet. Industrialization is responsible for increased events of corrosion loss of metals e.g., steel, brass etc. causing strain on natural resources e.g., ores thus affecting their sustainability. It may affect industrial operations if steps are not taken to control metal corrosion. For this purpose, the use of corrosion inhibitors is prevalent due to their cost effectiveness, retrofitting their application easily. Conventional inhibitors, hitherto used are good but are toxic. Recent Government regulations mandate use of green technology to avoid toxication and its ill effects. This initiated search for green inhibitors about two decades ago. Advancement in nanotechnology has seen the synthesis of carbon allotropes namely C_{60} fullerenes, graphene and carbon nanotubes which are more efficient inhibitors and non-toxic in nature. They interact strongly with metallic surfaces to provide holiday-free coats with polymers. They thus show better inhibition efficiency and may prove effective in maintaining sustainable environment from the natural resources of metals and pollution point of view.

***Keywords*:** Environment, sustainability, corrosion, metals, inhibitor, C^{60} fullerene, graphene, carbon nanotubes

Email: ajaysfpt@iitr.ac.in

Chandrabhan Verma and Chaudhery Mustansar Hussain (eds.) Carbon Allotropes and Composites: Materials for Environment Protection and Remediation, (337–382) © 2023 Scrivener Publishing LLC

14.1 Introduction

14.1.1 Basic Aspects of Sustainability

'Sustain' means to keep something/somebody available for doing work without any interruption. If we use a lamp of 40 watts for lighting a room, it spends 40 watts or 40 joules of electrical energy in 1 sec. and that same amount i.e. 40 joules is available immediately afterwards second after second without interruption, the lamp will continue to lighten the room uninterruptedly. Thus, electrical energy is available for lighting the room continuously. This is an example of 'sustainable operation'. If suppose this operation is not sustainable i.e. electrical energy is not available continuously, the lamp will also not lighten the room continuously and that situation will not be preferred, if somebody has to work at night for several minutes or hours. Much bigger problems arise in industries if we don't have a sustainable source of energy which is why the need for energy sustainability. Light energy from the sun is an example of a 'sustainable process of light emission. Because the nuclear process in sun undergoes fusion reactions and generates energy continuously producing light and another form of energy uninterruptedly. This is very important for the life of human beings, animals, plants etc. because this sunlight keeps the earth warm and helps one survive during winter. One may have noticed that days are warmer while nights are colder due to the absence of light during the night. Photosynthesis is one reaction in plants, which occurs under sunlight, which keeps them green and alive. Sun is also called a source of Renewable energy. When a source of energy can be replenished faster than it is used, it is termed a "renewable source" of energy. Other examples of sustainable operations or renewable energy are wind energy, water etc. Similarly, fossil fuels, e.g., coal, oil, etc. were also considered earlier as sources of renewable energy, but anymore now. Whatever we see on our planet and the hospitable conditions for life for the last hundreds of years is a consequence of the equilibrium that the living species have been able to maintain with nature so far.

On the other hand, the tendency of human nature is to pursue efforts to reduce the cases of their ailments to live life happily and for a longer time. There has seen a continuous increase in the maximum age limit of human beings. Consequently, in nature, humans have become the longest-living species on earth. Our ancestors first appeared on Earth approximately 6 million years ago, whilst the modern version of humans evolved approximately 200,000 years ago. Another trait of human beings is to continuously think and practice improving their lifestyle. In this process infrastructure development, industrialization etc. took place. With time, as we drew more and more from nature in the form of wood, coal, oil, metals, chemicals etc. and develop more and more industries, the ill effects of these changes started appearing. These changes could be detrimental for our planet as well as for humans and other species existing on our planet. Our planet may become uninhabitable for all of us without any plants, crops etc. It will thus body, marine life e.g., fish etc. and the planet will be a desert. If we have to keep our planet hospitable for livelihood, we need to pay attention and seek answers to (i) why these destructive changes are happening; (ii) what are the causative factors responsible for these changes; (iii) how can we avoid these changes by adopting newer technology/procedures, etc. Seeking about answering these questions and applying them in practice will lead to our goal of 'sustainability'.

14.2 Changes Being Observed in Nature and Their Effect on Our Planet

14.2.1 Water, Air, and Effect on Energy Generation

Due to urbanization, industrialization and the ever-increasing population on our planet, the sustenance of water and air has become a problem. Consequently, the amount of water in waterfalls and water flow in rivers is reduced thereby affecting turbine rotation which reduces hydroelectricity production. Wind energy production is getting affected by a larger number of buildings of greater heights coming up with each passing day so that they can accommodate the increased population. These buildings put obstacles in the free flow of air, thereby reducing the airspeed and rotation of wind turbines producing lesser electricity. A UN study estimates an annual increase in global population to ~ 83 million. So all the resources will be shared by 83 million additional people every year. Nature takes approximately one and a half years to restock and regenerate what the planet's inhabitants ingest in 1 year. This gives an idea about the strain on the resources due to the increased population.

14.2.2 Air Quality

Due to increased industrialisation, pollutants are entering the atmosphere making air polluted and unbreathable leading to respiratory problems in kids and older people. A clean air act (CAA) was agreed to sign in 1970 to accomplish national air quality levels that protect public health and welfare

Table 14.1 Air quality index vs air quality.

AQI	AQI conditions are
0–50	Good
51–100	Moderate
151–200	Unhealthy for sensitive Group
201–300	Very Unhealthy
301–500	Hazardous

by constructing air quality standards to control this. These aid in imposing limits on emissions of air pollutants from both stationary and mobile sources (e.g., industries, housing societies etc.) and mobile (e.g., railways, buses, ships etc.) sources. To achieve this, the CAA sets Air quality standards, with the help of EPA's (Environmental Protection Agency), for six air pollutants namely carbon Monoxide, lead, ground-level ozone, nitrogen dioxide, particulate matter and sulphur dioxide. The Clean Air Act has been a huge success. More than 200,000 early deaths and 18 million cases of respiration illness among children were avoided during its first 20 years. Table 14.1 depicts an idea of the quality of air classified in terms of the Air Quality Index (AQI).

14.2.3 Pollution (Air/Water)

This phenomenon is the introduction of unwanted, harmful to human life, chemicals through used/dirty water, aqueous media or air, and discharge of industry as gaseous emission or as wastewater. These could be carcinogenic, toxic chemicals e.g. Bisphenol A (BPA) an additive found in plastics and is toxic, used in making plastic bottles, cups etc., tetrachloroethylene found in dry cleaning solutions, reported to be cancerous and banned in the US since 2007, formaldehyde found in cleaners e.g. toilet bowl cleaner, rust removers, turpentine, non-stick cookware, fresheners etc., Chlorine, nitrous oxide, formaldehyde (carcinogen) found as preservatives in many household products, chlorine dioxide, perchlorate, reported being toxic, with some carcinogenic as well, used in bleach and some fertilizers.

Water is required for both domestic and industrial purposes. It is now infested with a variety of wastes, ranging from floating plastic bags to chemical waste, transforming our bodies of water into a poisonous pool. Water is a natural solvent, so most pollutants can easily dissolve in it and contaminate it. Water pollution has an impact on organisms, amphibians, and vegetation that live in or near water. Several people per day die as a result of drinking polluted and infected water. According to an Economist report (2008), over tens of thousands of children die each day from diarrheal illness, and the figures have been alarming in the last 5 years. Water pollution is caused by both natural and man-made activities. When particulates such as fertilizers, pesticide residues, and waste leach from garbage dumps and sewer tanks, they enter an aquifer and pollute it. According to the EPA of the United States, 50% of rivers and streams and one-third of lakes are polluted and unsafe for swimming, fishing, and drinking. Nutrient pollution, such as nitrates and phosphates, which are required by animals and plants to grow, causes significant pollution of freshwater sources as a result of farm waste and fertilizer runoff.

Each household's sewage and effluent are chemically treated and released into rivers/seas with fresh water. Sewage water contains pathogens, a common water pollutant, as well as other harmful bacteria and chemicals that can cause serious health problems and, as a result, diseases. Household garbage, such as paper, plastic, food, aluminium, rubber, and glass, is gathered and tossed into a river or sea, where it takes 2 weeks to 200 years to decay. They pollute the environment and endanger marine life.

Industries generate massive amounts of waste, much of which contains toxic chemicals and pollutants that pollute the air and water. They contain hazardous chemicals such as lead, mercury, sulphur, nitrates, asbestos, and others. Many industries with poor waste management discharge waste into freshwater, which flows into canals, rivers, and eventually into the sea. When coal and oil are burned in industry, they emit a significant amount of ash into the atmosphere. Acid rain is caused when toxic chemical particles combine with water vapor. Nuclear energy is generated through nuclear fission or fusion. Uranium, a highly toxic and radioactive element, is used to generate nuclear energy. To avoid a nuclear accident, the radioactive nuclear waste generated must be disposed of.

14.2.4 Carbon Footprint

Plastic is used in making various items for (i) industrial use e.g. large storage tanks, reaction chambers, pipes, valves, pumps, sheets to act as roof, barriers against corrosion of metals, (ii) domestic/transportation purpose e.g. tyres, tubes for vehicles, carry bags large and small, small/larger containers etc. Plastic is non-degradable (remains on earth for a very long time so a contributor of pollution). Plastic, being non-degradable, plastic waste is a big problem for normal activity. Danes generate 350K tonnes of plastic every year and only 60 K tonnes (~ 17%) is recycled for reuse. The rest remains as such and becomes a source of greenhouse gases i.e., carbon dioxide and methane – toxic gases. A measure of total amount of carbon dioxide (CO_2) and methane (CH_4) emissions/discharge from industries, due to various treatments/processing followed in them, is expressed as "carbon dioxide equivalent". Global average of annual carbon footprint per person in 2014 was 5 tonnes CO_2e (CO_2 equivalent). Average carbon footprint for a US citizen is 16 tons (3 times the global average). The Paris Agreement, between various countries on carbon footprints, is to limit it around 6 and ½ tonnes per person per year (an increase by only 1 and ½ tonne per year per person) by 2030, so that temperature of our planet does not increase by more than 1.5°C by then, with respect to preindustrial level temperature. The higher the carbon footprint generated the more polluting environment. These result in global warming leading to Human induced climate change now experienced as increased frequency of severe draught and flood, urban air pollution, toxic acid rain, coastal and ocean acidification.

14.2.5 Green House Effect

Green House Effect is an effect on the climate and hence on the temperature existing on earth because of the sun light passing through the gases present in the atmosphere above earth. Main greenhouse gases are CO_2, CH_4 and NOx. The greenhouse effect functions in two ways: to begin with, because these gases are poor absorbers, they are incapable of prohibiting solar radiation from entering the earth's surface. The temperature of the earth is enhanced by this phenomenon. After interacting earth's surface, they are reflected back and when returning away from earth get effectively absorbed by these gases and redirected back to earth's surface. This phenomenon further adds to the earth's temperature, and has been in play, since centuries, to maintain the earth's temperature so that it is conducive for existence of life on earth.

14.2.6 Ozone Layer Depletion

The ozone layer is a concentration of ozone gas. It shields us from the sun's damaging ultraviolet rays. CFCs (chlorofluorocarbons), which are used in industries and in everyday life, are destroying this vital layer (e.g., aerosol cans). These compounds' chlorine depletes the ozone layer. The gap in the ozone layer exposes humans and wildlife to harmful UV rays, resulting in a variety of skin diseases, including cancer.

14.2.7 Temperature

Temperature rises are caused by an increase in concentration of greenhouse gases in the atmosphere, primarily from human activities like the use of fossil fuels and farming, as well as industrial activity. Since with increase in population, these activities also increase, population increase is also one of the factors of global warming. A few examples of increase in climatic temperature are cited here. The Concordia research station, at 3,000m above sea level on the Antarctic Plateau, rarely experiences temperature rise above -25°C even in the summer. In midwinter it can fall to around -80°C. But this place, during winter, has recently experienced a heatwave, with the temperature reaching a high of -11.8°C. This is over 40°C warmer than the seasonal average. According to available records, since 1880, the earth's temperature has increased by 0.14° F (0.08° C) every decade; however, during the last 40 years, the pace of heating has been more than twice that, increasing by 0.32° F (0.18° C) per decade since 1981.2020 was the second-warmest year on record, with land areas reaching new highs. With rising temperatures, several changes occurred, the most concerning of which was the accumulation of CO_2, a greenhouse gas. When effective monitoring of the atmosphere's carbon-dioxide level began in the late 1950s, it was around 315 parts per million (ppm). It had reached 350ppm by that summer, coinciding with a heatwave that was going to bring record temperatures to much of North America.

14.2.8 Effect on Farm Products

Crops

Another cause of concern is effect on growth of crops of different types which provide Food, plants, wood etc. Since crops require water and fertilizer for growth, decrease in the amount of available water will affect crops. This will result in insufficient food, starvation, malnutrition and in extreme cases famine among the increasing population resulting in increased number of deaths.

Vegetation and fruits

The effects of temperature increases can be seen in vegetables, such as crops, fruits, and plants. They are as follows: (i) Because of the higher minimum temperatures, it has a greater impact on grain yield than on vegetative growth. These effects are visible in an increased rate of senescence, which reduces the crop's ability to fill the grain or fruit efficiently. (ii) Different plant seeds need different temperature for germination like broccoli and lettuce germinate in a temperature range from 55 to 77 °F. Once the optimum temperature is obtained, germination continues until a certain point beyond which it drops. (iii) Heat has a significant effect on photosynthesis and respiration. According to experts, both processes accelerate as temperatures rise. When temperatures rise to uncomfortably high levels, the two methods become inconsistent. Tomatoes, for instance, suffer when temperatures rise above 96 degrees Fahrenheit (36 °C).

The primary concern with these extreme temps has been to protect apples from sunstroke and bitter rot, but heat stress can also cause fruit size reduction. Early research indicates that temperatures above 30°C, as well as high night-time temperatures, can be harmful to growth. The ambient temperature all through fruit development is a significant determinant of fruit quality in fruit cultivation. High temperatures during fruit ripening inhibit the development of fruit color. Grapes, for example, have been reported to have poor coloration due to high temperatures.

Deforestation

Deforestation is taking place due to many reasons. Among them, the population explosion is the main cause. They are cutting down trees indiscriminately because people use wood for different purposes for providing shelter, agricultural land, fuel, mature, etc. If there is no tree, there will be no rain too. It will affect badly our farming. Consequently, the farmers will face a great crisis. Deforestation leads to many ecological damages resulting in affecting the flora and fauna of the area. Main effects of deforestation are as follows: (1) The existence of animals in the forest is going to be threatened. (2) Because of deforestation carbon dioxide is increasing worldwide day by day. Consequently, the world is becoming warmer resulting in increased global warming. (3) The sea level has been rising due to more and more melting of ice. In extreme cases, low-lying parts of the world will soon be engulfed by the sea. On the contrary, as a result of deforestation, new areas of the world are becoming deserts. It causes serious soil damage because trees protect the soil. Finally, the soil settles along the riverbank. Finally, frequent devastating floods are being caused. (4) The elimination of trees creates birds and other animals that live on them to flee. Because many living birds and animals live on trees, they will be unable to find food or shelter. Sixth, there will be no rain. As a result, our agriculture will face a severe drought. The planet will become unfit for human habitation.

Flooding

The increased use of fossil fuels (or the global increase in carbon footprint) is causing an increase in CO_2 in the atmosphere. This incredibly rapid increase in atmospheric CO_2 increases its ability to entrap energy, resulting in a greater amount of energy being returned to Earth. This phenomenon contributes to the earth's temperature gradually rising to hazardous levels. Furthermore, climate change is accelerating as a result of the greenhouse effect, such as global warming, melting of glaciers, and unpredictable temperatures, among other things.

Destruction of Marine Life

CO_2 absorption in sea water maintains its alkalinity. As carbon dioxide levels rise, a greater amount of it is being absorbed by oceans and other aquatic bodies. This will increase their alkalinity to dangerous levels thereby risking the marine life. This poses a significant risk to aquatic life, which faces extinction if alkalinity levels continue to rise.

14.2.9 Plastic

Plastics are polymers. They are light, less hard than metals, flexible so easy to bend and mould in various size and shapes, non-breakable and resistant to corrosion attack. Since they are corrosion resistant, they have been in use in industry since last many decades particularly to handle corrosive liquids and gases and also at places as electrical insulator e.g., electrical switches, connector covering as sleeve of wire/cable carrying electrical current in buildings, industry etc. However, the widespread use of plastics has resulted in severe plastic pollution. It's used to make a lot of our furniture, clothing, electronics, and food packaging, among other things. Natural materials used in manufacturing, such as paper, glass, and cloth, have been replaced by plastic in recent decades. Microplastics have been discovered everywhere in our environment, including the deepest ocean trenches, both poles, and the Himalayas. However, microplastics can be found in our homes as a result of the release of microplastics from sofas, carpets, drapes, and other synthesized textiles. Plastics can harm our health in three ways: (i) Every day, we eat, drink, and breathe microplastic particles that enter our bodies. (ii) Chemical additives are present in plastic products. Many of these chemicals have been linked to serious health issues, including hormone-related cancers, infertility, and neurodevelopmental disorders such as ADHD and autism. (iii) Plastics and microplastics attract microorganisms such as harmful bacteria when they end up in the environment (pathogens). Microplastics usually contain these pathogens may increase the risk of infection if they enter our bodies. Furthermore, plastics are not biodegradable, and bacteria are known to cause illnesses that grow on the surface of plastics.

14.2.10 Radiation Pollution

The definition of radiation pollution is that, while there are numerous sources of radiation, it is primarily high-energy nuclear radiation and high-energy particles that cause radiation pollution, posing serious health risks (such as cancer or death). Nuclear radiation and high energy particles are released during nuclear reactions triggered by human activity either by accident (Chernobyl disaster of 1986) or by design e.g. nuclear power plant etc. These are produced by radioactive isotopes of chemicals and disperse in the atmosphere, soil, etc. These materials pose a risk to living beings for decades (Atomic bombing of Hiroshima Nagasaki in 1945 during the 2nd world war). Radiation exposure to the atmosphere implies that it is present even in soils. Radioactive substances are still present in soil and react with the various nutrients, destroying them and rendering the soil infertile and highly toxic. Such soil produces crops that are contaminated with radiation and thus unfit for human and animal consumption.

14.3 Advantages of Green House Effect

The greenhouse effect imparts some advantages too e.g., it is a crucial component of sustainable life. Several advantages of greenhouse effects are listed below.

14.3.1 Supports and Promotes Life

It is not astonishing that only specific temperatures are necessary for human existence and life in general. While fluctuations occur in the earth's atmosphere, they are limited and not too erratic. Because greenhouse gases can trap solar radiation and reflect it back to the earth's surface, earth has not yet been frozen and the temperature is toasty enough for living species. As a result, the greenhouse effect assists the Earth in maintaining a reasonable temperature that allows it to be habitable. Furthermore, in the utter lack of greenhouse gas emissions, the ice regions would quickly melt, raising water levels to dangerous levels. Ozone, functions as a barrier against UV rays from the sun, which are harmful to humans by causing skin cancer.

14.3.2 Photosynthesis

This is the method by which plants produce their own food, thereby initiating the food chain's existence. It is dependent on three key elements: CO_2, water, and sunlight. CO_2, a significant greenhouse gas, is critical here. The greenhouse effect helps to raise CO_2 levels, which can result in increased food production. Plants grow larger and contribute more food in areas with higher CO_2 concentrations. In fact, doubling atmospheric levels of CO_2 can increase agricultural yields by 32%.

14.4 Industrial Sustainability

Efforts towards maintaining "industrial sustainability" or a sustainable environment in/due to industrial operations involve one unique aspect which has not been discussed so far in this chapter. This is related to machinery materials, engineering materials (raw materials), corrosion inhibitors. Different type of industries that exist presently are (i) Chemical or Process Industry (oil, refining, acid manufacturing, paper & pulp, printing, plastic and cloth manufacturing (ii) Transportation industry involving automobiles, Railways, ships (maritime industry), aeroplanes etc. (iii) Defence armaments e.g. tanks, armoured vehicles etc. (iv) Construction e.g. building of roads, bridges, railway tracks, housing for various offices, residential purpose etc.

Unique aspect, related to affecting industrial sustainability, is associated with the phenomenon of "Corrosion" which attacks metals largely but plastics also due to variety of chemicals which may be in use in the factory as process chemicals. Corrosion limits the useful life of the machinery materials, so called as 'metal loss' of mostly steels, aluminium, copper alloys etc. significantly resulting in either their maintenance or their premature replacement. This loss of metal is resulting in depletion of the natural resources (mines) in the form of ores from which respective metals are derived by appropriate metallurgy. So higher the corrosion rate or rate of maintenance/replacement of machinery material, higher will be the rate of depletion of ores of respective metals.

Alongside depletion of natural resources, corrosion imposes additional cost on the industry also. This corrosion cost is direct as well as indirect. The 'direct cost' is cost of replacement of corroded or failed part of machinery and/or structure (involves cost of material, its fabrication as machinery item, transportation to and installation at the factory site). The 'indirect' cost is due to plant shutdowns because of failure of machinery items, down time cost due to loss of production, additional energy cost in restarting the operation, legal proceedings in case of serious injuries or fatal accidents of the shop floor workers etc. Industrialised countries have conducted Corrosion surveys from time to time to check (Table 14.2) if corrosion

Table 14.2 Corrosion cost in billion US$.

USA	Year	1949	1978	1998	2016
	CC	5.5	70	276	1,100
Australia	Year	1972	1983		
	CC	0.699	1.6		
China and Saudi Arabia	Year	2003	2015		
	CC	0.9	110		
Global*	Year	2020			
	GCC	2505.40			
	PGDP	3.4%			
India	PGDP	5.7%			

CC – Corrosion Cost, GCC – Global Corrosion Cost, PGDP - % of GDP
* Mazumdar, M.A.J., Global Impact of Corrosion: Occurrence, Cost and Corrosion DOI:10.33552/GJES.2020.05.000618

cost is increasing so that precautionary measures may be taken to reduce the cost, which will also help in industrial sustainability.

In a review of 2020, annual global corrosion cost was estimated as US$ 2505.40 Billion, equivalent to 3.4% of global GDP. In comparison to this, India loses 5% to 7% of its GDP annually due to corrosion (Table 14.2). To estimate the overall effect of corrosion on industrial sustainability, corrosion cost should be linked with amount of metal lost on the basis of corrosion rate which in turn is related to loss of metal natural resources. This is so because metal natural resources or ore are non-renewable, since it takes millions of years for ore to form under earth. As the world is witnessing technological advancement with rapid pace, corrosion cost is expected to increase (Table 14.2). So increased efforts to negate the effect of corrosion will be required to sustain industrial environment. This can be achieved by taking appropriate steps of corrosion control. Here, one senses a positive aspect in this approach, whereas many industries see efforts to control pollution as merely extra expenditure on it. So they try to bypass the process of pollution control somehow. However, adopting pollution control measures gives an option of reducing corrosion cost through reduced corrosion attacks on machinery materials and so less expenditure on maintenance, replacement of machinery, downtime etc. Alongside, it also decreases the

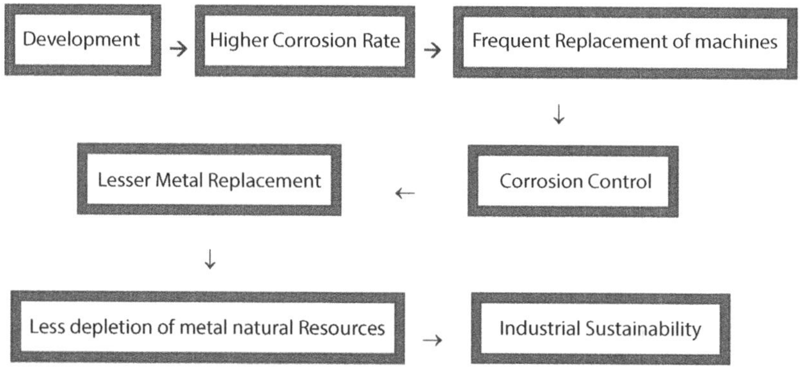

Figure 14.1 Block diagram of industrial development vis-à-vis industrial sustainability via corrosion control.

rate of depletion of metal resources and so help in maintaining industrial sustainability, as depicted in Figure 14.1.

The industry, of whatever type, uses water, air, chemicals, electrical energy, raw material etc. for producing the output be it metals and alloys, plastics, crude oil, petrol, paper etc. The industrial sustainability, therefore, is also affected by consumption of raw materials, chemicals, water and air, electrical energy etc. and so is required to be addressed to maintain sustainability. This is not just because use of these components will lead to their depletion and so will have to be addressed by their production in sufficient amount or reduce their usage somehow. But their use e.g. that of chemicals, raw material, polluted water/air, temperature of process conditions etc. also increases the aggressiveness of media hence corrosion of machinery and therefore loss of metals/alloys. This will affect sustainability of natural resources of materials. To overcome this problem, efforts will be required to control corrosion of machinery materials. This will help in making industrial environment sustainable. Forthcoming sections will discuss about corrosion, various means to control corrosion and amongst them a little more in detail about Corrosion inhibitors particularly the role of carbon allotropes in controlling corrosion.

14.5 Corrosion and Its Implications

14.5.1 Corrosion

Corrosion has a deteriorating effect on materials due to the chemicals present in liquified form in its affinity. This deteriorating effect weakens the machinery material which may cause leakage of chemical solutions present in the process equipment e.g. pipe, pumps, reaction chamber, storage tanks etc., maintenance of machinery by refurbishing it or in extreme cases may lead to its failure thereby requiring its complete replacement. Corrosion is caused by various electrochemical and chemical reactions between metal and chemicals in its surrounding both present in an electrically conducive medium, e.g., aqueous solutions having various chemicals in industrial machinery. The chemicals may be present for carrying out various process reactions but corrosion is a side effect the chemicals impart on the materials of machinery. Corrosion reactions are spontaneous in nature because their occurrence results in lowering the total energy of the system. In other words, occurrence of corrosion reactions changes a system, the machinery material e.g., steel, cast iron, brass etc. alongside the process liquors, from a high energy state to low energy without any external stimulus. Only condition is that both the components of the system should remain in electrical contact with each other, a condition which exist more or less in every industrial production unit, pipes, storage tanks, reaction chambers, structures e.g., underground tunnels, bridges, roads, transport vehicles e.g. ships, aeroplanes, trains, off shore buildings and structures etc.

14.5.2 Corrosion and Sustainable Environment

As described earlier, corrosion is natural phenomenon of spontaneous nature and it results in loss of useful life of metallic machinery and parts of various utility metallic objects. This in turn incurs economic losses to the industry and hence to the nation (resulting in corrosion cost. Pl, see Table 14.2). Also, it is responsible for loss of metals. As a simple example, one may consider a slab of steel of 1 mt^2 area and 2 cm thickness (This slab could be a part of outer surface of a cylindrical vessel).

Consider the surface (Figure 14.2) facing the arrow is exposed to corrosive media resulting in its corrosion along the direction of arrow uniformly through the thickness such that the upper 1 cm of thickness is corroded in 1 year of its operation as a part of machinery. Metal of this part of the

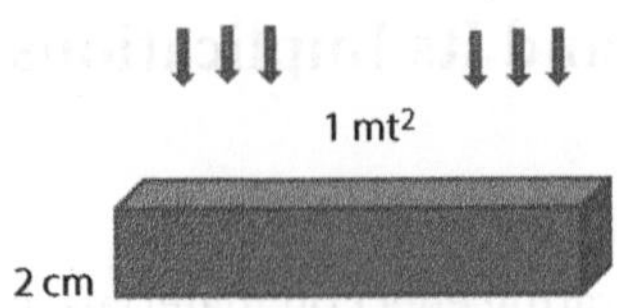

Figure 14.2 Schematic diagram of part of metal surface.

slab will be lost for bearing the load since corroded metal (e.g. steel will transform mainly as various type of iron oxides, hydroxides etc. which are weak and brittle) is weaker and is not ductile hence it can not bend and be given different shapes. It will thus be of no engineering use if the residual thickness of metal is less than the minimum required to bear the load as a part of machinery during mill operations. This then is termed as "Lost metal". This part needs to be replaced by a new metal piece of the same composition, size and shape.

Now, in the above example, how much amount of metal was lost due to corrosion. The case presented here is an example of 'uniform corrosion'. The argument and, therefore, the solution to overcome the corrosion effects will differ if type of corrosion responsible for failure of machinery or its part is 'pitting/crevice corrosion/stress corrosion cracking etc. It may be worth mentioning that uniform corrosion is one of the more frequently observed type of corrosion). Let steel corroded at a rate of 1 cm/year (known in corrosion parlance as 'corrosion rate'). That means thickness of the slab, useful for engineering purpose, lost due to corrosion was 1 cm in 1 year. This is equivalent to 78 kg of steel (of above shown dimension) lost due to corrosion in 1 year considering 7.8 gm/cc as density of steel. Its original weight was 156 kg. So roughly ½ of steel may be lost due to corrosion in 1 year. And this much amount of steel is to be replaced to keep the concerned machinery in operation. In other words, this means additional requirement of more metals in the form of ores from natural resources which are to be transformed into metals and alloys, the materials used to manufacture industrial machinery and various engineering parts etc., through metallurgical process before they are used in various industries in different size and shapes.

As per a recent estimate, present yearly requirement of steel in India is ~300 million tonnes. While a part of this steel will be used in new constructions of industry, rail network, bridges, buildings for offices, residential purpose etc. An approximate calculation, on the basis of yearly steel requirement in industry will be around 1/3rd of this amount i.e., 100 million tonnes. Considering steel installation in Indian industry in the form of pipes, storage tanks, reaction towers, pressure vessels etc., which are more vulnerable to corrosion owing to their higher surface area (surface/volume ratio), a good fraction of this steel may be required yearly to replace the corroded steel. So corrosion is considered as responsible for several tens of million tonnes of steel required as replacement in a corroded machinery equipment.

The above estimated amount of new steel required will need four to five times of this amount (i.e. a hundred million tonnes) as ores or natural resources alongside, off course, energy and chemicals required for treatment to change ores into commercial metals and alloys. Overall these operations will be responsible for depletion of natural resources of different metals or their ores. Thus, process of corrosion affects one aspect of environmental sustainability and so efforts are to be made or investigated for maintaining sustainable environment due to industrial operations. Considering different aspects which affect sustainability of environment, one of the types are those which are due to various activities performed by living species on earth mainly the human beings, the other aspect is due to industrial operations. The former aspect may be considered first in brief. The latter one is the subject matter of this chapter. However, there are some commonalities practiced, in general, in the two cases. These are described in the following lines.

14.5.3 Industrial Operations and Environmental Sustainability

Generally speaking, it means managing natural resources responsibly so that they are still be available for use in future. This can be practiced by adopting following approach:

- (i) Limiting use of natural resources e.g. fresh water can be sustained by curbing excessive use of water, even when not required.
- (ii) Preventing pollutants to enter water supply system
- (iii) Treating polluted water, from industrial output, water bodies open to atmosphere e.g. lakes, rivers etc. or other activities, to remove pollutants and recycle for use in the system. Can be used for drinking, cooking, agriculture, marine life sustenance etc.
- (iv) Educating mass for sustainable farming e.g. crop rotation, appropriate use of soil quality for a given crop, suitable fertilizer in optimum amount, timely watering crops etc. to get maximum yield.
- (v) Use of raw materials efficiently so that various items can be produced from minimum requirement of raw material per tonne/volume of product.
- (vi) Recycling of the used product so that fresh raw material is used to minimum possible extent, leading to its sustenance. Paper Industry is a good example in this direction.
- (vii) Develop the technology of production so as to make it more efficient. Thus one can reduce the input cost of raw

material, energy, chemicals, water etc. per unit of commercial product. This, in turn will help in saving these commodities thereby helping maintain environmental sustainability.

(viii) The industrial equipment should be designed such that they are more efficient, e.g., a boiler/high temperature reaction towers, digesters should be designed to minimize the heat losses, leakages of steam, vapors etc. This will help in energy, environmental and chemical sustainability.

The approach to practice environmental sustainability includes one or more of these steps (i) limiting use of natural resources e.g. freshwater can be sustained by curbing excess water use (ii) protecting them from becoming impure e.g. by preventing pollutants from entering the water supply (iii) by processing polluted water so as to remove pollutants to make it suitable for normal usage e.g. cooking, drinking, agriculture and as industrial process aqueous media for marine life sustenance etc.

For industrial sustainability cases, an additional nevertheless important example could be to avoid degradation of metals i.e. materials of machinery construction e.g. mild steel, stainless steel brass, aluminium etc., through corrosion. This is achieved by controlling/mitigating corrosion by several means (i) use of corrosion resistant materials (ii) use of coatings (iii) inhibitors (iv) cathodic protection system etc. Shortlisting of a procedure to control metallic corrosion is governed by the economics of corrosion control procedure vis-à-vis the corrosive environment and the process conditions which are responsible for corrosion attack on the machinery material.

14.5.4 Industrial Machinery Corrosion and Its Implications

Now we deal with main item of this chapter - Effect on sustainability due to corrosion and how to minimize it. Prior to deliberating on this aspect and considering that many of the readers of this book may not be aware much about "corrosion", it is suggested to refer to some text books on corrosion [1–3]. Books referred here are among the leading literature on corrosion and for most part they are written in easily understandable English. As we have written earlier that corrosion of industrial machinery, which is largely made of steel, leads to its either refurbishing or replacement by the same machinery fabricated of steel itself. Simply speaking, even if we don't consider any new fabrications of various industrial equipment/machinery, civil constructions, manufacturing of road, rail, air or water transport vehicles, one will need continuous supply of steel hence of iron from its natural resources. These natural resources, namely ores, will, therefore, go on depleting and the rate of depletion is faster than the formation of new ore deposits, since millions of years are required normally for formation of ore. This, in turn, will affect sustainability of metals. And since metals to a large extent are the material of constructions for various parts of industrial machinery, components etc., shortage of metal supply will affect production of industrial output which will affect the development process on our planet. Overall, life of the human beings and other living species on our planet will be affected severely.

Some estimates [4] indicating depletion in the iron ore reserves are given below.

Amount of iron in iron ore = 57% (considering 100% recovery of iron from iron ore)
Amount of ore required for 20 tonnes of iron = 20/0.57 = 35.08 tonnes ore
Loss of iron while going with slag in blast furnace = 5%
So to get 20 tonnes of iron, amount of ore required such that iron is = 20 + 5% of 20
= 20 + 1 = 21 tones
Amount of ore required to get 21 tonnes of iron = (100×21)/57 = 37 tonne iron ore

World resources of crude iron ore = 800 billion tonnes
Amount of iron = 216 billion tonnes (50% impurity, and 37 tonnes required for 20 tonnes iron.)

However, despite intensive extraction of iron throughout 20th century, its abundance has not reached alarming stage. Its depletion degree is 28% (Depletion degree of a mineral depends upon two factors – (i) its abundance in nature (ii) its production rate).

It was estimated that we would require 400 million tons of iron ore by 2020 to generate 200 million tonnes of steel. The current iron ore resources will not last more than 50 years. It may not last more than 15 to 20 years in some states. Due to preferable depletion of deposits with higher iron content, the potential reserve life of contained iron is only 19 years. The majority of active mines reported a decrease in ore reserves in 2018, owing primarily to mining depletion and changes in the mining area.

Another problem associated with increased mining of iron ore is due to decrease in surface area of earth. Long before having to worry about whether we can continue to make steel, one will have to concern about finding a place to stand. Iron ore deposits, for example, account for approximately 32% of the Earth's crust. The estimate given below can give you an idea of this.

The original radius of earth = R (say),
The radius of the earth surface, after removal of iron-containing earth's crust, which is 32%the of earth's crust = $R_1 = 0.68$ R
Surface of earth with iron crust = $S = 4\Pi R^2$
Surface of earth after removal of iron crust = $S_1 = 4\Pi R_1^2 = 0.46 (4\Pi R^2) = 0.46$ S

Thus, with the removal of iron from containing earth's crust, the surface of the earth will be reduced to 46% of the present surface [5]. In such a scenario, how people and various activities will be accommodated on earth unless there is a sharp reduction in the earth's population?.

14.6 Corrosion Control and Material Properties

To attain environmental sustainability due to industrial and related operations, one needs to control/mitigate/overcome corrosion of industrial machinery. There are several ways in which corrosion can be mitigated, these are (i) use of corrosion resistant materials (ii) use of insulating materials e.g. plastics/FRP's/Rubber/ceramics etc. either as material of construction or as coating (iii) Corrosion inhibitors (iv) cathodic/anodic protection (iv) galvanic compatibility etc.. Different methods may be effective in corrosion control in different environment and selection of a particular alternative depends upon its cost effectiveness so that as much as possible money is avoided from wasting due to corrosion resulting in minimizing the 'corrosion cost'. Before discussing these alternatives, it is equally important to discuss some related aspect of other properties of engineering materials to be used in constructing the industrial machinery. These are Mechanical properties, fabrication properties, high temperature stability etc.

14.6.1 Mechanical Properties

Tensile strength – This property represents maximum possible load that a material can bear before failing. It helps in determining the load bearing capacity of a materials sheet of a given thickness. This in turn determines the cost of material required in establishing a factory/plant. Obviously a stronger material is preferred, but these materials are costlier. Thus carbon steel are cheaper but have less strength that that of stainless steel which are costlier.

Hardness – It represents capability of materials to avoid surface changes due to wearing action. This could be required in cases where a material may come in contact with a moving liquid media having abrading particles e.g. silica, fibers etc. The surface becomes rough and in extreme cases, it may lead to loss of metal with time. If media is corrosive also, combined action of wearing and corrosion results in loss of material strength very rapidly, thus shortening its useful life as part of machinery. In general, a stronger material is harder and so has better resistance against wearing action.

Fabrication properties – These represents ease of a material to cut, bend, weld etc. A material which is ductile can be fabricated into different size and shapes easily as compared to a brittle material. It can also be joined, wherever required, with other parts of machinery, by welding, with ease.

High temperature stability – It is represented by melting point of a material. Usually metals and ceramics have high melting point so have better high temperature stability. Plastics have much lower melting point (normally less than ~ 50°C or so), so they can be used for handling liquid media only at around room temperature or so. Though by mixing with resins and fibers, plastics (fiber reinforced Plastics or FRP's) and composites (e.g. PVDF, Teflon etc.) have been developed which are stronger and better high temperature stability. In spite of this, for handling media beyond 50°-60°C, one uses metals and for temperatures beyond 100°C or so, it is economical to use ceramics. Thus on the basis of properties of various materials (Table 14.3) and their availability commercially in market, metals are used for most applications. But metals are prone to corrosion attack and corrosion resistant materials are costly, one looks for other alternatives for controlling corrosion so as to minimize its deleterious effects on the industrial operations and hence corrosion cost. In the process of minimizing corrosion cost, several alternatives of corrosion control are in vogue and continuous work on corrosion mitigation is still in progress. Following are mostly used alternatives:

(i) Corrosion Resistant Materials
(ii) Design consideration
(iii) Anodic/cathodic protection
(iv) Corrosion inhibitors

Table 14.3 Properties of materials.

S. no.	Material	Tensile strength (MPa)	Hardness (Brinell)	Ductility % elongation	Melting point °C
1	Iron	262	167	45	1538
2	C-steel (1020)	430	150	28	1400–1530
3	Stainless steel 304	515	250	21	-do-
4	Stainless steel 316	550	270	22	-do-
5	Brass (70Cu-30Zn)	300	60–70	68	1000
6	Plastics	~ 60–70	20	100–400	45–60
7	Ceramics Al_2O_3	340–1000	1900	- Brittle	2020

A brief discussion on these alternatives is given below.

14.6.2 Corrosion Resistant Materials

Carbon steel (simply speaking 'steel' and in engineering terminology termed 'mild steel') is the maximum used material of construction in the industrial operations. This is due to its various properties (discussed above) alongside ease of availability in different size and shapes, fabrication, compatibility with different materials and lower cost. Though its corrosion resistance is poor, it is still cost effective to use in handling mildly corrosive media by considering 'corrosion allowance' while finalizing its specifications. But when the machinery has to handle media which is highly acidic pH (~2 or less), have corrosive chemicals e.g. Cl^-, $S_2O_3^{--}$, sulfides, peroxide, O_2 etc., abrading chemicals e.g. silica and/or operating at high temperature, one uses corrosion resistant materials most common of these are stainless steels (martensitic, austenitic, ferritic, duplex stainless steel), brass, bronze, aluminum, Nickel etc. alloys. Plastics also handle corrosive media in the form of pipes, tanks, valves, pumps etc. if the operational conditions correspond to ~ room temperature and ceramics when operating temperatures are high e.g. boilers, condensers, evaporators etc.

14.6.3 Design Consideration

Pitting, crevice corrosion, galvanic corrosion, erosion corrosion etc. are types of corrosion which may occur due to some faulty design. Thus, a metal may experience (a) 'Galvanic corrosion'—if two dissimilar metals e.g., stainless steel and steel Figure 14.3(a) where bolt of steel experience excessive corrosion while stainless steel nut remains almost unaffected. This corrosion can be avoided if both metals are same or with least difference in their electrochemical potential. (b) 'Pitting' - if a metal is part of an equipment such that aqueous media gets trapped in some cavity or cavities [Figure 14.3(b)]. This can be overcome by selecting a metal with appropriate corrosion resistance. (c) Crevice corrosion – This type of corrosion is experienced when overlapping of two metallic plates or planar surfaces create an area where aqueous media can get trapped e.g. crevices [Figure 14.3(c)]. Metal part under the crevice experience high degree of corrosion in comparison to the rest of metal (Figure 14.4). This type of corrosion can be avoided by blocking these crevices so that aqueous media does not enter into them.

14.6.4 Erosion Corrosion

This type of corrosion occurs because of combined effect of corrosion and erosion due to solid particles in a liquid media. Thus a metal pipe may

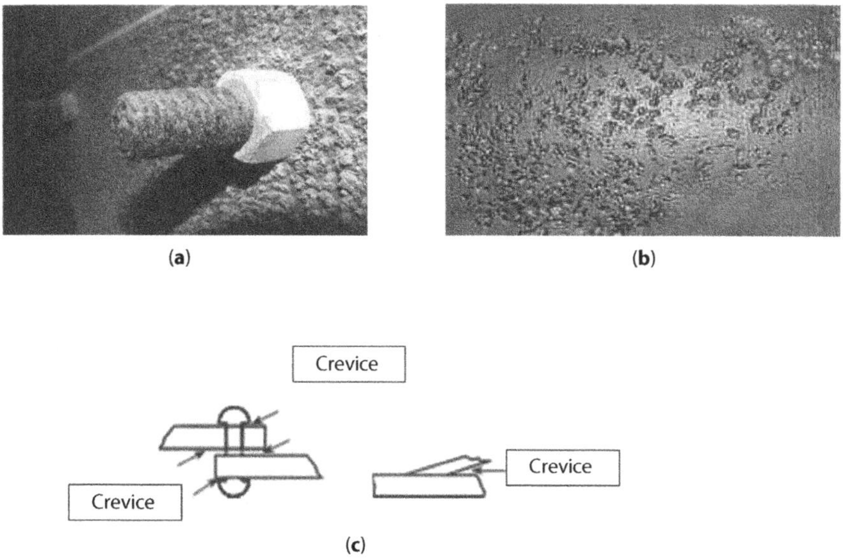

Figure 14.3 (a) Stainless steel (nut) and steel (bolt) galvanic corrosion. (b) Pitting corrosion. (c) Crevice corrosion.

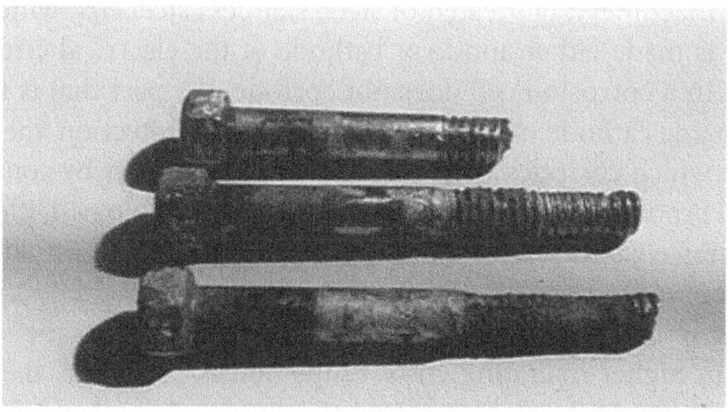

Figure 14.4 Crevice corrosion of through hull bolts.

be affected, particularly at bends, if it is handling aqueous media with dissolved solids. So will be fan of a pump being used as a component of the whole plant (Figure 14.5) which revolves around while transferring media from one place to another. Erosion corrosion can be avoided by (i) reducing the flow velocity of media (ii) by using harder material of pipe, pump etc. (iii) increasing the cross-section of pipe at the bends.

Figure 14.5 Erosion corrosion of fan of a pump operating in a media with suspended solids.

14.6.5 Cathodic/Anodic Protection

In cases where corrosion is very high, it becomes more cost effective to apply this method of corrosion control. In this method, the part of the industrial machinery, fabricated of steel, stainless steel etc., which is to be protected is made either anode or cathode of the electrical circuit. Since a cathode in a corrosion cell does not corrode, the part that is to be protected is made cathode of the circuit in 'cathodic protection' method. It is the anode which corrodes. This method is applied either by considering a 'Sacrificial Anode' or an 'Impressed current' methodology. Different type of structures e.g. working platform, bridges/tunnels etc. off shore or on shore, underground gas/oil pipelines etc. can be protected by this type of protection system. Corrosion increases frictional drag on the ship's hull, resulting in higher operating costs. Ship's hull is protected, therefore, by this technique. A high-quality coating gives a smooth hull and, when combined with a well-engineered cathodic protective device, aids in ensuring that the surface is kept in good condition while operating costs are kept to a minimum. In case of 'sacrificial anode' technology, a material which is active type e.g. Mg, Al or Zn etc. is made to corrode in a galvanic cell formulation and saves the structure (cathode) that is to be protected from the vagaries of corrosion. This technology is applied to protect hulls of ships, water heaters, pipelines, above/underground tanks, refineries etc. GI (galvanised iron/steel) pipes, sheets etc. are among the popular domestic products using principle of sacrificial anode.

14.6.6 Corrosion Inhibitors

Controlling corrosion by using corrosion inhibitors is among the more popular corrosion protection methods, in cases where their addition does not affect chemical reactions meant for producing the desired product. Corrosion inhibitors are a class of chemicals which when added to the corroding medium inhibit corrosion to significant extent. The extent of inhibition is estimated using following equation:

$$\text{Inhibition efficiency} = \eta\% = [1 - (i_{corr}'/i_{corr})]x100$$

i_{corr}' - corrosion rate with inhibitor, i_{corr} - corrosion rate without inhibitor

and its value may sometimes reach as high as 90% or even higher. This is equivalent to $i_{corr}' = 0.1\ i_{corr}$ or corrosion current, on adding inhibitor, reduced to $1/10^{th}$ of corrosion current without inhibitor. Inhibitors might also be used as coating over a metal which is to be protected. In many cases, the extent of corrosion inhibition obtained, for a given inhibitor, depends upon amount of inhibitor added. But this is not true for all and it is advisable that laboratory tests must be done to determine the optimum concentration of the inhibitor to be added to the process solution in order to attain maximum corrosion inhibition. Inhibitors reduce corrosion by either delaying the cathodic/the anodic or both processes and so termed as (i) cathodic (ii) anodic and (iii) mixed inhibitor respectively. Some prevent corrosion by producing an adsorbed coating that serves as a barrier. 'Cathodic inhibitors' are those which impart corrosion inhibition by slowing down the reduction (cathodic) reaction and thus increasing cathodic polarization and shifting corrosion potential to more cathodic or lesser values. As, Bi, and Sb are examples of these inhibitors which decrease hydrogen reduction reaction and thus overall corrosion rate. 'Anodic inhibitors' act by enhancing anodic polarisation and thus shifts the corrosion potential to higher values which may fall in the passive region of anode. They are therefore also called passivating inhibitor. Examples are chromate, nitrates, phosphate etc. Use of anodic inhibitor may be dangerous if it shifts corrosion potential in the trans-passive region beyond passivation range of anode metal. This leads to observance of localised corrosion. 'Mixed inhibitors' delay both anodic and cathodic reactions of corrosion process.

Conventionally used inhibitors are found to affect 'environment' which affect 'human health' and needs to assess 'safety considerations'. So, corrosion inhibitor studies have become focussed on these two aspects now-a-days. Accordingly, researchers have been working increasingly on the utilization of 'environment friendly' or 'Green' corrosion inhibitors. As a result, many 'organic compounds' have been developed as Green inhibitors, such as extracts, elapsed nontoxic medicines, and so on. Depending on how they respond to the metal's surface and potential, these organic inhibitors can be classified as anodic, cathodic, or mixed [6–8]. Because of their association with E4, organic compounds are now recognized among the most efficient and lucrative methodologies of corrosion inhibition. This is backed up by the fact that the worldwide economy for corrosion inhibitors had been $7.2 billion in 2019 and is expected to grow by 3.6% to $9.6 billion by 2026 [9]. Because they are environmentally friendly, the market for plants retrieved and organic green corrosion inhibitors is expected to grow by 7.0% by 2026. Keeping the environment in mind, newer areas of focus include the use of rare earth elements (REM), reactive functional groups compounds, the synergistic of organic/inorganic compounds using REM, and the encapsulation of inhibitors [10, 11]. These alternatives, as expected, exhibit very minimal side effects and strong corrosion inhibition. Despite having good properties, as mentioned above, organic inhibitors have one major drawback and that is very low corrosion inhibition efficiency which is limited due to two factors (i) less surface coverage of metal which is to be protected. This is due to large particle size of organic molecules (~ microns), which reduces covered surface of metal. Further interaction between organic inhibitor molecule and metal is physicochemical adsorption. Of the two, physical adsorption is through Van der Waal forces, which are very weak and non-directional. Overall the inhibitor's adsorbed surface does not provide good protection. (ii) These compounds show less solubility and electrolyte catalysed rearrangements responsible for donor-acceptor interaction between free electron pairs of charged inhibitor molecule and vacant low energy d-electron orbitals of metal [12].

14.6.7 Nanomaterials

As discussed above that although organic inhibitors have negligible toxicity and other advantages e.g. cost-effectiveness, its use is restricted when one needs to opt for a corrosion inhibitor which gives a robust protection against corrosion to the industrial machinery materials so as to minimise the corrosion cost. One needs to find an alternative material which has acceptable properties and additionally should cover metal surface more completely (higher surface coverage of the base material) so as to achieve the goal of robust corrosion protection. Among various candidate materials, nanomaterials (or nanostructured materials) fulfil most of the requirements required for acting as efficient inhibitor. A major requirement, for attaining higher inhibition efficiency, is existence of large surface area of the particles/grains of chosen material. Nanomaterials fall in that category since they have finer grains (1–100 nm). Owing to tremendously fine grain size, their grains have higher surface area to volume ratio. Consequently, the surface coverage by nanomaterials, of metal to be protected, is quite high as compared to the normal materials and they behave as excellent corrosion resistant material. Additionally, nanomaterials have outstanding strength, hardness and other physical properties, because of their high grain boundary to volume ratio [13, 14]. However, not all nanomaterials are corrosion resistant. So, one considers application of functionalised nanomaterials.

Functionalized Nanomaterials

The process of adding additional functionality, features, capabilities, or characteristics to a substance by modifying the surface chemistry of the material is known as functionalization. This may also include enhancement of certain properties which otherwise are existing in the materials but not to the desired level/magnitude. Thus the process of functionalisation of a material, which has desirable properties say for the purpose of fabricating industrial machinery, involves the modification of surface properties required for a particular purpose e.g. in present case, it is improvement of corrosion inhibition etc. which otherwise do not exist in the material. Significant advances have been made in the synthesis of nanoscale materials recently. It was followed by efforts from synthesis to produce useful structures and coatings with increased wear and corrosion resistance. By changing the process parameters or using nanograined feedstock powders, existing PVD and CVD procedures for making preparations microcrystalline coatings can be utilized to produce nanostructured coatings. Thermally activated coating deposition has been used successfully to produce nanocrystalline (nc) coatings. Nanostructures encourage selective oxidation, resulting in the formation of a protective oxide magnitude with superior bonding to the substrate. The advantages of organic polymers, such as deformability and water resistance, can be effectively combined with the benefits of advanced inorganic molecules, such as hardness and permeability, in a polymer nanocomposite coating.

14.7 Carbon Allotropes and Corrosion Inhibition

Carbon is the second most abundant mineral in the human body after oxygen, accounting for approximately 18% of a human's body weight. Carbon's function is critical because it can form a bond with other light elements as well as with itself. As a result, carbon is able to link together to form a connected series of various carbon units e.g. ethylene, propylene, vinyl chloride etc. which paves the way to successfully formulate the structure of organic polymers which are observed in biological molecules and in chemistry. So when one is looking at the possibility of synthesising green corrosion inhibitors, it is quite natural to consider carbon polymers as candidate material. As a result, carbon science is quite popular today, and carbon nanostructures of various low-dimension allotropic forms are observed to play an active role in the fields of nanoscience, materials engineering, engineering, and technology. These are called 'Carbon allotropes 'and are low dimensional (~ nm) namely C_{60} family of buckyballs, carbon nanotubes (CNT), graphite, graphene etc. [15]. Since focus in this chapter is on recently suggested corrosion inhibitors which are environmental friendly and have higher inhibition efficiency than conventional inhibitors through utilisation of carbon allotropes, only such carbon allotropes will be discussed. Following are the types of carbon allotropes useful as corrosion inhibitors

0D Carbon Dots	0D
Buckminster fullerene C_{60}	0D
Graphite → Graphene (G)	2D
Carbon nanotubes (CNT).	1D

14.7.1 Carbon Dots (CD) or Carbon Quantum Dots (CQD)

Carbon-based quantum dots are a new type of carbon nanomaterial with dimensions less than 10 nm. They are made up of graphene quantum dots (QGDs) and carbon quantum dots (QCDs) (CQDs, C-dots or CDs). They were discovered in 2004 during the preparative electrophoresis purification of single-walled carbon nanotubes. CDs have been synthesized using two methods: (i) the top-down method and (ii) the bottom-up method. A more massive carbon structure, such as CNTs or graphite, is broken down into nanoparticles (10 nm or less) utilizing arc discharge, laser ablation, and electrochemical methods in the top down approach, whereas CDs are synthesized from smaller carbon units (small organic molecules) in the bottom up approach using electrochemical/chemical oxidation, laser ablation, ultra - sonic treatment, thermal decomposition, and so on [16]. The reader is advised to consult the reference cited here for more information on the methods.

Carbon Dots as Corrosion Inhibitors

Because of their unusual properties, such as decent water solubilization, biocompatibility-compatibility, low cytotoxicity, remarkable antimicrobial properties, chemical resistance, superior temperature activity, and nonflammability, carbon dots (CDs) have piqued the interest of corrosion inhibitors. CDs are useful as corrosion inhibitors and even for microbial corrosion mitigation due to these properties. One study [17] used environmentally friendly, highly penetrable, and cost-effective carbon dots as a corrosion inhibitor on mild steel in a 15% HCL solution. Although mild steel is commonly employed in manufacturing sectors, it is highly susceptible to corrosion, resulting in significant corrosion costs to the industry. The aggressiveness of a 15% HCl solution causes significant corrosion and degradation to steel, reducing its capabilities and limiting its use. To avoid it, corrosion control alternatives must be found. One such alternative is the use of inhibitors.

The hydrothermal method was used to create a novel corrosion inhibitor N (nitrogen doped)-, S (sulfur doped)-CDs. FTIR, UV-vis, photoluminescence (PL), Raman spectroscopy, XPS, and HRTEM techniques were used. Weight loss analysis, electrochemical impedance spectroscopy (EIS), and potentiodynamic polarization methods were used to assess the inhibitory effect of N, S-CDs on mild steel corrosion in a 15% HCl [18].

Work has also been done to investigate the effect of doping on the enhancement of anticorrosion protection of carbon dots. The functionalized carbon dots (FCDs) were synthesized by conjugation of citric acid carbon dots (CA-CDs) and imidazole and their corrosion behaviour was tested in 1 M HCl solution. FCDs were synthesized. The electrochemical characterizations showed that the FCDs can effectively reduce the corrosion rate of steel in 1 M HCl medium with an inhibition efficiency up to 90% in case of FCDs content of 100 mg/L. According to the Langmuir adsorption model, the excellent protection property could be attributed to the good coverage of the formed adsorption film of FCDs on the metal surface. Nyquist plots obtained from electrochemical impedance spectroscopy (EIS) showed that the diameter of capacitive reactance arc increased upon the addition of the different compositions (imidazole, CA-CDs and FCDs), confirming the corrosion inhibition of steel in presence of these solutions. This diameter increased with the increase of the inhibitor dose in case of imidazole and FCDs, while it was stable in case of CA-CDs which means that the barrier adsorbed film over the steel metal becomes denser with the increase of imidazole and FCDs concentrations. Inhibition efficiency calculated from EIS date revealed significant difference in efficiency between the different inhibitors with the increase of inhibitor concentration.

14.7.2 Buckminster Fullerene C^{60}

It is a 0D nanomaterial carbon allotrope in the form of a sphere (Figure 14.6). It has 6 proton and 54 neutrons. It's the tiniest fullerene molecule with no two pentagons sharing an edge (which can be destabilizing) (Table 14.4). In terms of its natural occurrence, it is also the most common. C_{60} has the shape of a soccer ball, with twenty hexagons and twelve pentagons and a carbon atom at each polygon's vertices and a bond along each polygon's edge. A C_{60} molecule has a van der Waals diameter of about 1 nanometer (nm) and a nucleus-to-nucleus diameter of about 0.7 nm.

Buckminster Fullerene as Corrosion Inhibitor
Carbon-based nanoparticles with unique properties for use as nanofillers are used to create multifunctional nanomaterials with dramatically improved performance. These functionalised carbon nanocomposites, due to their properties and sizes, are found useful in different field of interest. Here we discuss about their application as corrosion inhibitors.

In comparison to neat polymer materials, fullerene-C60 nanomaterials exhibit reduced tension, hardened mechanical characteristics, and improved anti-corrosion behavior [23]. Because of their unique spherical structure with nanoparticles diameter, fullerene-C60 (C60) nanomaterials can serve as nanofillers to strengthen the polymer matrix. C60 nanoparticles exhibit enhanced anti-corrosion and mechanical properties when compared to CNTs and GNPs. Liu *et al.* [40] examined fullerene-C60–reinforced epoxy with C60 concentrations ranging from 0.25 to 1.0 wt.%. The preparation of the coating is described in detail in ref. [34]. Employing wire bar coaters with a width of 30m, the final mixture of EP/C60 epoxy composites was painted on the cast iron's surface. The cast iron test samples were electrochemically tested in 3.5wt% NaCl solution with EP/C60

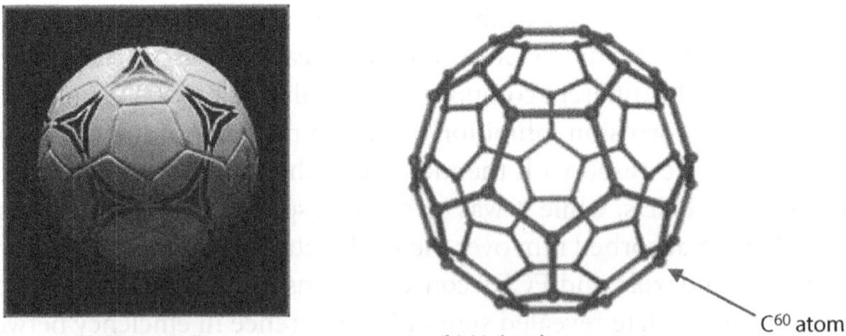

(**a**) Soccer Ball (**b**) Molecular structure

Figure 14.6 Buckminster fullerene C^{60}.

Table 14.4 Properties comparison of various carbon allotropes (nanomaterials).

Carbon nanomaterials	Dimension	Hybridization	Surface (m^2/gm)	Thermal conductivity (W/m/K)	Tenacity	Hardness
Fullerene	0	sp^2	80–90	0.4	Elastic	High
Graphite	3	sp^2	~ 10–20	1500–2000	Flexible Nonelastic	High
Graphene	2	sp^2	~ 1500	4840–5300	Flexible Elastic	Max (for Single layer
CNT	1	sp^2	~ 1300	3500	Flexible Elastic	High

composite coatings. Tafel plot measurements were taken at a scan rate of 5mV/s in the potential range of -2.5 to 1 V of corrosion potential. The coated electrodes were measured after 24 hours of presoaking in 3.5wt% NaCl solution in the test cell, while the bare iron electrode was evaluated immediately. The results establish corrosion inhibition being demonstrated by EP which is further increased by addition of fullereneC60. Maximum inhibition is provided by the coating when amount of fullereneC60 is increased to 0.50 wt%. Further addition of the nanomaterial reduces the inhibition efficiency.

Magnesium (Mg) alloys have beneficial properties such as good mechanical performance, little density, better damping capacity, very max tensile ratio, high castability, and excellent recyclability. Their widespread industrial applications are limited by serious corrosion problems they face due to their low corrosion resistance and so, appropriate technology is desirable to regulate the corrosion. With this aim, oxidized fullerene/sol-gel nanocomposite was synthesised and tested for corrosion defense of Mg alloy [38]. For reasons for selecting nanocomposite of oxidised fullerene, readers may refer to the reference cited Samadianfard *et al.* [41]. Furthermore, oxidized fullerene was used in this case because I functionalization of fullerene nanoparticles is generally thought to be advantageous in terms of corrosion protection due to their better even dispersion and compatibility with the coating, and (ii) fullerene can be effortlessly functionalized through numerous ways. In the range of 0 to 500 mg/l, oxidised Fullerene nanoparticles were added to the sol. Following that, the pretreated alloy specimens were immersed for 10 minutes in sol solutions containing oxidized fullerene nanoparticles. It was ensured that the coating would not crack. The EIS assessed the corrosion behavior of the coatings. The Nyquist and Bode plots depicted the EIS response of the coatings (with varying contents of OF nanoparticles) after 15, 30, 60, 120, 240, 480, and 1440 minutes. The values of the Rct gradually reduced as the immersion time increased from 15 to 480 minutes, which can be ascribed to an increase in the corrosion rate. Furthermore, the relative increase in Rct at the completion of the immersion could be elucidated by the gathering of corrosion products on the alloy surface.

To fit the experimental data, a three-time constant electrical circuit with four impedance elements (Rox-hyd, Ri, Rf, and Rs) and three CPE (CPEf, CPEi, and CPEox-hyd) was chosen. Changes in the polarization resistance (Rp= Rf + Ri + Rox-hyd), which is inversely related to the rate of corrosion, can be used to estimate improvements in the corrosion protection capacity of the applied coatings. The changes in Rp for the neat and OF-containing sol-gel coatings were plotted in semi-logarithmic (log Rp versus immersion time) form as a function of the immersion time in the corrosive solution.

14.7.3 Graphene

Graphene is obtained from graphite. Graphite is formed due to sp^2 hybridization with C replacing H and it has honeycomb structure (Figure 14.7). Graphite consists of carbon atoms bonded together in huge networks (layers) that are piled one over the other (Table 4.4). It is opaque, very flexible but non-elastic, has high thermal conductivity and chemically inert. The layers are weakly bonded to each other by van der Waal's force This eases the movement between layers causing them to be slippery and soft [15]. From graphite (being a multilayer structure where layers are held together by the weak van der Waals force), one can obtain graphene (being a single layer, one of the many layers forming graphite, with nanometer thickness) from simple techniques e.g. (i) Exfoliation - In 2007, adhesive tape was used to split graphite into graphene [19]. It took several exfoliation steps, each resulting in a slice, until the only one layer of 'graphene' remained. The flakes are then deposited on a silicon wafer after exfoliation. This method can produce crystallites larger than 1 mm in size and visible to the naked eye. In a similar vein, a sharp diamond wedge was used in 2011 to penetrate the graphite source and exfoliate layers in order to obtain graphene. (ii) by mixing graphite in a proper dispersing medium and sonicating/centrifuging the resulting medium for separating graphene layer [20]. But these methods are either time consuming and/or not efficient producing lesser concentration of graphene (~0.01 mg/ml) in N-methylpyrrolidone (NMP). Later many other methods have been developed which produce graphene in higher concentration and more quickly. The reader is referred to literature available elsewhere [21]. A report entitled "The Global Market

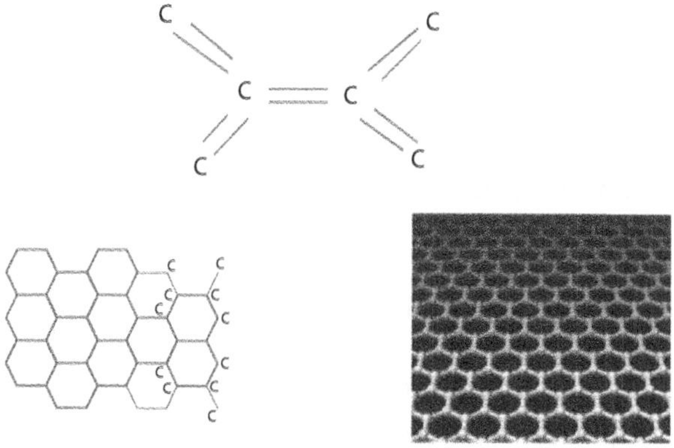

Figure 14.7 Honeycomb structure from sp^2 hybridization. In graphene H is replaced by carbon.

for Graphene to 2017" by Future Markets, Inc. (Edinburgh, UK) 2012 revealed that graphene production volume in 2010 was 28 tons and is expected to grow to 573 tons by the end of 2017 [22].

Graphene as Corrosion Inhibitor

When exposed to 3.5% NaCl, the corrosion rate for graphene coated carbon steel is much lower (0.05 mm per year) than for uncoated carbon steel (0.09 mm per year). Thus although, pure graphene shows corrosion inhibition on mild steel through its coat, the quality of coating observed is much better if chemically modified graphene/GO is coated to mild steel. To test corrosion inhibition capability of graphene, a coating of epoxy with nanofillers was considered. Graphene was dispersed in epoxy resin using a high-speed disk (HSD) disperser and ultrasonication to form this coating. EIS testing was used to assess the corrosion evaluating the progress of the coatings. The coating films were cast on S-36 steel panels, and the EIS was measured using a three-electrode system. The coating specimen's steel panel served as the working electrode, with a platinum and saturated calomel electrode serving as the counter and reference electrodes, respectively. All three electrodes were submerged in a solution of 1.0% NaCl. The EIS data was collected using an EIS spectrometer and displayed as impedance vs. frequency curves. Furthermore, salt spray visibility (ASTM B117) was used as an expedited durability test to assess the long-term durability of nanofiller-reinforced coatings [23].

Chemically Modified Graphene Oxide (GO)

Graphene, graphene oxide, and their composites have been evaluated for use as corrosion protection coatings for various alloys and metals in various electrolytes. However, due to their lower solubility in polar electrolytes, literatures on the corrosion protection effect of chemically altered GO are relatively scarce. Because of their nanosize, chemically modified GO have high surface area and therefore they cover metal surface effectively and high protection against corrosion to metal. In one test, the corrosion inhibition effect of one of the tested functionalized GO namely, aminoazobenzene-GO (AAB-GO), for mild steel corrosion in acidic solution of 1 M HCl. FT-IR, TEM, XPS, and XRD techniques were used to characterize these GO composites. The analysis of open circuit potential (OCP) versus time curves indicates that AAB-GO becomes effective by adsorbing on the metallic surface. AAB-GO acted as mixed corrosion inhibitors, according to a potentiodynamic study. The inhibitor AAB-GO adsorbs at the metal-electrolyte interface and acts as an interface-type corrosion inhibitor. The presence of this composite in a corrosive medium increases the magnitude of resistance to charge transfer and, as a result, corrosion resistance.

Polypyrrole (PPy) has been widely recognized as one of the top candidates between many conductive polymers for an anti-corrosion coating since its anti-corrosion ability was discovered. Several studies have been conducted in which the polypyrrole-based composite coating has demonstrated appealing anti-corrosion properties due to its easy polymerizing nature and mechanical stability. Nickel-based composite coatings have recently gained popularity in the automotive and aerospace industries due to their corrosion resistance. However, numerous studies have suggested that this coating is ineffective in chloride-containing environments. A GO/PPy composite with nickel was suggested as a remediation as an effective corrosion protective layer for bare mild steel. Their commercial applications, however, have been hampered by their high cost and the lack of instruments in bulk production of high-quality coating. Furthermore, pure graphene coatings are susceptible to surface defects when exposed to corrosive environments over time, resulting in localized corrosion of the metal substrate. In comparison, graphene oxide (GO) is easily produced by chemically oxidizing graphite in bulk and has the same corrosion resistance as graphene. As a result, it has a great deal of potential for use as a corrosion inhibitor. Because of the functional groups on GO, it is hydrophilic, which allows for better dispersion in the electrolyte and may be advantageous for anti-corrosion coating. Using the previously mentioned information, a nickel graphene oxide polypyrrole composite coating was deposited for corrosion protection in a recent work [24]. Because both GO and polypyrrole are anti-corrosive, polypyrrole-based composite coatings were created using the electrodeposition method by integrating GO into the polymer matrices with nickel, and their anti-corrosion characteristics were studied. The constructed Nickel-graphene oxide-polypyrrole (Ni-GO-PPy) composite coatings were implemented to mild steel using a low-cost electrodeposition process. These coatings were characterized using XRD, FE-SEM, EDAX, and FT-ATR. After dipping coated steel samples in a 3.5% NaCl solution, potentiodynamic polarization experiments were carried out to investigate the coatings' anti-corrosion properties. Long-term immersion tests were also carried out to investigate the long-term corrosion behavior of the coated steel in corrosive media and compare the results with those obtained from electrochemical tests. Material and solution, synthesis of graphene oxide, preparation of graphene oxide, polypyrrole (GO-PPY) composite, electrode fabrication by electrodeposition, characterization methods and corrosive environment tests, and other experimental details are described in reference [24]. Characterization techniques used included

XRD, UV, FTAR, FTIR, EDAX, FESEM, and TGA. Long-term (6 months) immersion tests were used to determine the corrosion rates and extent of highly localized corrosion encountered by uncoated and covered mild steel samples in NaCl solution.

These tests show a gradual decrease in the corrosion rate and extent of localized corrosion from uncoated steel to Ni, Ni-GO, and Ni-GO-PPy coated steel. This reduction can be attributed to the composite coating's well-developed, less porous, and more hydrophobic nature, which helped to slow the movement of corrosive ions towards the metal surface. The fracture in the coating on the steel specimen was more noticeable in nickel coatings than in other coatings. As a result of the coating's more hydrophobic, compact nature, and less porous structure, overall localized and uniform corrosion is noted less in composite (Ni-GO-PPy) coatings. Electrochemical polarisation tests were also performed on the uncoated/coated steel samples to further check the corrosion inhibition ability of the different coatings. Various parameters related to corrosion of metal samples. With the help of cyclic polarization measurements, one can estimate the tendency of metal to experience localised corrosion.

Measurements on uncoated mild steel were not taken because it does not undergo localised corrosion. In case of coated samples, increasing value of pitting and repassivation potentials indicate higher resistance of the coated steel towards both pitting and crevice corrosion respectively. Thus one observes maximum inhibition of localised corrosion also on mild steel when it is coated with Ni-GO-Ppy coating. This work demonstrate that coating of graphene

Graphene, in its pure or derivative form has been a topic of increasing importance in the scientific community for around more than two decades. Most of this work has been done in laboratories where conditions are controlled and purified forms of chemicals are used in strict environmental conditions. Due to graphene's unique chemical, structural, electrical, and mechanical properties, it shows applicability for many areas within nearly almost every industry be it chemical industry e.g. Pulp and Paper industry etc., the oil and gas industry, marine industry, transportation which includes shipping, trains, motor vehicles etc. However, to study application of graphene in industry is a tedious task due to varying conditions in terms of temperature, chemical concentration and purity level, mixture of chemicals, flow of liquid media, high pressure conditions etc. The author has found it very difficult to find work on corrosion in industry using carbon nanoallotropes.

The application of graphene in the oil and gas industry, which is one of the more corrosion-prone industries, has gained popularity in recent years, with the majority of research going to take place in the last ten years or less [25]. Corrosion affects pipes, production tubing, casing, and virtually any type of downhole hardware that is not made of plastic in this industry. Graphene and its derivatives have been studied for their potential use as a corrosion-resistant coating for downhole metal equipment. The benefits of using graphene as a covering include its strong mechanical and electrical properties, impermeability to gases and liquids, low high reactivity, and the ability for the coating to be significantly thinner than a conventional coating whilst also exhibiting high results. A 200nm thick graphene oxide hybrid membrane was implemented to stainless steel inside one study. The study concluded that while the GO coating did not provide much corrosion resistance on its own, it performed admirably as part of the hybrid blend [26]. In a subsequent study, an oil-based graphene oxide base was used as a coating for steel in a seawater atmosphere [27]. The study concluded that their graphene oxide coating was highly effective in a high salt environment, reducing corrosion rate by over 10,000 times when compared with the non-coated control group steel. In the final study, pure graphene nanoflakes were risen directly on a stainless steel substrate by the Dumee group [28]. This same group has successfully grew among 3 and 15 layers on the steel, resulting in improved corrosion resistance and electrical characteristics without affecting the stainless steel substrate's properties.

14.7.4 Carbon Nanotubes (CNTs)

These are carbon allotropes (Figure 14.8) with nanostructures that can have a length-to-diameter ratio of up to 28,000,000:1, far greater than any other material (Table 4.4). CNT's are formed by rolling graphene nanosheets [Figure 14.8(a)] and so exhibit cylindrical structure [(Figure 14.8(b)] with

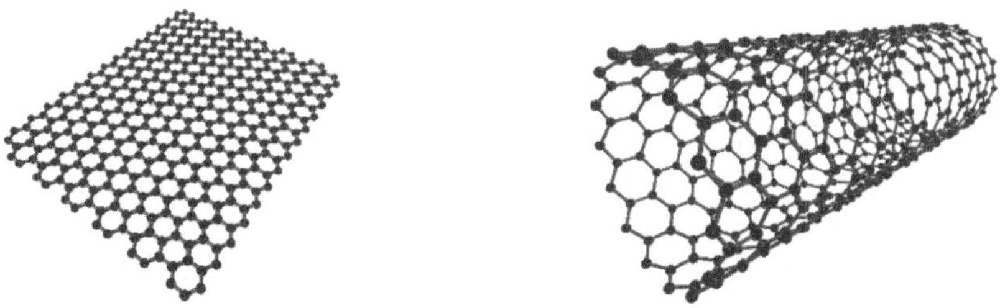

(a) (b)

Figure 14.8 Structure of (a) graphene nanoplatelet, (b) carbon nanotube.

a nanoscale diameter. There are two types of CNT's namely SWNT (single wall nanotube) as shown in Figure 14.8(b) having ~ 1 nanometer (nm) diameter and MWNT (multi wall nanotubes). The MWNTs are made up of only two coaxial cylinders. The multi-walled nanotubes have an outer diameter of 5.5 nm and an inner diameter of 2.3 nm. Carbon nanotubes have a thickness or diameter of a few nanometers, which is approximately 50,000 times smaller than that of the width of a human hair, and their length can be many centimeters [23].

CNT and its derivatives are widely used as corrosion inhibitors due to a variety of fascinating properties such as excellent mechanical strength, good mechanical, high thermal and chemical resistance, high specific surface area ratio, and high dispersibility, as well as an excellent ability to interact with the metallic surface. The properties of these cylindrical carbon molecules are novel. In terms of tensile and elastic modulus, carbon nanotubes are the greatest and stiffest materials available. The covalent sp2 bonds formed between both the personal carbon atoms contribute to this strength. A multi-walled carbon nanotube with a tensile strength of 63 gigapascals was tested in 2000 (equivalent to its ability to withstand 6300 kg of weight on a cable with a cross-section of 1 mm^2). CNTs are much greater than stainless steel and have a higher toughness than both stainless steel and Kevlar. This makes carbon nanotubes ideal for industrial applications. CNTs, on the other hand, are not as strong when compressed. This is due to their hollow construction and high length/diameter ratio. When subjected to compressive, torsional, or bending stress, they tend to buckle. All nanotubes are expected to be excellent thermal conductors along the tube while being excellent insulators laterally to the tube axis. At room temperature, carbon nanotubes can transmit up to 6000 watts per meter per Kelvin. Copper, a metal known for its high thermal conductivity, transmits only 385 watts per meter per K. Carbon nanotubes are thought to be temperature stable up to 2800°C in vacuum and 750°C in air.

Synthesis

To make CNT, a one-atom-thick sheet of graphene is rolled into a tube. This results in the formation of a single-walled carbon nanotube (SWNT). Layers of these graphene sheets can be rolled to form multi-walled carbon nanotubes (MWNT), which have slightly different properties than single-walled carbon nanotubes. These CNTs have been created in a variety of diameters and lengths so that they can be used with the specific properties required for various applications.

There are presently three main methods for producing carbon nanotubes: arc discharge, laser ablation of graphite, and chemical vapor deposition (CVD). Arc discharge and laser ablation of graphite are two techniques that begin with graphite and end with a gas from which carbon nanotubes are formed. The most widely used method for producing carbon nanotubes is CVD, which can produce the nanotubes in greater quantities, at a lower cost, and under easier-to-manage conditions. The process includes combining a heterogeneous catalyst with fuel reaction gases, which results in the formation of carbon nanotubes on the catalyst within a furnace.

To improve the purity of the created nanotubes, methods for creating nanotubes require a filtration process. Because impurities cause structural defects, they alter the physicochemical properties of nanotubes. Metal encapsulated nanoparticles, metal particles, and amorphous carbon can pollute nanotubes, causing structural defects and altering the physicochemical properties of the nanotubes. As a result, after the tubes have been formed, a purification step is required. Purification can be accomplished at the end of the process through either acid hydrolysis or ultrasound interaction.

Carbon Nanotubes as Corrosion Inhibitor

CNTs are carbon nanotubes with a diameter of one nanometer (nm). CNTs and their derivative products (functionalized CNTs) are widely used in material science for a variety of industrial applications due to their exceptionally tensile strength, high mechanical power, and nanosized structure, thermal and chemical resistance, high surface to volume ratio, and excellent ability to interact with metals [13]. Due to their high electrical conductivity and so show lesser corrosion resistance. Thus they are not used directly for corrosion inhibition but used as corrosion inhibitors, especially in aqueous phase, after covalently functionalising it to gain the desirable anticorrosive properties. Covalent functionalization of SWCNTs and MWCNTs can be achieved using various chemical transformations. Covalent functionalisation increases (i) its dispersion, (ii) interfacial adhesion ability and decreases (i) hydrophobicity, (ii) electrical conductance. There are few shortcomings also while intending to use nanomaterials for corrosion protection e.g. their tendency to agglomerate in aqueous media so as to form bigger molecules (agglomeration). This in turn (i) lowers surface to volume ratio (ii) reduces nanomaterials solubility in aqueous electrolyte. But nanomaterials can be used as anticorrosive material when they are mixed with polymer matrix e.g. nylon, polyamide, polystyrene, epoxy resin etc. To improve corrosion resistance, 0.5% to 5% of nanomaterials are used as nanofillers in the polymer matrix. Nanomaterials, obviously, block the surface micropores of the polymer matrix, preventing the diffusion of corrosive species such as water, oxygen, and corrosive ions.

CNT derivatives, particularly polymer composites, have received a lot of attention as anticorrosive materials for ferrous and non-ferrous alloys. As anticorrosive composite coatings for carbon steel in 3.5% NaCl, PPy film, CNT-PABS (PPY/polyaminobenzene sulfonic acid-functionalized single-walled carbon nanotubes), and CNT-CA (PPy/carboxylic acid-functionalized single-walled carbon nanotubes) were used. The anti-corrosive effect of various formulations was measured using PDP, SEM, and TEM techniques. The results showed that CNT-PABS and CNT-CA distribute uniformly in the composites and provide better anticorrosive effectiveness than pure PPy. The corrosion inhibition effect of the various formulations was also found to be in the following order: PPy CNT-CA CNT-PABS. The anticorrosive effect of a PPy/MWCNTs nanocomposite on 304 stainless steel in a 3% NaCl solution has also been studied.

Fukuda *et al.* [37] studied corrosion inhibition on magnesium alloy in NaCl due to CNT's. Magnesium (Mg) alloys are used in a variety of structural and mechanical components due to their low density of 1.738 g/cm3. It is widely acknowledged that replacing traditional structural materials such as steels or aluminium alloys with magnesium alloys can significantly reduce transportation vehicle fuel consumption. This work was carried to basically address the problem associated with high corrosion tendency of Mg alloys. Secondly, they are weak and instead of depending on improvement on mechanical strength by using Mg alloys, CNT's have been considered as more suitable for improving the strength of these alloys. In addition, the effect of CNTs on Mg corrosion is expected to improve the resistance to corrosion of the Mg alloy CNT composite. A previous study found that AZ91D Mg-MMC with CNTs (AZ91D/CNT composite) had better corrosion resistance than pristine AZ91D alloy. According to the report, CNTs act as a water repellent and reinforce the surface insulating barrier, and both of these changes improve corrosion resistance. More studies were done to further investigate the role of CNT's in improving corrosion resistance of Az31BMg alloy/CNT composite in 0.51 M NaCl at pH = 6.2. Details of the preparation methods for forming Az31B/CNT composite may be seen elsewhere [37]. For corrosion testing, both immersion and electrochemical polarization tests were conducted. Three samples were tested (a) Az31B without CNT, Az31B/CNT composite with (b) CNT = 0.89 vol% and (c) CNT = 2.66 vol%. 12 hours long immersion test showed following weight loss, due to corrosion, in the three samples :

Weight loss observed in case of sample without CNT is least while it is maximum for AZB/CNT composite having higher amount of CNT. It indicates role of CNT in enhancing corrosion loss. Potentiodynamic polarisation measurement of sample 'a' shows passivation behaviour, indicating protection properties of the corrosion product Mg(OH)2 formed on the sample and lower corrosion potential responsible for lesser degree of mass loss, while in other two cases (sample b and c), passivation has ceased to exist resulting in higher degree of corrosion of Mg than in sample (a). Corrosion potential in these cases is around 0.7 V higher than that in case of sample 'a'. Surface morphology of sample 'c' after polarization test show specimen surface locally damaged at primary particle boundary where lots of CNT's are present. Surface potential measurement show lower potential at this point. Surface potential difference, between three Mg matrix and the dark areas, were calculated 1.06, 1.09, and 1.14 V. Thus area near particle boundary, where CNTs are accumulated acts as anodic and Mg matrix area acts as cathodic site of the galvanic cell. This leads to faster corrosion loss of Mg areas near particle boundary and much lesser corrosion loss in Mg matrix in case of sample 'b' and 'c' of the AZB/CNT composite. Overall corrosion loss is much smaller in case of pristine Mg alloy sample due to absence of Mg-CNT galvanic cell. Thus CNT in such cases does not help in enhancing corrosion resistance of Mg alloys. This is one of the eye opening examples indicating that whereas in case of steel CNT's provide corrosion inhibition but not in case of Mg alloys. So while considering any carbon nanoallotropes for inhibition purpose, one should check to avoid the possibility of galvanic corrosion [29–36, 38, 39].

14.8 Conclusion

Thus we see that carbon nanoallotropes possess extra-ordinary properties either as fullerene/carbon dots, graphite, graphene or CNT's or when they form composite with polymers. These properties e.g. high mechanical strength, hardness, high temperature stability, high surface/volume ratio providing higher surface coverage (per unit mass) for metal protection, barrier properties for chemicals and electrical resistance, thermal conductivity, amenability to interact and bond with metal surfaces, flexibility and easy surface functionalization makes them efficient, cost effective and eco-friendly alternatives to the harmful inorganic and organic corrosion inhibitors. Incidentally graphite and carbon derivatives are abundant, cost-effective and light materials, so they don't add much weight to the metal surface which is to be protected, being abundant will not lead to their scarcity and off-course their use will reduce the carbon prints and lower the corrosion cost significantly. Although graphene, fullerene, carbon dots, etc. have been investigated in the last decade as anticorrosive materials, and some of them have already been included in industrial coatings, there are still some challenges and possible perspectives that have to be addressed in finite research. These are (i) Upscaling – related to work out suitable technology so that carbon base nanomaterials may be scaled up to the industry size level. This is due to the fact that optimal parameters for large-scale synthesis of carbon components and testing their inhibitory activity on an industrial scale may differ from optimized conditions on a lab scale. So, in addition to laboratory-based tests, efficient coatings should be developed and applied at a larger scale. (ii) Safety concerns - Because these materials are nanosized, they can penetrate into human body cells and may be hazardous to work with, much like working in smoke/dust. Carbon nanomaterial safety risks are entirely dependent on the type of carbon nanostructure, surface composition, shape, size, and application. Research should be conducted to investigate the long-term health effects of these materials as well as their fate in the environment. (iii) Graphene quality - The conductivity of graphene and graphite films varies greatly depending on surface defects, orientation, synthesis method, and purity. This variation will result in different surface reaction rates. As a result, quality control of graphene and carbon nanotube materials, as well as their influence on inhibition efficiency, must be investigated further and linked to variations in carbon nanomaterial properties.

14.8.1 Commercialization

The commercialization of graphene materials has been steady, and the graphene market is expected to grow substantially over the decade 2021 to 2031. Graphene was initially oversold to the industry as a wonder material that would revolutionize nearly every industry overnight. Today, graphene platelets are progressively and correctly regarded as a component of the vast carbon additive materials. Also a consumer realizes that all are not same, may have different properties and so perform differently. So one has to procure a given graphene material depending upon his requirement. Graphene powders and platelets, like carbon nanotubes, are primarily a substitute material that outperforms carbon black, graphite, and other additives. As a result, it must compete with incumbent solutions on both price and performance. Graphene suffered as a new specialty material from high and differing prices and pricing models. This, however, has changed. For the time being, graphene platelet prices have fallen and are beginning to converge. Prices, however, will not settle around with a single point, reflecting the variety of graphene types and giving it a unique chemical character. According to current data, graphene company income has been steadily increasing since 2013. This rise will continue at a similar rate until

2021/22, when the model predicts an inflexion point, putting the industry into its rapid volume growth stage.

14.8.2 Synergy in Mixed Nanohybrids

Carbon nanoallotropes are thought to perform better when combined with other anti-corrosive materials. Graphene layers, for example, serve primarily as a barrier and provide mechanical stability for the coating in such hybrids, as in the case of graphene oxide reinforced epoxy resins. As a result, the development of hybrid materials composed of multiple carbon nanoallotropes provides practical opportunities to create materials with combined properties of the constituents via synergistic and cooperative effects. As a result, hybrid carbon-based nanomaterials can now find applications in fields where traditional materials were uncompetitive. As a result, more research into more efficient and environmentally friendly combinations for sustainable coatings is required.

References

1. Fontana, M.G., *Corrosion engineering*, pp. 1–576, McGraw Hill Book, Co., Singapore, 1987.
2. Uhlig, H.H. and Revie, R.W., *Corrosion and corrosion control: An introduction to corrosion science and engineering*, pp. 1–490, John Wiley & Sons, Inc., Hoboken, New Jersey, 2008.
3. Singh, A.K., *Microbially induced corrosion and its mitigation*, pp. 27–128, Springer, Singapore, 2020.
4. Valero, A. and Valero, A., Physical genomics combining the energy and hubert peak analysis for predicting mineral resources depletion. *Resour. Conserv. Recycl.*, 54, 1074, 2010.
5. https://www.fortunebusinessinsights.com/industry-reports/high-strength-steel-market-101854, 2020.
6. Ita, B.I. and Offiong, O.E., Adsorption studies on the corrosion inhibition properties of 2-acetylpyrole and 2-acetylpyrole-(2-acetylthiosemicarbazone) on mild steel in hydrochloric acid medium. *Global J. Pure Appl. Sci.*, 5, 497, 1999.
7. Megalai, S.M., Manjula, P., Manonmani, K.N., Kavithaa, A., Babyd, N., Metronidazole, a corrosion inhibitor for mild steel in aqueous environment. *Portugaliae Electrochimica Acta*, 30, 395, 2012.
8. Atmani, F., Lahem, D., Poelman, M., Buess-Herman, C., Olivier, M., Mild steel corrosion in chloride environment: Effect of surface preparation and influence of inorganic inhibitors. *Corros. Eng. Sci. Technol.*, 48, 9–18, 2013.

9. Luo, X., Pan, X., Yuan, S., Du, S., Zhang, C., Liu, Y., Corrosion inhibition of mild steel in simulated seawater solution by a green eco-friendly mixture of glucomannan (GL) and bisquaternary ammonium salt (BQAS). *Corros. Sci.*, 125, 139, 2017.
10. Verma, C., Ebenso, E.E., Quraishi, M.A., Hussain, C.M., Recent developments in sustainable corrosion inhibitors: Design, performance and industrial scale applications. *Mater. Adv.*, 2, 3806, 2021.
11. Sorensen, P.A., Kiil, S., Dam-Johansen, K., Weinell, C.E., Anti-corrosive coatings: A review. *J. Coat. Technol. Res.*, 6, 135, 2009.
12. Brycki, B.E., Kowalczyk, I.H., Szulc, A., Kaczerewska, O., Packiet, M., *Corrosion inhibitors, principles and recent applications*, M. Aliofkhazraei (Ed.), pp. 1–33, Intechopen (UK), London, 2017.
13. Verma, C., Quraishi, M.A., Ebenso, E.E., Hussain, C.M., Recent advancements in corrosion inhibitor systems through carbon allotropes: Past, present and future. *Nano Select*, 2, 2237, 2021.
14. Saji, V.S. and Thomas, J., Nanomaterials for corrosion control. *Curr. Sci.*, 92, 51, 2007.
15. Nasir, S., Hussein, M.Z., Zainal, Z., Yusof, N.A., Carbon based nanomaterials/allotropes: A glimps of their synthesis, properties and some applications. *Materials*, 11, 295, 2018.
16. Sharma, A. and Das, J., Small molecules derived carbon dots: Synthesis and applications in sensing, catalysing, imaging and biomedicine. *J. Nanobiotechnology*, 17, 24, 2019.
17. Saraswat, V., Kumari, R., Yadav, M., Novel carbon dots as efficient green corrosion inhibitors for mild steel in HCl solution: Electrochemical, gravimetric and XPS studies. *J. Phys. Chem. Solids*, 160, 110341, 2022.
18. Berdimurodov, E., Varma, D.K., Kholikov, A., Akbarov, K., Guo, L., The recent development of carbon dots as powerful green corrosion inhibitors: A prospective review. *J. Mol. Liq.*, 349, 118124, 2022.
19. Geim, A.K. and Mcdonald, A.H., Graphene: Exploring carbon flatland. *Phys. Today*, 60, 35, 2007.
20. Alzari, V., Nuvoli, D., Scognamillo, S., Piccinini, M., Gioffredi, E., Malucelli, G., Marceddu, S., Sechi, M., Sanna, V., Mariani, A., Graphene-containing thermoresponsive nanocomposite hydrogels of poly(N-isopropylacrylamide) prepared by frontal polarization. *J. Mater. Chem.*, 21, 8727, 2011.
21. *Graphene production and application*, S. Ameen, M.S. Akhtar, H. Shik shin (Eds.), p. 106, Intechopen London, UK, 2020.
22. Future Markets, *The global market for graphene 2010-2017*, Future Markets, Inc., Edinburgh, London, UK, 2012.
23. Wang, X., Tang, F., Cao, Q., Qi, X., Pearson, M., Li, M., Pan, H., Zhang, Z., Lin, Z., Comparative study of three carbon additives carbon nanotubes, graphene and fullerene C-60 for synthesizing enhanced polymer nanocomposites. *Nanomaterials*, 10, 838, 2020.

24. Pandey, U., Singh, A.K., Sharma, C., Development of anticorrosive nickel graphene oxide polypyrrole composite coatings on mild steel employing electrodeposition technique. *Synth. Met.*, 290, 117135, 2022.
25. Neuberger, N., Adidharma, H., Fan, M., Graphene: A review of applications in the petroleum industry. *J. Pet. Sci. Eng.*, 167, 152–159, 2018.
26. Mondal, J., Aarik, L., Kozlova, J., Niilisk, A., Mander, H., Maeorg, U., Graphene nanoplates based protective and functionalising coating for stainless steel. *J. Nanosci. Nanotechnol.*, 15, a, 6747–50, Sept. 2015.
27. Singbabu, Y.N., Shivkumar, B., Singh, J.K., Bapari, H., Pramanik, A.K., Sahu, R.K., Efficient anti-corrosive coating of cold rolled steel in a sea water environment using an oil based graphene oxide ink. *Nanoscale*, 7, 17, 8035–47, 2015.
28. Dumee, L.F., He, L., Kong, L., Growth of nano-textured graphene coatings across highly porous stainless steel supports towards corrosion resistant coatings. *Carbon*, 2, 042, 2015.
29. Wen, G., Bai, P.P., Tian, Y., A review of graphene based materials for marine corrosion protection. *J. Bio Tribo Corros.*, 7, 27, 2021.
30. Raman, R.K.S. and Tiwari, A., Graphene: The thinnest known coating for corrosion protection. *JOM*, 66, 637, 2014.
31. Tong, Y., Bohm, S., Song, M., The capability of graphene on improving the electrical conductivity and anti-corrosion properties of polyurethane coatings. *Appl. Surf. Sci.*, Part 1, 72, 424, 2017. http://dx.doi.org/doi:10.1016/j.apsusc.2017.02.081.
32. Aneja, K.S., Böhm, H.L.M., Böhm, S., Khana, A.S., Functionalised graphene as a barrier against corrosion. *FlatChem*, 1, 11, 2017.
33. Yu, Z., Di, H., Lv, L., Ma, Y., Pan, Y., Zhang, C., He, Y., Fabrication of graphene oxide-alumina hybrids to reinforce the anti-corrosion performance of composite epoxy coatings. *Appl. Surf. Sci.*, http://dx.doi.org/doi:10.1016/j.apsusc.2015, 351, 986, 2015.
34. Othman., N.H., Ismail, M.C., Mustapha, M., Sallih, N., Kee, K.E., Jaal, R.A., Graphene-based polymer nanocomposites as barrier coatings for corrosion protection. *Prog. Org. Coat.*, 135, 82, 2019.
35. Hayatdavoudi, H. and Rahsepar, M., A mechanistic study of the enhanced cathodic protection performance of graphene-reinforced zinc rich nano-composite coating for corrosion protection of carbon steel substrate. *J. Alloys Compd.*, 727, 1148, 2017.
36. Chang, C.H., Huang, T.-C., Peng, C.-W., Yeh, T.C., Lu, H.I., Hung, W.I., Weng, C.J., Yang, T.I., Yeh, J.M., Novel anticorrosion coatings prepared from polyaniline/ graphene composites. *Carbon*, 50, 400, 2012.
37. Fukuda *et al.*, H., Szpunar, J.A., Kondoh, K., Chromik, R., The influence of carbon nanotubes on the corrosion behavior of AZ31B magnesium alloy. *Corros. Sci.*, 52, 3917, 2010.
38. Cui, G., Bi, Z., Zhang, R., Liu, J., Yu, X., Li, Z., A comprehensive review on graphene-based anti-corrosive coatings. *Chem. Eng. J.*, 373, 104, 2019.

39. Teng, S., Gao, Y., Cao, F., Kong, D., Zheng, X., Ma, X., Zhi, L., Zinc-reduced grapheme oxide for enhanced corrosion protection of zinc-rich epoxy coatings. *Prog. Org. Coat.*, 123, 185, 2018.
40. Liu *et al.*, S., Gu, L., Zhao, H., Cheng, J., Yu, H., Corrosion resistance of graphene-reinforced waterborne epoxy coatings. *J. Mater. Sci. Technol.*, 12, 17, 2015.
41. Samadianfard, R., Seifzadeh, D., Habibi-Yangjeh, A., Jafari, T.Y., Oxidizedfullerene/sol-gel nanocomposite for corrosion protection of AM60B magnesium alloy. *Surf. Coat. Technol.*, 385, 125400, 2020.

Index

Printed and bound by CPI Group (UK) Ltd, Croydon, CR0 4YY
17/09/2026
14964277-0002